Leckie
the education publisher
for Scotland

Primary **Maths** for **Scotland**

Early Level Maths

Teacher Guide

Series Editor: Craig Lowther

Authors: Julie Brewer, Sheena Dunlop and Lesley Ferguson

001/04122019

10 9

ISBN 9780008359690

Published by
Leckie
An imprint of HarperCollinsPublishers
1 Robroyston Gate, Robroyston, Glasgow, G33 1JN
T: 0844 576 8126   F: 0844 576 8131
leckiescotland@harpercollins.co.uk   leckiescotland.co.uk

HarperCollins *Publishers*
Macken House, 39/40 Mayor Street Upper, Dublin, 1 D01 C9W8, Ireland

Publisher: Fiona McGlade
Project editor: Julianna Dunn

Special thanks
Copy editor: Mitch Fitton
Illustrator: Ann Paganuzzi
Layout and typesetting: Siliconchips Services Ltd
Proofreader: Lucy Hollingworth

A CIP Catalogue record for this book is available from the British Library.

Acknowledgements

Printed in the United Kingdom.

FSC www.fsc.org
MIX
Paper | Supporting responsible forestry
FSC® C011748

Contents

Introduction **1**

Primary Maths for Scotland Programme Overview **1**

Primary Maths for Scotland Programme Components **2**

Primary Maths for Scotland Early Level Teacher Guide **2**

Primary Maths for Scotland Early Level Record Book **7**

Primary Maths for Scotland Early Level Digital Package **8**

Suggested Teaching Plans **11**

1. Estimation and Rounding **27**

Estimation and Rounding A: By the end of Nursery **27**

Estimation and Rounding B: By the end of Primary 1 .... **37**

2. Number and Number Processes – Number Recognition and Place Value **50**

Number Recognition and Place Value A: By the end of Nursery **50**

Number Recognition and Place Value B: By the end of Primary 1 .... **80**

3. Number and Number Processes – Four Operations **110**

An Introduction to Addition and Subtraction A: By the end of Nursery **110**

Addition and Subtraction B: By the end of Primary 1 .... **121**

4. Fractions **144**

An Introduction to Fractions A: By the end of Nursery **144**

Fractions and Fair Shares B: By the end of Primary 1 .... **150**

5. Money **161**

An Introduction to Money A: By the end of Nursery **161**

Money and Shopping B: By the end of Primary 1 .... **166**

6. Time **172**

An Introduction to Time A: By the end of Nursery **172**

Telling the Time B: By the end of Primary 1 .... **181**

7. Measurement **189**

An Introduction to Measurement A: By the end of Nursery **189**

Early Measurement B: By the end of Primary 1 .... **196**

8. Patterns and Relationships **202**

Patterns and Relationships A: By the end of Nursery **202**

Patterns and Relationships B: By the end of Primary 1 .... **210**

9. 2D Shapes and 3D Objects **218**

An Introduction to 2D shapes and 3D objects A: By the end of Nursery **218**

Name and Discuss the Properties of 2D shapes and 3D objects B: By the end of P1 .... **224**

10. Angle, Symmetry and Transformation **232**
Directions, Reflections and Symmetry A: By the end of Nursery **232**
Directions, Reflections and Symmetry B: By the end of Primary 1 .... **241**
11. Data and Analysis **252**
Data and Analysis A: By the end of Nursery **252**
Data and Analysis B: By the end of Primary 1 .... **262**

Introduction

Primary Maths for Scotland Programme Overview

This Primary Maths for Scotland Early Level Teacher Guide has been published in conjunction with the Primary Maths for Scotland Early Level Record Book; Digital Package; Early Level Problem Solving Pack and Early Level Assessment Pack. These titles complete a whole-school/nursery mathematics programme which also includes First Level and Second Level.

This series has been written by a team of Scottish numeracy experts specially for Scottish schools and has been written to the Curriculum for Excellence Benchmarks to plan routes through the Experiences and Outcomes for Numeracy and Mathematics. They have been designed to develop conceptual understanding alongside procedural fluency whilst being easy to implement alongside existing plans and resources.

When the team created these resources, we use researched-informed pedagogy to guide us. We have used the principle of social constructivism to inform the way we teach and how we have phrased the resources and teaching ideas. We believe that pupils should be active participants in the learning process and be challenged and encouraged to talk about their ideas and share their strategies for solving mathematical problems. We also want our, and your, pupils to enjoy learning mathematics. This mind-set and motivation makes learning and teaching a truly positive experience for all.

At Early Level active participation is supported by learning through play. This could be play initiated and structured by an adult, or free-play directed by the child, with the adult providing a suitable environment and the tools or props to help learning flourish. The Early nursery setting and classroom will need the space, environment and resources to support this play.

Both the inside and outside environment can be filled with objects to support counting, construction, sand and water play, games, measurement, making collections and so on. These objects can be 'found' or sourced cheaply by using items from nature; stones, shells, dried beans, nuts and autumn items or recycled items such as plastic bottles and their tops, straws, ribbon, fabric or old clothes (like socks for matching and sorting), cups and pots. Many items can have numerals painted onto them and covered with varnish so that they last. 'Bought' resources like counters, 2D shapes and 3D objects are usually very robust, so they can be used for construction, added to sand trays and used for printing with paints. By making resources freely available children will be encouraged to play, discuss and be creative with mathematical ideas.

Structured discussion and play can be supported through interest tables and displays. The number of items on the interest table can increase over time, e.g. by counting up to 100 over a hundred days. Wall displays can reflect the seasons and track the day, date, month and weather. Displays of numerals in the room (even as labels on storage trays) and outside as well as displays of tens frames with magnetic counters, number lines with movable arrows can support children to make connections between collections, counting and the symbols we use in mathematics.

This 'rich' mathematical environment, tied with active play, discussion and enjoyment will support children to build inquisitive and positive attitudes to learning about the connected world of mathematics.

Primary Maths for Scotland Programme Components

Primary Maths for Scotland Early Level Teacher Guide

The Teacher Guide is a planning and teaching support manual for delivering the Early Level Numeracy and Mathematics Curriculum for Excellence. It is a flexible resource which can be used alongside teachers' existing planning, resources and scheme of work. The purpose of this book is to provide Early Level teachers with:

- lots of support for planning and delivering the Early Level curriculum;
- creative ideas for numeracy stations and numerate play as well as group and whole-class work;
- support to develop their own CPD;
- guidance on questioning and observing assists teachers to make judgements about pupils' achievement;
- time saving support.

At the beginning of the Teacher Guide is a suggested teaching plan for Early Level. We have organised the curriculum into two parts; A and B, which correspond to where we would expect children's mathematical knowledge and skills to be by the end of Nursery (level A) and by the end of P1 (level B). The teaching plan can be used by teachers as their planning document or can be ignored in favour of an existing or different plan. The Teacher Guide has been organised by Experiences and Outcomes, and not by suggested teaching plan, so that each topic is easily found and teachers can use the resources as they see fit.

Suggested teaching plan

Book	Chapter	Content	Units	Benchmark/Es&Os
Early	A1. Estimation and Rounding	By the end of Nursery • Uses subitising to compare the size of objects in a group. • Use observation/ comparison skills to	• 1.1 Use 'bigger' 'greater', 'smaller' and 'the same as', 'more than', 'less than' 'fewer than' to describe differences between small collections of similar objects and between easily	I am developing a sense of size and amount by observing, exploring, using and communicating with others about things in the world around me.

This provides a suggested order of topics to follow. The resources are designed to be flexible and will fit with an established curriculum design, but if you are looking for a revised order then this is your starting point. The order is designed to support the pupils to progressively build knowledge and skills, whist being able to return to previously learned topics as part of a spiral curriculum of deliberate spaced practice. This helps pupils to remember and practise old skills that they need to continue to learn. If you are working with a multi-stage class, it would be prudent to choose one sequence of topics for the class and then to differentiate to meet the needs of groups and individuals.

Chapters

Each chapter in the Teacher Guide is divided into two sections which correspond to levels A and B. For each of the two sections within a topic, the following guidance is provided:

Section overview

Units of learning	Review units	Prior learning	Benchmarks/Es&Os
Unit 1.1 **Comparing collections visually** Comparing sets visually if there is a large numerical difference.		• In familiar settings do they understand what the concept 'more' means? For example, to ask for more food if they are hungry. • Experience of making collections.	**MNU 0-01a** • Recognises the number of objects in a group, without counting (subitising) and uses this

This table shows the Curriculum for Excellence Experiences and Outcomes and Benchmarks covered by each unit in this section. It is not our expectation that a child who successfully completes a single unit has demonstrated completion of a benchmark, rather it is through regularly returning to ideas and building skills and knowledge that ultimately a child will be able to demonstrate their understanding of each benchmark in new and unfamiliar contexts. This is when they, and you, know they are secure.

- Column 1 lists the Teacher Guide units in this section, which correspond to the Textbook units.
- Column 2 directs teachers to units that children can review if they are not yet ready to complete the work in the unit and do not possess the required pre-requisite knowledge.
- Column 3 describes the prior learning that children are expected to have mastered before embarking on the unit.
- Column 4 shows the coverage of the CfE and Benchmarks within this section.

Benchmarks and sections of benchmarks that are not relevant or covered within the section have been greyed out.

Conceptual context

Conceptual context

If we hope for reasonable responses, rather than wild guesses, it is important to nurture children's understanding of estimation from a very young age.

Comparing two concrete collections which are very obviously different in size should be encouraged even before a child can use counting to check. Strategies such as physically pairing and lining items up help develop an understanding of the concepts of 'more', 'less' and 'the same'. Using benchmarks such as subitising or groupings of regular patterns, e.g. five frames, are very valuable tools which also encourage comparisons without counting. It is vital that children experience collections that have 'like' and 'unlike' items so that the numerosity of the

This section provides an overview of the topic. It will include a brief explanation of key aspects of learning covered, including prior knowledge, calculation methods and next steps in learning.

Resources

Resources

- Interlocking cubes
- Five frames – different sizes
- Sorting hoops and hula hoops
- Building bricks
- Coloured counters
- Number/numeral tracks
- Posting boxes
- Numeral cards

This section lists the physical resources which can be used to support the teaching and learning of the units in this topic. A Glossary of Physical Resources can be found and downloaded at collins.co.uk/PrimaryMathsforScotland/freeresources.

We have listed resources that we use regularly with our own classes. We do not expect you to buy specific items of equipment; many objects; like counters or cubes will be available in a number of different forms. We do believe strongly that when pupils first encounter new ideas that they should have access to and be encouraged to use concrete materials. As the pupils become more confident they can move onto imaging the materials in their minds and then move onto using the number facts and skills without the help of manipulatives.

Digital

This section shows which of the games, songs and digital mathematical tools available in the Early Level Digital Package are the most suitable to support teaching and learning within this topic.

Key vocabulary

Key vocabulary

- **More than/a little bit more**
- **Fewer than** – for countable quantities, e.g. There are **fewer** bananas **than** apples.
- **Less than** – for non-countable quantities, e.g. There is **less** water in my glass **than** in the pond.

The main mathematical terms relating to the units in this section are listed.

A full glossary of all the key mathematical terms and definitions can be found online at this link collins.co.uk/PrimaryMathsforScotland/freeresources.

The key vocabulary is listed and defined to promote the use of correct mathematical words and consistency across classes.

Misconceptions

Potential misconception	Addressing the misconception
Unit 1.3 They may not realise that the final number that they say represents the final set. When pointing to 4 items they may have to start their count from one again. They think they need to say the string of numbers 1 to 4, not just the final count of 4.	• Lots of opportunities to collect and count small collections using a wide range of objects. Model the final count. • Evidence in environment and displays of value associated with a number. • Practise forward number sequence up to 10.

We have listed many of the common misconceptions that pupils will reveal to you when they have an opportunity to talk about their ideas. Column 1 indicates the units where the potential misconceptions are most likely to occur.

Where errors or misconceptions are highlighted it is important to address these immediately to support the pupil to change their thinking.

Teaching ideas

These suggested activities can be used to introduce a topic prior to pupils practising new skills or as part of the reinforcement of learning. The teaching ideas are often presented in a context, or as a game or song, but do adapt the context for the children to take advantage of local events, activities and facilities that they will be more familiar with. There are ideas for books to use as an initial stimulus, but again use the books that you have available to you and adapt the activity.

The team have highlighted key questions in bold that you can use with individuals or groups of pupils. These questions are carefully constructed to provoke thought and to help reveal the pupils' current understanding and any misconceptions. The answers and explanations that the pupils offer, will provide feedback that will help you and them to decide on the next steps in their learning.

We have provided instructions for the activities, sometimes with pictures to clarify how the activity can run. There are also some worksheet resources that you can use to quickly get set up. Sometimes the activity will work as planned, other times the children will take you in an unexpected direction, such is the nature of learning, so we do encourage you to keep notes in order to improve on these activities, much as we have developed these ideas ourselves.

Whole class

Whole Class

- **Songs** – For example, 'Five Little Speckled Frogs'. Choose five children to be the frogs and ask them to sit on a bench (the log). Place a blue blanket beneath the log and ask the children to notice what happens to the two groups when a frog jumps off the log into the pond. Sing the song and stop to compare the two groups. **What happens if one frog jumps off the log?** Ensure the frogs are lined up underneath the log but paired with the frogs still on the log. **Will there be more or fewer frogs left on the log? How do you know? Are there more/fewer frogs on the log than in the pond?**

Where suitable, we have included ideas for teaching an objective to a whole class or a group. Often these can also be adapted as a station activity, but the idea of large group tasks is to ensure that all the children have had the same introduction as well as the opportunity to build the learning community.

Stations

Stations

- **Farm Set** – Set out the play farm and different pens/fields. Arrange groups of animals into dice patterns around the farm. **Which field has the most animals in? Which pen has the fewest animals?** Take one animal out of the field. **Are there more or fewer animals left in the field?**

Stations or invitations to play support a self-directed and play-based approach to learning mathematics, and can help to support or embed mathematical concepts for children. At Early Level children will still need some adult support or supervision, but once they become confident with the station activity the adult can observe and record the child's learning.

What to observe

This section provides guidance for practitioners on making teacher judgements about a child's mastery of each objective within the Education Scotland Benchmarks.

Photocopiable resources

Many activities are accompanied by Photocopiable resources which are referenced throughout this Teacher Guide and are available to download from collins.co.uk/PrimaryMathsforScotland/freeresources.

Glossary of key vocabulary

A full glossary of all the key mathematical terms and definitions from each topic can be found online and can be downloaded from collins.co.uk/PrimaryMathsforScotland/freeresources.

Primary Maths for Scotland Early Level Record Book

The Primary Maths for Scotland Early Level Record Book can be used to record achievement of the Early Level benchmarks and to demonstrate mastery of Early Level.

It is a booklet for each child which can be used as a formative or summative assessment of a child's mastery of the Early Level benchmarks. Each page requires the child to complete an exercise to demonstrate mastery of an aspect of the Early Level curriculum. Mastery of the entire Early Level curriculum can be demonstrated by completing the Record Book. It is expected that many children will complete the Record Book during the summer term of P1 as a move towards more formal learning and to assess or evidence attainment of early level.

Record book

Pages 3–5

The contents page provides a grid showing the coverage of the Early Level benchmarks which practitioners can use as a checklist.

Curriculum organiser	Curriculum coverage	Page name	Page	Completed
Estimation and rounding	MNU 0-01a	More or fewer?	3	☐
		Colour match	4	☐
		Regular patterns	5	☐
Number and number processes	MNU 0-02a	Join me up	6	☐
	MNU 0-04	Me first!	7	☐
		Higher and faster	8	☐
		Collections	9	☐
		How many?	10	☐
		Party time	11	☐
		Colour the facts	12	☐
		Finding totals	13	☐
		Something is missing	14	☐
		Double up	15	☐
		Farm calculations	16	☐

Pages are simple, with minimal text, colourful and engaging. Teacher's Notes at the bottom of the page provide guidance for supporting the child to complete the exercise.

Primary Maths for Scotland Early Level Digital Package

The Primary Maths for Scotland Early Level Digital Pack is an online pack which supports teachers in delivering the Early Level numeracy and mathematics Curriculum for Excellence in nursery and P1.

- 26 maths tools support teachers in demonstrating tricky mathematical concepts using a wide variety of representations.
- 9 maths games can be played at different levels to begin to develop mathematical fluency.
- 6 animated maths songs embed key skills such as counting back from 10, increasing a number sequence by 1, etc.

These can be used with the whole class to demonstrate concepts, to embed knowledge and to practise key skills or can be set up on tablets, computers or whiteboards for group work, as invitations to play or as part of a learning carousel. They are fun and engaging, with a rigorous pedagogy underneath, and can support children to master key mathematical concepts and skills.

Games

Title	Objective	Level 1	Level 2	Level 3
Tree Ants	Count items to 10	Up to 4	Up to 6	Up to 10
Fishy Facts	Recall addition and subtraction facts to 20	Add and subtract from 1-9 where the addend or subtrahend is 4 or less	Add and subtract any two single digit numbers (with positive answers)	Add and subtract any two single digit numbers (with positive answers), using the + and - signs
Ants on the bus	Recall addition and subtraction facts to 20	Numbers up to 10	Numbers up to 15	Numbers up to 20
Double the fun	Solve problems involving doubling, halving, grouping and sharing	Doubles to 5	Halves to 10	Doubles to 10 & Halves to 20
Coconut calculations	Recall addition and subtraction facts to 20	Basic addition and subtraction facts to 20	Missing numbers in addition and subtraction facts to 20	More missing numbers in addition and subtraction facts to 20
Tangrams	Move and rotate shapes to solve puzzles	Rabbit, dancer, duck	Cat, swan, boat	Spinning top, house, candle
Ant crossing	Provide positional directions	6x6 grid	8x8 grid	10x10 grid
Falling shapes	Identify and match named shapes	2D and 3D shapes	3D shapes	More 3D shapes
Treasure Measure	Order objects by weight and capacity	Order 2 shapes/ containers	Order 3 shapes/ containers	Order 4 shapes/ containers

Animated songs

Title	Objective
Hoopy Zoo	Adding one more to 5 (increases a number sequence by 1)
Fingers and toes	Counting to 10
Wobbly tower	Counting back from 10
Draw some numbers with me	Writing numbers 1 to 10
Shapes song	Naming shapes
My favourite toy	Using position words including in, on, under and behind

Mathematical tools

Tool	Description
Counting tool	Use the counting tool to assist with counting from 1-20.
Tree tool	Use real-life objects to practice addition and subtraction.
Dice tool	Demonstrate the concept of probability or use the tool in conjunction with activities which require dice.
Number line	Use the number line to assist with counting and calculations.
Spinner	Demonstrate the concept of probability and making a calculated estimate.
Number cards	Display sets of numbers and calculations on movable number cards.
Place value	Explore how multiple-digit numbers are made up of millionths, hundreds of thousands, tens of thousands, thousands, hundreds, tens, units, tenths, hundredths and thousandths.
Base ten	Demonstrate the relationship between units, tens, hundreds and thousands.
Bead sticks	Assist with counting, and show the relationship between thousands, hundreds, tens, units, tenths, hundredths and thousandths.
Fraction wall	Demonstrate fractions visually with the fully customisable fraction wall.
Snake fraction tool	Demonstrate fractions visually with the fully customisable fraction snake.
Number square	Demonstrate and explore counting and number patterns.
Function machine	Demonstrate one- and two-step calculations with this animated tool.
Beads and laces	Create repeating patterns with the beads and laces tool.
Pattern tool	Create patterns and sequences of shapes and design a jumper.
Shape set	Use the shape set to compare 2D and 3D shapes.
Carroll diagram	Demonstrate how to classify and group data with this sorting tool.
Venn diagram	Demonstrate how to classify and group data with this sorting tool. Import images to give a real-life context.

Tool	Description
Pictogram	Demonstrate the creation and interpretation of pictograms.
Money	Practise counting money, or doing calculations involving money.
Clock	Demonstrate the features of analogue and digital clocks, and explore time.
Thermometer	Demonstrate how to measure and record temperature, using either Fahrenheit and Celsius scales.
Capacity	Demonstrate how to measure capacity with this range of animated water containers.
Weighing	Demonstrate weight by weighing objects of different masses on a range of animated weighing scales.
Nets	Explore the nets of 3D shapes such as cones, cylinders, cubes, cuboids, prisms, pyramids and prisms.
Rotate and reflect	Demonstrate the rotation and reflection across a horizontal, vertical and diagonal mirror line of a range of 2D shapes.

Further resources and ideas are available from collins.co.uk/PrimaryMathsforScotland/freeresources.

Primary Maths for Scotland Early Level Problem Solving Pack

This pack is available to support your pupils to develop the numeracy and mathematics skills from the Education Scotland Benchmarks. We advise that after a topic has been covered with a class or group that you then use the relevant problem with that class/group a week or so after the topic has ended. The Problem Solving Pack is designed to be used flexibly to fit with the curriculum order the pupils are following.

Primary Maths for Scotland Early Level Assessment Pack

Assessment Packs are available to support you to make assessment decisions on each pupil's progress through the numeracy and mathematics curriculum. The Early Level Assessment pack includes Yearly Progress checks that can be used throughout the year to formatively assess ongoing progress and End of Level assessments that cover all the Experiences and Outcomes of the Broad General Education Numeracy and Mathematics curriculum.

We welcome your thoughts and feedback, please contact the team here: leckiescotland@harpercollins.co.uk

Suggested Teaching Plans

In the early years children will direct the learning through their play choices, so there is no sequence given here. Rather the practitioner should look to introduce the teaching ideas as part of play and during themes or events that arise as the year unfolds.

Below is the scope and sequence for Primary 1. This sequence is designed so that the pupil progressively builds the skills and knowledge required. Skills learned early in the year will support later topics. As new topics are introduced prior skills and knowledge will need to be revised and misconceptions exposed through discussion and diagnostic tasks. As the year continues pupils will benefit from regular spaced practice to aid memorisation of techniques, vocabulary and facts.

Primary Maths for Scotland Early Level P1 Suggested Teaching Sequence

Chapter	Block	Units
Data and Analysis	1	B11.1
Number Recognition and Place Value	2	B2.1 - B2.6
Time	3	B6.1 - B6.2
Number Recognition and Place Value	4	B2.7 - B2.9
Number and Number Processes – Four Operations	5	B3.1 - B3.2
Estimation and Rounding	6	B1.1 - B1.4
Number Recognition and Place Value	7	B2.10 - B2.17
Patterns and Relationships	8	B8.1 - B8.2
Time	9	B6.3
Measurement	10	B7.1 - B7.2
Data and Analysis	11	B11.2 - B11.4
Money	12	B5.1 - B5.2
Fractions	13	B4.1 - B4.4
Number Recognition and Place Value	14	B2.18 - B2.21
Number and Number Processes – Four Operations	15	B3.3 - B3.6
Angle, Symmetry and Transformation	16	B10.1 - B10.2
Properties of 2D Shapes and 3D Objects	17	B9.1 - B9.5
Estimation and Rounding	18	B1.5 - B1.6
Fractions	19	B4.5 - B4.9
Number and Number Processes – Four Operations	20	B3.7 - B3.12

Chapter	Block	Units
Number and Number Processes – Four Operations	21	B3.13 - B3.15
Number and Number Processes – Four Operations	22	B3.16 - B3.17
Money	23	B5.3 - B5.4
Number and Number Processes – Four Operations	24	B3.18 - B3.19
Time	25	B6.4 - B6.6
Fractions	26	B4.10 - B4.14

Full Early Level Scope and Sequence

Book	Chapter	Content	Units	Benchmark/Es&Os
Early	A1. Estimation and Rounding	By the end of Nursery • Uses subitising to compare the size of objects in a group. • Use observation/ comparison skills to compare without counting – e.g. area taken up/ amount jar filled. • Uses language to compare the size of groups of objects. • In play and other contexts talk about number problems throughout.	• 1.1 Use 'bigger' 'greater', 'smaller' and 'the same as', 'more than', 'less than' 'fewer than' to describe differences between small collections of similar objects and between easily compared amounts. • 1.2 Encourage children to use numbers to compare sets rather than just matching items one to one. • 1.3 Make predictions to say if a group will get bigger or smaller depending on the change proposed. • 1.4 Create collections of equal size by one-to-one matching. • 1.5 Make collections equal sized by moving objects from one group to another. • 1.6 Compare collections by pairing if close in size and amount of items is out of their range. • 1.7 Use idea of 'stopping numbers' to work out how many things are missing/ extra in a collection. E.g. If we're tidying away 5 trucks and we have 4, we still have to find one more.	**MNU 0-01a** • Recognises the number of objects in a group, without counting (subitising) and uses this information to estimate the number of objects in other groups. • Checks estimates by counting. • Demonstrates skills of estimation in the contexts of number and measure using relevant vocabulary, including less than, longer than, more than and the same. I am developing a sense of size and amount by observing, exploring, using and communicating with others about things in the world around me.
	B1. Estimation and Rounding	By the end of P1 • Uses subitising to compare the size of objects in a group. • Uses language to compare the size of groups of objects. • Check estimates by counting or by using benchmarks. • Apply skills in estimation in other contexts and use appropriate language.	• 1.1 Make predictions/ estimates based on the size of one group to say how many are in another group. • 1.2 Estimate the size of one collection by counting or comparing it to the size of a known group (subitising). • 1.3 Apply estimation skills in other contexts using appropriate language. • 1.4 Use who has the most, least fewest, same as, is equal to. • 1.5 Use idea of 'stopping numbers' to work out how many things are missing/ extra in a collection.	

Book	Chapter	Content	Units	Benchmark/Es&Os
	A2. Number and Number Processes – Number Recognition and Place Value	By the end of Nursery • Introduced to the concept of zero. • Recognise and use marks to represent the symbol for zero. • Recite the forwards number word sequence (FNWS). • Recite the backwards number word sequence (BNWS). • Beginning to identify the number before and after. • Ordinal numbers. • Use language to identify before, after in-between. • Order numbers forwards and backwards. • Identify and recognise numerals. • Use one-to-one correspondence to count a collection of objects. • Know that the total is unaffected by the order in which we count a collection of objects, or the place at which we start (conservation of number). • Know that the total is unaffected by the way in which the objects are arranged (conservation of number). • Know that the last number said is different to the others since it both 'labels' the last object and tells how many there are in the whole group. • They also know that the number in a set will remain constant as long as no items are added to the set, or taken from the set.	• 2.1 FNWS to 10 • 2.2 BNWS to 10 • 2.3 FNWS to 20 • 2.4 Begin to identify the number before (0-10) and may use a drop back to one strategy. • 2.5 Begin to identify the number after (0-10) and may use a drop back to one strategy. • 2.6 Identify and recognise numerals including zero (0-10). • 2.7 Mark make to represent numbers, know how to record some numbers 0-10. • 2.8 Sequence and order numerals (0-10) forwards and backwards. • 2.9 Identify missing numbers in a counting sequence (0-10). • 2.10 Use language to identify before, after in-between. • 2.11 Know that the total is unaffected by the order in which we count a collection of objects, or the place at which we start (conservation of number). • 2.12 Know that the total is unaffected by the way in which the objects are arranged or if they are hidden behind or covered up (not imaging just -conservation of number). • 2.13 Understands the abstract principle, e.g. the '3ness' of a number. • 2.14 Beginning to understand a group gets bigger when we add more objects to it.	**MNU 0-02a** • Explains that zero means there is none of a particular quantity and is represented by the numeral 0. • Recalls the number sequence forwards within the range 0 - 30, from any given number. • Recalls the number sequence backwards from 20. • Identifies and recognises numbers from 0 to 20. • Orders all numbers forwards and backwards within the range 0 - 20. • Identifies the number before, the number after and missing numbers in a sequence within 20. • Uses one-to-one correspondence to count a given number of objects to 20. • Identifies 'how many?' in regular dot patterns, for example, arrays, five frames, ten frames, dice and irregular dot patterns, without having to count (subitising). • Groups items recognising that the appearance of the group has no effect on the overall total (conservation of number). • Uses ordinal numbers in real life contexts, for example, 'I am third in the line'.

Book	Chapter	Content	Units	Benchmark/Es&Os
		• Understand that a quantity can be represented by different things (abstract).	• 2.15 Beginning to understand a group gets smaller when we take objects away. • 2.16 Use one-to-one correspondence to count a collection of objects and to count out a required number of items. ○ Different sizes, colours/ sets ○ Actions, sounds, words ○ Things that cannot be moved, i.e. pictures in book • 2.17 Can count items in a row forwards and backwards. • 2.18 Can join or separate two collections in the 1-10 range. • 2.19 Can use finger patterns to keep track of temporal patterns. • 2.20 First 'throw' then 'grow' finger patterns to 1-5, 10 and some small doubles, 1 + 1,2+2. • 2.21 Subitise regular and irregular dot patterns to 5, dice patterns and dominoes. • 2.22 Know that the last number said is different to the others since it both 'labels' the last object and tells how many there are in the whole group. • 2.23 Ordinal numbers, 1st, 2nd, 3rd.	• Uses the language of before, after and in-between. • When counting objects, understands that the number name of the last object counted is the name given to the total number of objects in the group. I have explored numbers, understanding that they represent quantities, and I can use them to count, create sequences and describe order.

Book	Chapter	Content	Units	Benchmark/Es&Os
	B2. Number and Number Processes – Number Recognition and Place Value	By the end of P1 • Understand the concept of zero. • Recognise and use the symbol for zero. • Recite the forwards number word sequence (FNWS). • Recite the backwards number word sequence (BNWS). • Identify the number before and after. • Ordinal numbers. • Use language to identify before, after in-between. • Order numbers forwards and backwards. • Identify, recognise number names and numerals and records them. • Learn to skip count, 2s, 5s and 10s. • Use skip counting to work out how many 2s, 5s and 10s. • Use one-to-one correspondence to count a collection of objects.	• 2.1 BNWS to 20 (alter start no.) • 2.2 FNWS to 30 (alter start no.) • 2.3 Recognise, identify and record numerals including zero (0-20). • 2.4 Identify the number before (0-20) without dropping back to one. • 2.5 Identify the number after (0-20) without dropping back to one. • 2.6 Explain what zero represents. • 2.7 Sequence and order numerals (0-20) forwards and backwards. • 2.8 Introduce regular groupings to 10. Dice and domino patterns, irregular patterns and five frames, ten frames- five wise and pair wise and conceptual subitising patterns to 10. • 2.9 'Throw' finger patterns to 1-5, 10 and doubles up to 5 + 5 and 'grow' patterns 6-9. • 2.10 Partition 'to 5' and 'within 5' as number facts. • 2.11 Show '5 and' facts on concrete materials. • 2.12 Identify missing numbers in a counting sequence (0-20). • 2.13 Ordinal numbers, 1st – 10th. • 2.14 Begin Skip count in 2s, 5s and 10s up to 20, 50 and 100 respectively. • 2.15 Know a group gets bigger when we add more objects to it.	

Book	Chapter	Content	Units	Benchmark/Es&Os
			• 2.16 Know a group gets smaller when we take objects away. • 2.17 Understands that as one moves up the counting sequence, the quantity increases by one and the quantity decreases by one when one moves down or backwards in a sequence. • 2.18 Use one-to-one correspondence to count a collection of objects.	
	A3. Number and Number Processes – Four Operations	By the end of Nursery • Subitise (regular and irregular patterns). • Doubles patterns. • Learn to explore and talk about quantity using appropriate vocabulary, e.g. more, less, makes, add, take away. • Use concrete materials to represent addition and subtraction. • Solve story problems – throughout whole organiser.	• 3.1 Subitise flashed patterns to 5 (dice, five frames and irregular/ random). • 3.2 Learn to explore and talk about quantity using appropriate vocabulary, e.g. more, less, makes, add, take away, altogether. • 3.3 Addition and subtraction by joining sets using concrete materials in range 0-10 starting count at one. • 3.4 Know that the last number said is different to the others since it both 'labels' the last object and tells how many there are in the whole group. • 3.5 Doubles patterns 1+1,2+2.	**MNU 0-03a** • Partitions quantities to 10 into two or more parts and recognises that this does not affect the total. • Adds and subtracts mentally to 10. • Uses appropriately the mathematical symbols +, – and =. • Solves simple missing number problems. • Counts on and back in ones to add and subtract. • Doubles numbers to a total of 10 mentally. I use practical materials and can 'count on and back' to help me understand addition and subtraction, recording my ideas and solutions in different ways.

Book	Chapter	Content	Units	Benchmark/Es&Os
	B3. Number and Number Processes – Four Operations	By the end of P1 • Counting on and back (prior to addition and subtraction). • Subitise (regular and irregular patterns). • Doubles (patterns then facts). • Addition and subtraction using: count all on concrete materials; count all using imaging; count on and count back. • Partition numbers (within 5 then up to 10). • Know all partitions to 10. • Learn to explore and talk about quantity using appropriate vocabulary, e.g. more, less, makes, add, take away. • Use mathematical symbols +, -, = • Use concrete materials to represent addition and subtraction. • Fact families. • Add and subtract mentally using known facts. • Solve story problems. • Solve missing number problems.	Throughout use story problems represented in different ways • 3.1 Conceptual Subitising patterns to 10 (dice, 5-wise, pair-wise, and irregular). • 3.2 Odd and even numbers based on patterns. • 3.3 Doubles number facts to 10. • 3.4 Fact families 'to five' and 'within 5'. • 3.5 Use mathematical symbols +, -, = • 3.6 Addition and subtraction using: count all on concrete materials count all using imaging. • 3.7 Say the next 2,3 or 4 numbers in a sequence (FNWs – 30 and BNWS -20). • 3.8 Can count on/back from an unscreened numeral to locate a numeral on a screened numeral track. • 3.9 Can count the jumps from one unscreened number to a screened numeral on a number track. • 3.10 Addition using: count on, using concrete materials. • 3.11 Addition using: count on, using screened concrete materials. • 3.12 Addition using: count on, by imaging. • 3.13 Subtraction using: count back using concrete materials.	

Book	Chapter	Content	Units	Benchmark/Es&Os
			• 3.14 Subtraction using: count back, using screened concrete materials. • 3.15 Subtraction using: count back, by imaging. • 3.16 Fact families to 5 (then 10). • 3.17 Know partitions to 10. • 3.18 Add and subtract mentally using known facts (facts to 5, doubles to 10). • 3.19 Solve very simple missing number problems.	
	A4. Fractions	By the end of Nursery • Sharing collections and objects (whole). • 'Fair' shares. • Sharing into halves. • Language of fractions.	• 4.1 Equal/fair shares. • 4.2 Can break a whole part into parts and describe how they made it fair. • 4.3 Understands that for shares to be equal a quantity may remain. • 4.4 Begin to use the language of fractions in everyday situations (e.g. an equal piece, an equal part, a fair share). This is largely modelled by an adult. • 4.5 Begin to recognise that when an object is divided into two equal parts each part is called one-half.	**MNU 0-07a** • Splits a whole into smaller parts and explains that equal parts are the same size. • Uses appropriate vocabulary to describe halves. • Shares out a group of items equally into smaller groups. I can share out a group of items by making smaller groups and can split a whole object into smaller parts.

Book	Chapter	Content	Units	Benchmark/Es&Os
	B4. Fractions	By the end of P1 • Sharing collections and objects (whole). • 'Fair' shares. • Sharing into halves. • Sharing into quarters. • Combining halves to make a whole. • Combining quarters to make a whole.	• 4.1 Begin to use the language of fractions in everyday situations (e.g. an equal piece, an equal part, an equal share, half, quarter). • 4.2 Can break a whole into equal parts. • 4.3 Make a whole from halves. • 4.4 Can understand the term 'half way' and 'half of' and find by practical means halves and quarters of basic shapes/lengths. • 4.5 Identify half of a whole region/set/length and can 'see' by matching if both parts are equal in size or can use a counting strategy to check. • 4.6 Knows if the 2 parts are not equal, they cannot be called halves. • 4.7 Recognises the notation for one half and one quarter. • 4.8 Share into quarters. • 4.9 Find the quarter of a whole. • 4.10 Understands that one quarter can be made by halving the half. • 4.11 Begin to make a whole from quarters. • 4.12 Find half of a quantity. • 4.13 Find a quarter of a quantity. • 4.14 Problems where the group of items does not share equally and there are 'left-overs'.	

Book	Chapter	Content	Units	Benchmark/Es&Os
	A5. Money	By the end of Nursery • Money play. • Identifies coins to at least £2. • Appropriate vocabulary, e.g. total, cost, amount, change.	• 5.1 Money play situations. • 5.2 Name coins 1-10p. • 5.3 Order coins by value.	**MNU 0-09a** • Identifies all coins to £2. • Applies addition and subtraction skills and uses 1p, 2p, 5p and 10p coins to pay the exact value for items to 10p. I am developing my awareness of how money is used and can recognise and use a range of coins.
	B5. Money	By the end of P1 • Money play. • Identifies coins to at least £2. • Addition and subtraction using coins to 10p. • Appropriate vocabulary, e.g. total, cost, amount, change.	• 5.1 Name all coins. • 5.2 Know the value of all coins. • 5.3 Use 1-10p coins to buy items (applying correct language for money). • 5.4 Use addition and subtractions for money purchases.	
	A6. Time	By the end of Nursery • Language of time for everyday events. • Link events in their daily routines to specific times, e.g. morning, afternoon, lunchtime, night-time. • Days of the week. • Months. • Seasons of the year. • Time devices, e.g. visual timetables, hand claps, sand timers, clocks, calendars, etc. • Clock faces.	• 6.1 Daily routines. • 6.2 Use language to describe time, e.g. before, after, earlier, later. • 6.3 Days of the week introduction. • 6.4 Seasons. • 6.5 Months introduction (birthdays). • 6.6 Describe and use time devices.	**MNU 0-10a** • Links daily routines and personal events to time sequences. • Names the days of the week in sequence, knows the months of the year and talks about features of the four seasons in relevant contexts. I am aware of how routines and events in my world link with times and seasons, and have explored ways to record and display these using clocks, calendars and other methods.

Book	Chapter	Content	Units	Benchmark/Es&Os
	B6. Time	By the end of P1 • Language of time for everyday events. • Link events in their daily routines to specific times, e.g. morning, afternoon, lunchtime, night-time. • Days of the week. • Months. • Seasons of the year. • Time devices, e.g. visual timetables, hand claps, sand timers, clocks, calendars, etc. • Analogue time hours. • Analogue time half hours. • Clock faces - Digital equivalents.	• 6.1 Know sequence of days of the week. • 6.2 Know sequence of the months of the year. • 6.3 Time to the hour. • 6.4 Time to the half hour. • 6.5 Drawing time on an analogue clock. • 6.6 Digital equivalents.	• Recognises, talks about and where appropriate, engages with everyday devices used to measure or display time, including clocks, calendars, sand timers and visual timetables. • Reads analogue and digital o'clock times (12 hour only) and represents this on a digital display or clock face. • Uses appropriate language when discussing time, including before, after, o'clock, hour hand and minute hand.
	A7. Measurement	By the end of Nursery • Language of measurement for size. • Comparative language. • Apply language for different forms of length. • Opposites. • Orders objects by size, length, mass, capacity. • Describes objects size, length, mass, capacity. • Experiences situations where measurement is used. • Use non-standards units of measurement for objects size, length, mass, capacity.	For length, mass and capacity: • 7.1 Experiences situations where measurement is used, e.g. baking. • 7.2 Comparative language. • 7.3 Opposites. • 7.4 Use non-standard units.	**MNU 0-11a** • Shares relevant experiences in which measurements of lengths, heights, mass and capacities are used, for example, in baking. • Describes common objects using appropriate measurement language, including tall, heavy and empty.

Book	Chapter	Content	Units	Benchmark/Es&Os
	B7. Measurement	By the end of P1 • Language of measurement for size. • Comparative language. • Apply language for different forms of length. • Opposites. • Orders objects by size, length, mass, capacity. • Describes objects size, length, mass, capacity. • Experiences situations where measurement is used. • Use non-standards units of measurement for objects size, length, mass, capacity. • Estimates and measures using non-standards units of measurement for objects size, length, mass, capacity.	• 7.1 Use descriptive language for measurement. • 7.2 Estimates and measures using non-standards units of measurement.	• Compares and describes lengths, heights, mass and capacities using everyday language, including longer, shorter, taller, heavier, lighter, more and less. • Estimates, then measures, the length, height, mass and capacity of familiar objects using a range of appropriate non-standard units. I have experimented with everyday items as units of measure to investigate and compare sizes and amounts in my environment, sharing my findings with others.
	A8. Patterns and Relationships	By the end of Nursery • Continues patterns. • Uses language to describe patterns. • Reproduce the same patterns in different media or with different objects. • Recognise and continue simple number patterns. • Recognise, discuss, duplicate and create simple numeric sequences.	For objects, shapes, numbers: • 8.1 Copy simple patterns. • 8.2 Continue simple patterns. • 8.3 Create simple patterns. • 8.4 Develops the language to describe patterns.	**MTH 0-13a** • Copies, continues and creates simple patterns involving objects, shapes and numbers. • Explores, recognises and continues simple number patterns. • Finds missing numbers on a number line within the range 0 - 20. I have spotted and explored patterns in my own and the wider environment and can copy and continue these and create my own patterns.

Book	Chapter	Content	Units	Benchmark/Es&Os
	B8. Patterns and Relationships	By the end of P1 • Continues patterns. • Uses language to describe patterns. • Reproduce the same patterns in different media or with different objects. • Recognise and continue simple number patterns. • Recognise, discuss, duplicate and create simple numeric sequences. • Find missing numbers on a number line within the range 0-20.	• 8.1 Create and describe patterns. • 8.2 Find missing number on a number line.	
	A9. Properties of 2D Shapes and 3D Objects	By the end of Nursery • Describe the properties of shapes and objects using informal language. • Use the names of shapes and objects. • Describe the properties of shapes and objects using mathematical language. • Find similarities between shapes.	• 9.1 Describe the properties of shapes and objects using informal language. • 9.2 Sort shapes and objects by describing features.	**MTH 0-16a** • Recognises, describes and sorts common 2D shapes and 3D objects according to various criteria, for example, straight, round, flat and curved. I enjoy investigating objects and shapes and can sort, describe and be creative with them.
	B9. Properties of 2D Shapes and 3D Objects	By the end of P1 • Describe the properties of shapes and objects using informal language. • Use the names of shapes and objects. • Describe the properties of shapes and objects using mathematical language. • Find similarities between shapes. • Use properties to describe and match shapes and objects and justify choice.	• 9.1 Recognise and name common shapes. • 9.2 Recognise and name common objects. • 9.3 Recognise the properties of shapes and objects using mathematical language. • 9.4 Describe the properties of shapes and objects using mathematical language. • 9.5 Use properties to describe, sort and/or match shapes and objects and justify choice.	

Book	Chapter	Content	Units	Benchmark/Es&Os
	A10. Angle, Symmetry and Transformation	By the end of Nursery • Give simple directions. • Use more complex language to give directions. • Use language to describe position. • Use language to describe the relative position of objects. • Identify reflections. • Identify/describe symmetry in the environment. • Create and identify symmetrical pictures or arrangements of concrete materials.	• 10.1 Give basic direction, 'the box is in the corner'. • 10.2 Begin to understand the language of position and direction, including in front, behind, above, below, left, right, forwards and backwards in movement games. • 10.3 Identify reflections. • 10.4 Identify/describe symmetry in the environment.	**MTH 0-17a** I have had fun creating a range of symmetrical pictures and patterns using a range of media. **MTH 0-19a** • Understands and correctly uses the language of position and direction, including in front, behind, above, below, left, right, forwards and backwards, to solve simple problems in movement games. • Identifies, describes and creates symmetrical pictures with one line of symmetry. In movement, games, and using technology I can use simple directions and describe positions.
	B10. Angle, Symmetry and Transformation	By the end of P1 • Give simple directions. • Use more complex language to give directions. • Use language to describe position. • Use language to describe the relative position of objects. • Identify reflections. • Identify/describe symmetry in the environment. • Create and identify symmetrical pictures or arrangements of concrete materials.	• 10.1 Use language to describe the relative position of objects. • 10.2 Create and identify symmetrical pictures or arrangements of concrete materials.	

Book	Chapter	Content	Units	Benchmark/Es&Os
	A11. Data and Analysis	By the end of Nursery • Collect objects and sort them. • Match objects and talk about the similarities and differences. • Collect data. • Pose questions to answer a problem or situation. • Display data. • Use technology to display data. • Recognise signs and charts in a familiar context. • Recognise signs and charts outside the immediate environment. • Use signs and charts to plan and make decisions.	• 11.1 Collect objects and sort them. • 11.2 Match objects and talk about the similarities and differences. • 11.3 Recognise signs and charts in a familiar context. • 11.4 Collect data by grouping items/objects together. • 11.5 Display data by creating a picture using the concrete materials collected. • 11.6 Ask simple questions to collect data. • 11.7 Recognise signs and charts outside the immediate environment.	**MNU 0-20a** I can match objects, and sort using my own and others' criteria, sharing my ideas with others. **MNU 0-20b** • Asks simple questions to collect data for a specific purpose. • Collects and organises objects for a specific purpose. • Applies counting skills to ask and answer questions and makes relevant choices and decisions based on the data. • Contributes to concrete or pictorial displays where one object or drawing represents one data value, using digital technologies as appropriate. • Uses knowledge of colour, shape, size and other properties to match and sort items in a variety of different ways. • Interprets simple graphs, charts and signs and demonstrates how they support planning, choices and decision making. I can collect objects and ask questions to gather information, organising and displaying my findings in different ways.
	B11. Data and Analysis	By the end of P1 • Collect objects and sort them. • Match objects and talk about the similarities and differences. • Collect data. • Pose questions to answer a problem or situation. • Display data. • Use technology to display data. • Recognise signs and charts in a familiar context. • Recognise signs and charts outside the immediate environment. • Use signs and charts to plan and make decisions.	• 11.1 Use signs and charts to plan, choose and make decisions. • 11.2 Collect data by counting one-to-one. • 11.3 Display data by creating a picture or using technology. • 11.4 Ask simple questions to collect data and make suggestions how to do this.	

1 Estimation and Rounding

Estimation and Rounding A

Section overview

Units of learning	Review units	Prior learning	Benchmarks/Es&Os
Unit 1.1 **Comparing collections visually** Comparing sets visually if there is a large numerical difference.		• In familiar settings do they understand what the concept 'more' means? For example, to ask for more food if they are hungry. • Experience of making collections.	**MNU 0-01a** • Recognises the number of objects in a group, without counting (subitising) and uses this information to estimate the number of objects in other groups. • Checks estimates by counting. • Demonstrates skills of estimation in the contexts of number and measure using relevant vocabulary, including 'less than', 'longer than', 'more than' and 'the same'. I am developing a sense of size and amount by observing, exploring, using and communicating with others about things in the world around me.
Unit 1.2 **Comparing and describing collections by matching** Comparing concrete materials by pairing or lining up one-to-one.	**Estimation and Rounding A** Units 1.1, 1.2, 1.3, 1.4, 1.5, 1.6 and 1.7	• One-to-one correspondence when counting objects. • Experience of making collections. • Understanding of the concept 'more than'.	
Unit 1.3 **Comparing small sets by grouping and counting** Comparing concrete sets (1-5 range) by using subitising patterns and starting to use counting to check estimates.	**Estimation and Rounding A** Units 1.1, 1.2, 1.3, 1.4, 1.5, 1.6, 1.7, 1.9 and 1.17	• One-to-one correspondence when counting objects. • Experience of making collections. • Understanding of the concept 'more than' and 'the same'. • Can subitise up to 5 (e.g. regular dice patterns). • Confidently count and order numbers to 5.	
Unit 1.4 **Making equal collections** Comparing two concrete sets and moving or adding objects to make them equal.	**Estimation and Rounding A** Units 1.1, 1.2, 1.3, 1.4, 1.5, 1.6, 1.7, 1.9 and 1.17	• One-to-one correspondence when counting objects. • Experience of making collections and understanding that numbers can count anything. • Understanding of the concept 'more than' and 'the same'. • Can subitise up to 5 (e.g. regular dice patterns). • Confidently count and order numbers to 5.	
Unit 1.5 **Comparing and describing measures** Visually notice the differences between measures.	**Estimation and Rounding A** Unit 1.1 **An Introduction to Measurement A** Units 7.1, 7.2, 7.3 and 7.4	• Understanding of the concepts 'more than', 'less than' and 'the same'. • Understanding of big, small and little.	
Unit 1.6 **Reasonable estimates** Begin to understand that some things are reasonable and others are not.	**Estimation and Rounding A** Units 1.1, 1.2, 1.3, 1.4, 1.5, 1.6, 1.7, 1.9 and 1.17	• Understanding of the concept 'more than' and 'the same'. • Make observations based on their experiences, e.g. granny is older than they are, mummy is not smaller than a baby.	

Conceptual context

If we hope for reasonable responses, rather than wild guesses, it is important to nurture children's understanding of estimation from a very young age.

Comparing two concrete collections which are very obviously different in size should be encouraged even before a child can use counting to check. Strategies such as physically pairing and lining items up help develop an understanding of the concepts of 'more', 'less' and 'the same'. Using benchmarks such as subitising or groupings of regular patterns, e.g. five frames, are very valuable tools which also encourage comparisons without counting. It is vital that children experience collections that have 'like' and 'unlike' items so that the numerosity of the collection is the focus and not other attributes such as size, colour or themes. For example, you should be using different fruits not just apples. The children need to be given lots of opportunities to collect and describe concrete collections.

When children compare the groups by matching objects on a one-to-one basis, they should:

- Physically pair items up by taking an item from each collection one at a time. Then encourage them to consider what it means to have some left over in one set. Point this out to them, e.g. This means this collection is bigger as it has more counters left over.

or

- Place the items in a line or on tens frames next to each other so that the size of each collection can be compared. It is important that one line is not more spaced out than another.

When considering the collections for comparing make sure there are some that are equal.

Consider collections with exactly the same attributes and others that are mixed in terms of:

- size
- colour
- pattern
- mixed within the same theme, e.g. different animals not just cows.

Real life and familiar situations for the children are an excellent way to introduce the language of comparisons and estimation in context. Most children at nursery age are familiar with the concept of wanting more, having enough or having a little bit of something, e.g. food or toys. To help develop the trickier concept of 'less/fewer than' the relationship between the two should always be modelled clearly, e.g. 'John, you have more grapes than Ben so that means Ben has fewer grapes than you, John'.

Using the correct vocabulary is an important aspect of this chapter and children need varied experiences estimating and comparing both numbers and measures. When comparing measures, it is important to choose the correct vocabulary. They are developing their comparing skills, being able to say 'This one has more than/is taller than/is lighter than that one'.

Opportunities for 'having a good guess' and asking for their opinions will also help develop their sense of something being reasonable.

Asking a child to estimate and compare collections by counting is only useful if they can actually count up to the number in the collection. For some children this will be a small number, while for others it may be up to 30. They must also have a strong understanding of the cardinal value of numbers, otherwise they will not realise that the last number said determines the number in the set.

Resources

- Interlocking cubes
- Five frames – different sizes
- Sorting hoops and hula hoops
- Building bricks
- Sorting bears
- Finger pattern cards
- Regular dice pattern cards
- Spoons – various sizes
- Coloured counters
- Number/numeral tracks
- Posting boxes
- Numeral cards
- Collections, e.g. dinosaurs, toy cars
- Estimation jars
- Containers – different sizes
- Irregular dot cards for matching

Digital

- Fishy Facts

Record book

Pages 3–5

Key vocabulary

- **More than/a little bit more**
- **Fewer than** – for countable quantities, e.g. There are **fewer** bananas **than** apples.
- **Less than** – for non-countable quantities, e.g. There is **less** water in my glass **than** in the pond.
- **The same as**
- **How many?**

Misconceptions

Potential misconception	Addressing the misconception
Unit 1.3 They may not realise that the final number that they say represents the final set. When pointing to 4 items they may have to start their count from one again. They think they need to say the string of numbers 1 to 4, not just the final count of 4.	• Lots of opportunities to collect and count small collections using a wide range of objects. Model the final count. • Evidence in environment and displays of value associated with a number. • Practise forward number sequence up to 10. • One-to-one correspondence practice – they touch, say and move each item in turn and only once!
Units 1.3 Many children do not understand that each counting number is one more than the previous number, e.g. they guess that if 6 bears go into cave and then 2 more, there are 3 altogether. They do not understand that numbers get larger the further you go up in order.	• Building staircases using cubes or number rods can help develop this understanding. • Opportunities to help children recognise that if they add one more they will get the next number in the sequence and if they take one away they will get the previous number.
Units 1.1, 1.2, 1.3, 1.4, 1.5 and 1.6 The language used can be confusing for children.	• Lots of practical experiences to practise describing the variety of different contexts of length comparisons. • Adults modelling the appropriate language and asking children to copy the statement, e.g. 'The box is heavier than the pencil. Now you say it.' • Correct inaccurate use of language, e.g. 'The man is longer than the boy.'
Units 1.1, 1.2, 1.3, 1.4 and 1.5 Children of this age may not understand that two groups can be compared	• Children will need lots of concrete experience of physically manipulating objects into collections and sub-sets and combining small sets to make larger ones. • Concrete materials that can be manipulated must be used before visuals that cannot be moved, i.e. comparing pictures in a book.
Units 1.1, 1.2, 1.3 and 1.4 Some children consider the size of an item instead of the number of items when they are comparing the numerosity of the group.	• Pair up the items so children can see that it is the number of items that is important, not the size, e.g. one tennis ball is the same quantity as one golf ball.
Unit 1.3 They may not understand that when counting items, each item can only be counted once.	• Use practical resources and demonstrate moving items from an uncounted place to a counted place, to ensure that they are not counted more than once. Move each item as it is counted. Encourage the child to describe the completed count, e.g. '1, 2, 3, 4, 5. There are 5 cubes'.
Units 1.1, 1.2, 1.3, 1.4, 1.5 and 1.6 Look out for children who do not change their first estimate even after checks have been made to check for reasonableness.	• Wild guesses should be explored and the reasonableness of the answer discussed. Remember that children cannot be expected to give reasonable estimates above the range that they can count. Reduce the size of the collection to within the child's capabilities.
Units 1.1, 1.2, 1.3, 1.4, 1.5 and 1.6 They may label the set with the wrong number, but this does not mean they cannot recognise the comparisons between the sets.	• Ask them whether the collection is more than the other one instead of asking for an estimate of 'how many' there are in the collection. Familiar things are a good place to start, food in particular.

Teaching ideas

Unit 1.1 Comparing collections visually

Whole class

- While singing the song 'Ten in the Bed', ask the children to act out the song by getting them to jump off the pretend 'bed' (a blanket on the floor) and onto the bare floor. Stop at different points in the song and ask the children to describe what they notice. Ask them if there are more children in the bed or on the floor. If they answer, 'The bed', model the statement, 'There are more children in the bed than on the floor.' Also model the accompanying statement, 'That means there are fewer children on the floor.' **How do they know? How could they check?** They could line up against each other to see which group is bigger. Repeat throughout the song and remember to include when there are an equal amount of children on the floor and on the bed. **What do they notice? What word is best to describe this situation?**
- **Role Playing at Snack Time: Café** – Give the children a tray of small sandwiches (real or pretend) and ask them to place them on the paper plates. Explain that some plates should be for very hungry customers and some for not so hungry. Get them to open the café and ask their customers to choose which plate they would like. If the children are all sitting at one table, it will be easier to ask questions that enable them to compare each other's plates. Ask: **Do you have more than your friend? How do you know? Which plate had the most sandwiches on? Who got the plate with the fewer/fewest on?** Do not expect an amount, more an observation of the space taken on the plate.

Stations

Estimation Station – Here are three different ways to set up an estimation station.

- 1 – Place identical sized cubes into six identical transparent containers. Vary the amount in each container, ensuring the difference between them is noticeable visually. Let the children pick two containers to compare and allow them to pick them up to get a good look. **Which jar has the most cubes? Which jar has the fewest cubes? Can you see a jar with a small number of cubes in? How do you know?** Tell them, 'This jar has the fewest cubes in and this jar has more in.' **Is there another jar that almost has the same amount of cubes in as this one?**
- 2 – Place collections with different themes and varying numbers into identical transparent containers, e.g. 8 pom-poms in one, 2 teddy bears in another, 30 tiny beads in the next, etc. Let the children pick the containers up to allow them to get a good look. Ask: **Which jar has the most things in? Which jar holds the fewest things in?** Explain that 'the fewest' means 'has a smaller number of items'.
- 3 – Keep the collections the same theme throughout or a different theme in each container but change the size of the transparent containers. Ask: **Which container has the most things in? Which container holds the fewest things? What would happen if we put the biggest collection into the smallest container?**

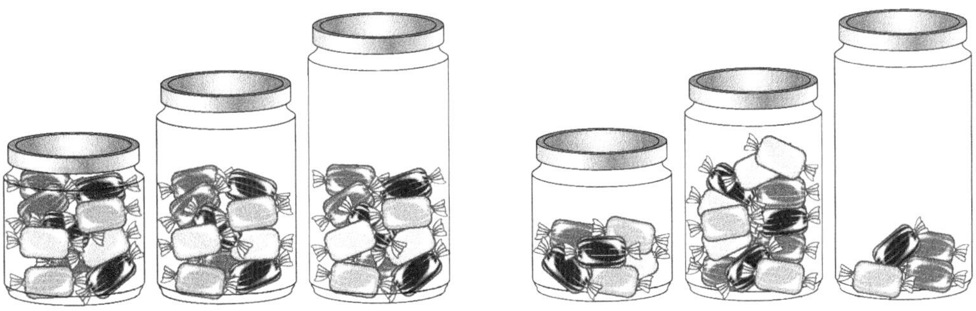

- **Dinosaurs** – Make two piles of dinosaurs in the sand which have an obvious difference in the size of the collection. Ask the children: **Which pile of dinosaurs shows the biggest group? Which is the smallest group? Is one group of dinosaurs bigger than the other? Do both groups have the same number of dinosaurs?** Change the size of both collections and repeat. Sometimes make sure the groups are equal. Can they notice visually without matching that they 'look the same'?

Unit 1.2 Comparing and describing collections by matching

Whole Class

- **Songs** – For example, 'Five Little Speckled Frogs'. Choose five children to be the frogs and ask them to sit on a bench (the log). Place a blue blanket beneath the log and ask the children to notice what happens to the two groups when a frog jumps off the log into the pond. Sing the song and stop to compare the two groups. **What happens if one frog jumps off the log?** Ensure the frogs are lined up underneath the log but paired with the frogs still on the log. **Will there be more or fewer frogs left on the log? How do you know? Are there more/fewer frogs on the log than in the pond?**
- **Stories** – Stories such as *The Very Hungry Caterpillar* by Eric Carle are useful to compare the number of items mentioned in a story. It is best done with hoops or sorting rings and examples of the foods on individual cards (so the three plums will be on three separate pieces of card). Share the food cards out amongst the class and get the children to come out as each day's food is mentioned and place their cards in the appropriate hoop. Once the story is finished the children can be asked to compare the number of items eaten on certain days by taking them out of the hoop and lining them up against each other. Ask: **Which hoop has the most/fewest items of food? Which day did the caterpillar eat more items of food?** Watch out for the children that concentrate on the size of the food not the number. Rich discussion can be had about the number of items compared to the size of the portion.
- **Outdoor Learning: Collecting Fir Cones** – Give the children an equal-sized bucket or bag and tell them they are going to collect as many fir cones as they can in one minute. Once the time is up, ask the children to find a partner. Get them to tip their fir cones onto their own large sheet of paper. **Who do you think has got the most/fewest fir cones? How could we check?** Get your empty buckets ready. Both of you pick up one fir cone. Say, 'I have a fir cone' and then drop it back in your bucket. Repeat until one person has no more cones left. The person that still has some left must have had more to start with.

Stations

- **The Garage** – Have a pile of cars next to the car park (cardboard box with spaces marked out) and the road mat. Ask the children to park some cars in the garage and some on the road. Ask: **Are there more cars in the garage or on the road mat? How could you check?** Bring them down the ramp and line them up next to each other. **What do you notice? Were you correct?**

- **Train Set** – Set up engines with different numbers of carriages. Ask: **Which engine has the most carriages? How do you know?** Take one carriage from each engine and pair them up. Repeat until one engine has no more carriages. Whoever still has some left, had more in the first place. Say, 'This engine had more carriages because there are still more left.'
- **Sandpit** – Hide collections of different coloured jewels in the sand tray. Tell the children they are going to be pirates. Give each pirate a treasure box with an appropriate coloured sticker on the top. Tell them if they can find the same coloured jewels as the sticker on their box, they can put it in their box. After collecting, the children estimate who has the most/fewest by looking in everyone's treasure box. **Who has more treasure than their friend? Do you have lots more or a few more than him? How could you check?** Then line the jewels up in the sand to check.

Unit 1.3 Comparing small sets by grouping and counting

Whole class

- Set out four hula hoops in the hall/playground (two opposite each other). Place 10 bean bags into two of the hoops. Arrange for half the class to stand behind each empty hula hoop. Tell the children they are going to run a relay race to see how many bean bags they can collect in two minutes. Tell them they must take it in turns to run down the hall, pick up a bean bag, then run back and place it into their team's hoop. The next child repeats this action. Once the time is up, the bean bags in both hoops are arranged into dice patterns. Show them a 6 dice pattern and ask them if they have more in the finishing or starting hoop? **How many bean bags did you collect? How many are left in the other hoop? Are there more or fewer in the finishing hoop?** Repeat with different dice patterns.
- **Outdoor Learning** – Hang 10 cm pieces of coloured wool (worms) around the nursery garden. Tell the children they are mummy and daddy birds who are going to find worms for their hungry babies. Ask them to place each worm onto a square on a large laminated five frame. Tell them only one worm can go in each square. **How many worms did you collect? How do you know? Does your friend have more or fewer worms than you?**
- **Tidy Up Time** – Have large painted/laminated five frames on the floor in the nursery to help the children organise the toys and equipment used. Tell them there is space for five toy trucks on the large five frame. Ask the children to collect the trucks and place each one in its own square. **We have two trucks, how many more do we need to find? How many spaces are there left? Have we nearly got all of them? Have we got lots of trucks still to find?** Organise familiar items used in the nursery using regular dot patterns or five frames. **Do we have the correct number of items in each box? Are there not enough? More than enough?**

Stations

- **Farm Set** – Set out the play farm and different pens/fields. Arrange groups of animals into dice patterns around the farm. **Which field has the most animals in? Which pen has the fewest animals?** Take one animal out of the field. **Are there more or fewer animals left in the field?**
- **Water Table** – Have a collection of laminated large dice patterns or irregular dot patterns within 1–5 range. Place a collection of magnetic fish in the water table and ask the children to go fishing with their rods. Tell the children to place their fish on the dot cards. **How many have you caught? Have you caught more than your friend? Who caught the most/fewest? How do you know?** Encourage them to notice whether one collection is more than the other, e.g. 4 is a bigger number than 2, so you have caught the most.

Unit 1.4 Making equal collections

Whole class

- **Making Equal Groups with Small Numbers** – Put the children into groups of three or four. Give each group a small bag/box with some stones in (assorted 1–5 range). Ask the children to tip the stones out and decide whether they have enough for them to each have one. Ask how many stones there are. Ask how many children there are in the group. Say, 'We need the same amount of stones as children.' **Do we have enough? Do we have more or fewer than we need? How will you know if you have enough?** Encourage the children to use counting as a strategy. If there are 4 of you and only 2 stones, you will need 2 more stones.
- **Making Equal Groups Using a Lining Up Strategy** – Place a bundle of sticks on the floor and tell the children they all need to take one stick on their walk. **Do we have enough? Do we need lots more? Do we need a few more? How can we make sure?** Place the sticks out in a line and ask each child to stand next to a stick. Repeat the questions. **Have we got the correct amount now? How do we know when we have enough?** We have the exact number of sticks that we need. The number of sticks we have equals the number of people we have. Make sure that you sometimes have too many/few sticks as well as the exact amount.

Stations

- **Making Equal Groups Using a Pairing Up Strategy** – Place out eight teddies and three toy cars. **Will all the teddies get a car?** Ask the children to pair up each teddy with a car. Ask them to notice how many teddies do not have a car. **How many more cars will we need so that every teddy gets a car? How do you know?** Some children will be able to count the teddies and then go and find that amount of cars, other children will keep adding cars until each teddy has one. Try it with fewer teddies than cars or equal amounts. **Do we have the same amount of teddies as cars?** Say, 'This means they can all drive a car.' **Do we have more teddies than cars?** Say, 'We have fewer cars than teddies.' Ask: **Do we have a lot more teddies than cars?**
- **Making Equal Groups Using Subitising Patterns – dice patterns**. Explain that they are going to feed two rabbits but they must make sure each rabbit gets the same amount of food. Place six lettuce leaves in one bag and two in another. Ask the children to look in the bags and check if they think the bags have the same amount of food in each one. Ask the children to place the lettuce out in a dice pattern. **How can we make the bags equal for the two hungry rabbits? What do we need to do? How can we check?** They could take some lettuce from one group and give it to the other and on other occasions they could add some extra lettuce from a spare bag.

Unit 1.5 Comparing and describing measures

Whole class

- **Height** – Tell the children that you are investigating who is the tallest and shortest person in the nursery. Compare two people at a time. **Who do you think is the tallest/shortest? How can we check?** Make them stand next to each other. **Were you right?** Model the correct statements, e.g. 'Mrs Jones is taller than Alex so that means Alex is shorter than Mrs Jones.' Explain that we do not say 'longer/smaller than'. Repeat with two new people. **Who is the tallest person/child in the room?**

- **Drawing** – Set different sizes of equipment out for a drawing session. Explain to the children they are going to explore the length of things while they are drawing. Ask: **Which pencil is the longest? Which one is the shortest?** Hold up a crayon and ask if they can find one that is longer/shorter. Place them next to each other. **Were you right?** Say, 'This crayon is longer than this one.' Ask if the rubber is smaller or longer than the pencil. Ask them to find a pencil which is the same size as the one you are holding. **How can you check?** Encourage them to place them next to each other. Model the statement, 'These two pencils are the same length.'
- **Capacity** – At the snack table provide the water/juice in two identical sized jugs. **Which jug has the least juice in it? Which jug has the most juice?** Place them next to each other. **Does this jug have a lot more juice or a little more?** The conversation can be held at different times during snack time as the level of the juice changes. This activity could be done with different sized jugs and comparisons made by how many cups can be filled. Ask the children: **Which jug do you think has the most water in?** Ask them to guess how many cups can be filled from each jug. Ask them to test their estimates. **Were you right?** Explain that the cups need to have roughly the same amount of water in them for this to be fair.

Stations

- **Foam Bricks** – Tell the children they are going to be in two teams and they will help one of the Little Pigs build his house of bricks. Explain that the naughty Wolf has hidden the foam bricks around the nursery and they are going to find them and bring them back to their team's pile on the carpet. Let's see who can collect the most bricks. Tell them not to pile them on top of each other. When there are no bricks left to find, ask the children to look at both collections of bricks and guess who has collected the most. Now ask them to pile them up one on top of the other. **Who has the taller pile of bricks? Were you right? How can you check? Which pile has more/fewer?** Compare the height of the two piles by moving them next to each other.
- **Making Collections** – Ask two children to help you bring the snack items into the kitchen. Give them an identical shopping bag each, but make sure each bag is filled with a marked difference in quantity, e.g. apples for snack. One has 4 in and the other has 12. Let the children feel the weight of them both. **Which bag has more in? Which bag is heavier?** Always model the opposite statement, 'This bag has fewer in.' 'This bag is lighter than the other bag.'
- **Capacity** – Fill six transparent plastic bottles of the same size with varying amounts of coloured water; make sure you have an empty, almost empty, nearly half, half, nearly full and full bottle. Mark one of the bottles with a star on the lid. One bottle is chosen as the star bottle. Ask the children to find the full bottle, the almost empty bottle, etc. Ask them to find the bottle with the star on it. **If this is the star bottle, which bottle has lots more water? Which is nearly the same?** Tell them they have to place the bottles next to each other and look at eye level to see the water line.

Unit 1.6 Reasonable estimates

Whole class

- **Story Time** – When reading stories, make a point of asking the children if something sounds reasonable. Read, *The Tiger Who Came to Tea* by Judith Kerr. **Could you eat as much food as the tiger? Could you eat more than him?** Also include questions where the answers are reasonable. Stories by Shirley Hughes, for example, have many situations which are familiar to the children's own lives. Using *Dogger* for example, ask: **Do you have a favourite teddy like Dave or do you have lots of favourite toys? Could you eat an ice cream like Dave's? Could you eat 8 of them?**
- **General Conversations About Volume** – Set the home corner up with visuals that suggest how many children can fit into particular areas of the nursery. Explain that a naughty teddy bear came in during the night and made signs to show how many children can play at each area; however, some of the signs make sense and some are silly suggestions. Accompany the children as they find them and discuss whether the suggested number of people is appropriate. **Can 8 children fit into the toy bed? Can 4 children sit at the snack table? Can only 1 child play on the climbing frame? How do you know? Why do you think this?**

Stations

- **Snack Time** – Set the snack table up with either far too little food or way too much. Ask the children's opinions on whether the food in the container/on the plate is going to be too much, too little or just the right amount. Discuss how much is enough food. **I only have two bananas; can all 8 of you have some banana? Do we have enough bananas? What can we do? What if I only have two grapes/a large bunch of grapes?**
- **Gardening** – Lay all the gardening tools, compost and plants on the floor next to a small and a large wheelbarrow. Tell the children that they are going to be gardening and you will need their help to move all the equipment. Tell them that you think they will be able to get everything into the smallest wheelbarrow. **Can we fit all the gardening tools into this smaller wheelbarrow? Do I need to use both wheelbarrows? Will this big garden fork go in the small wheelbarrow?** Once at the garden, continue asking their opinions. Show them a tray with 12 plants. **Can I put all of these into one plant pot? How many more pots will I need? Will I need lots more or just a few more?** Match each plant with a plant pot to check there are enough. Repeat this with too many pots and not enough plants. **Do I have enough plant pots? What should I do with the rest of the plant pots? Do I need lots more plants or a few more?**

What to observe

- Children can state which collection has 'more' when there is a large visual difference.
- Children are beginning to use pairing as a strategy for comparing two sets.
- Children are beginning to use lining up as a strategy for comparing two sets by looking at the difference in size.
- Do children successfully use a one-to-one correspondence to count a small collection?
- Children may be able to compare 2 numbers (within 1–5 range), e.g. 4 is more than 2, because 4 is a bigger number.
- Children may be able to predict how many there will be if you add or take one away (in 0–5 range) by using subitising rather than counting.

Estimation and Rounding B

Section overview

Units of learning	Review units	Prior learning	Benchmarks/Es&Os
Unit 1.1 **Estimating and comparing collections visually or by matching**	**Estimation and Rounding A** Units 1.1 and 1.2	• Experience of comparing two collections that are obviously very different in size. • Use everyday language specific to comparisons, e.g. more than, bigger than, smaller than. • Experience of using a matching strategy, i.e. pairing or lining up to compare two collections. • Experience describing the smallest collection as having 'fewer than' but not using it independently.	**MNU 0-01a** • Recognises the number of objects in a group, without counting (subitising) and uses this information to estimate the number of objects in other groups. • Checks estimates by counting. • Demonstrates skills of estimation in the contexts of number and measures using relevant vocabulary, including 'less than', 'longer than', 'more than' and 'the same'. I am developing a sense of size and amount by observing, exploring, using and communicating with others about things in the world around me.
Unit 1.2 **Estimating and comparing collections by grouping and counting** Estimate the size of one collection by comparing it to the size of a known group (subitising) or counting.	**Estimation and Rounding A** Unit 1.3	• Secure one-to-one correspondence when counting objects. • Understand the final number said means the number in the collection. • Experience of using a grouping strategy to compare two collections. • Understanding of the concept 'more than', 'the same.' • Experience describing the smallest collection as having 'fewer than' but not using it independently. • Confidently count and order numbers to 10.	
Unit 1.3 **Estimating and comparing collections by using benchmarks** Make predictions/estimates based on the size of one collection to say how many are in another collection.	**Estimation and Rounding A** Units 1.4 and 1.6	• Secure one-to-one correspondence when counting objects. • Experience of making and comparing collections with different attributes including size. • Understanding of the concept 'more than' and 'the same'. • Can subitise up to 5 (e.g. regular dice patterns and irregular) • Experience describing the smallest collection as having 'fewer than' but not using it independently. • Confidently count and order numbers to 10. • Recognise numerals up to 10.	

Units of learning	Review units	Prior learning	Benchmarks/Es&Os
Unit 1.4 **Making equal collections** Comparing two concrete collections and moving or adding objects to make them equal.		• Secure one-to-one correspondence when counting objects. • Experience of making collections and understanding that numbers can count anything. • Understanding of the concept 'more than' and 'the same'. • Experience describing the smallest collection as having 'fewer than' but not using it independently. • Can subitise up to 5 (e.g. regular dice patterns and irregular) • Confidently count and order numbers to 10. • Recognise numerals up to 10.	
Unit 1.5 **Estimating and comparing measures** Estimating and comparing measures visually or by matching.	**Estimation and Rounding A** Units 1.5 and 1.6	• Understanding of the concepts 'more than', 'less than' and 'the same'. • Understanding of big, small and little. • Experience of comparing measures by pairing or lining up. • Can note when there is an obvious difference in the volume, height, length of something and has everyday language to describe this.	
Unit 1.6 **Reasonable estimates** Understanding that some things are reasonable and others are not	**Estimation and Rounding A** Unit 1.6	• Understanding of the concept 'more than' and 'the same.' • Experience describing the smallest collection as having 'fewer than' but not using it independently. • Can make observations based on their own experiences, e.g. My daddy is taller than me.	

Conceptual context

Giving children plenty of opportunities to estimate and compare collections enables them to make decisions in the future on the reasonableness of their answers.

At this stage it is very important that concrete materials are used at all times. Children need opportunities to make collections and compare them against other collections. They will develop their use of matching strategies to confidently prove their initial estimates were correct. They understand that items from each collection need to be paired up with each other or lined up opposite each other, no matter their difference in size, colour or theme. They will recognise if there are any left in one of the collections then this means it is the 'bigger' collection.

They will also understand that comparing collections by using a familiar subitising pattern (within 1–10 range) will enable them to predict whether a collection has more or fewer than another collection.

Working on equivalence (same as) with very young children lays important foundations for later work on calculations and algebra. For children to truly understand equivalence they also have to explore what it isn't! Sometimes the result of the comparison should be that they are 'the same' and children should also be given opportunities to make collections equal.

Once children have a strong understanding of the principles of counting, they will be ready to estimate and compare sets by counting. They will need to:

- Know the 1–10 number sequence and then to 30.
- Understand that an item should be counted once and once only.
- Know how many items they have and that the last number they said means the total number of items (cardinal principle).
- Begin to understand that it doesn't matter which order you count the items – you will always get the same amount if none are missed out (order irrelevance).
- Understand that numbers can be used to count anything, e.g. 3 can mean 3 dogs, 3 apples and 3 people.
- Understand that each counting number is always one more than the previous number.

When confident at this level, children may be able to compare collections by counting how many more or fewer there are in a set, but this still does not mean they will naturally link these concepts to addition and subtraction.

Resources

- Interlocking cubes
- Tens frames – different sizes
- Sorting hoops and hula hoops
- Building bricks
- Sorting bears
- Finger pattern cards
- Regular dice pattern cards
- Spoons – various sizes
- Coloured counters
- Double-sided counters
- Number/numeral tracks
- Posting boxes
- Numeral cards up to 30
- Collections, e.g. dinosaurs, toy cars
- Estimation jars
- Containers – different sizes
- Irregular dot cards for matching
- Dice – large and small
- True and false cards
- Pegs
- 'Same as', 'more than,' and 'fewer than' cards
- Beads and laces
- Pompoms
- Balance scales

Key vocabulary

- **Nearly, close to, in-between, almost, just about/over, not quite**
- **More than/a little bit more**
- **Fewer than** – for countable quantities, e.g. there are **fewer** bananas **than** apples
- **Less than** – for non-countable quantities, e.g. there is **less** water in my glass **than** in the pond
- **Most/least**
- **Greater than**
- **Taller than/shorter than** – for measuring height
- **The same as/is equal to**
- **Half of/double**
- **How many?**
- **Compare**
- **Estimate**

Misconceptions

Potential misconception	Addressing the misconception
Units 1.2 and 1.4 Children may not realise that the final number that they say represents the final set. When pointing to 4 items they may have to start their count from one again. They think they need to say the string of numbers 1 to 4, not just the final count of 4.	• Lots of opportunities to collect and count small collections using a wide range of objects. Model the final count. • Evidence in environment and displays of value associated with a number. • Practise forward number sequence up to 10. • One-to-one correspondence practice – they touch, say and move each item in turn and only once!
Units 1.2, 1.3, 1.4 and 1.6 Many children do not understand that each counting number is one more than the previous number, e.g. they guess that if 6 bears go into a cave and then 2 more, there are 3 altogether. They do not understand that numbers get larger the further you go up in order.	• Building staircases using cubes or number rods can help develop this understanding. • Opportunities to help children recognise that if they add one more, they will get the next number in the sequence and if they take one away, they will get the previous number.
Units 1.1, 1.2, 1.3, 1.4, 1.5 and 1.6 The language used can be confusing for children.	• Lots of practical experiences to practise describing the variety of different contexts of length comparisons. • Adults modelling the appropriate language and asking children to copy the statement, e.g. 'The box is heavier than the pencil. Now you say it.' • Correct inaccurate use of language, e.g. 'The man is longer than the boy.' to 'The man is taller than the boy.'

Potential misconception	Addressing the misconception
Units 1.1, 1.2, 1.3, 1.4, 1.5 and 1.6 Children of this age may not understand that two groups can be compared.	• Children will need lots of concrete experience of physically manipulating objects into collections and sub-sets and combining small sets to make larger ones. • Concrete materials that can be manipulated must be used before visuals that cannot be moved, i.e. comparing pictures in a book.
Units 1.1, 1.2, 1.3, 1.4, 1.5 and 1.6 Some children consider the size of an item instead of the number of items when they are comparing the numerosity of the group. They may not understand that when counting items each item can only be counted once.	• Pair up the items so children can see that it is the number of items that is important not the size, e.g. one tennis ball is the same quantity as one golf ball. • Use practical resources and demonstrate moving items from an uncounted place to a counted place, to ensure that they are not counted more than once. Move each item as it is counted. Encourage the child to describe the completed count e.g. 1, 2, 3, 4, 5. There are 5 cubes.
Units 1.1, 1.2, 1.3, 1.4, 1.5 and 1.6 Look out for children who do not change their first estimate even after checks have been made to check for reasonableness.	• Wild guesses should be explored and the reasonableness of the answer discussed. • Remember that children cannot be expected to give reasonable estimates above the range that they can count. • Reduce the size of the collection to within the child's capabilities.
Units 1.1, 1.2 and 1.6 They may label the set with the wrong number, but this does not mean they cannot recognise the comparisons between the sets.	• Ask them whether the collection is more than the other one instead of asking for an estimate of 'how many' there are in the collection. Familiar things are a good place to start, food in particular.
Units 1.1, 1.2, 1.3, 1.4, 1.5 and 1.6 Some children think the quantity has changed just because it now takes up more space, e.g. 4 counters in a line, how many do you see? Then space them out, how many now? These children will often say a higher number than before.	• Lots of practical opportunities to explore the concept that if you don't add or take anything away you've not changed the amount, e.g. give them one lump of play dough. Then break it into pieces. Ask if they have more now or the same? Listen carefully to what they say. • Snack time can provide good examples to show that it looks different but is actually the same, e.g. cut a banana up and then ask if they have more or the same. Put it back together to show that it is the same.

Teaching ideas

Unit 1.1 Estimating and comparing collections visually or by matching

Whole class

- **Cars** – They will need two coloured A4 mats (paper) and up to 20 toy cars each. They will also need a set of sorting cards; two cards with 'same as', two cards with 'more than' and two cards with 'fewer than' written on them (see Photocopiable resource 1.1). In partners, the children take it in turns to be the 'starter' and the 'follower'. They place the cards face down on the table. The starter places a collection of cars on their mat. The 'follower' looks at the collection whilst the 'starter' counts to 5. The collection is then covered. One card is turned over. The 'follower' must do as the card instructs and attempt to make a collection on their mat, which is either, 'more than', 'fewer than' or 'the same as.' Explain how difficult it is to compare when the cars are spread out. **How can we organise the cars to make it easier to compare the collections?** Encourage the children to check by pairing up the cars, one from each collection at a time. **Did your partner follow the instructions on the card? Were they close? How do you know? Once all the cars have been matched up, who still has cars left? What does this mean?**

is the same as	is the same as
is more than	is more than
is fewer than	is fewer than

Photocopiable resource 1.1 – Sorting Cards.

- **Towers** – Ask four children to come to the front and turn their backs to the class. Give them 15 interlocking cubes each and ask them to make a tower of any size. They then turn round and show them to the class at the same time. **Who has the most cubes? Who has the fewest cubes? Does anyone have the same amount? How can we check to make sure?** Encourage the children to put them next to each other; resting on a table will help to keep the comparison fair. Model the relevant statements; 'You have more cubes than he does but fewer than she does.' 'You two have the same amount of cubes, they are the same height.'

Stations

- **Who Has More?** – Have 10 double sided counters in a cup. Find a partner to play this game, and decide who is going to be red and who is going to be yellow. Tip them out onto the carpet. Player 1 has to make a statement. 'There are more red counters than yellow counters.' Pair them up to prove it. Pair one red counter up with a yellow counter until you run out of one colour. 'There are some yellow counters without a partner.' **What does this mean?** The winner has to say, 'I have more counters than you', the loser has to say, 'I have fewer counters than you'. Repeat, the person with the most gets a point; the first person to 5 is the winner. The number of counters used could be changed within a 5 to 30 range.
- Put a collection of sorting bears out and some cards with different amounts of randomly placed spots on (see Photocopiable resource 1.2). Place the cards face down on the table. Choose two cards and compare them. **Guess which card has the fewest dots?** Place a plastic bear on top of each spot. **Which card has the most bears?** Lay the bears from the first card up one side of the 2 × 20 grid (see Photocopiable resource 1.3), one to each square. Place the bears from the second card on the column next to them. Whichever column has the most bears gets the point. **Were you right? How do you know? Can you see that one card has more than the other?**

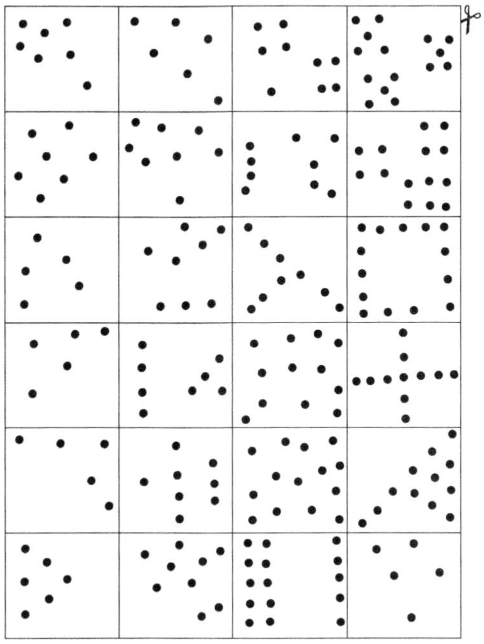

Photocopiable resource 1.2 – Random Dots to 20.

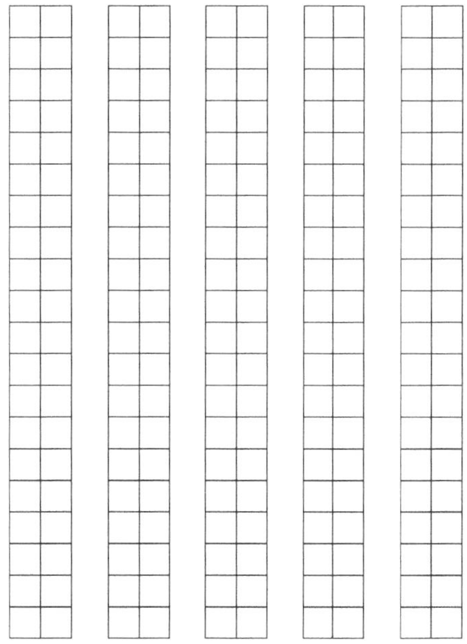

Photocopiable resource 1.3 – 20 x 20 grid.

Unit 1.2 Estimating and comparing collections by grouping and counting

Whole class

- **Story Time –** Take a familiar story and tell the children you are going to play 'Spot the Mistake'. When recapping deliberately change the amount mentioned in the story, e.g. 'I think there are four dogs on the farm'. **Am I correct? Were there more or fewer dogs? How can we check?** Go back and find the picture that will help. Put up fingers for each dog. Get the children to put up four fingers. **What do you think?** This is good for numbers up to ten. Encourage the children to say the final true statement, e.g. 'There were more dogs in the book than the teacher remembered.'
- **Comparing Numbers** – Take a pack of numbers up to 20 (or 30) (see Photocopiable resource 1.4). Hold two cards up. **Which number is bigger than the other? Which number is smaller? How can we prove it?** Ask them to peg both numbers on a large numeral track in range of 0–20 (or 0–30). This has more because it has 8 and this only has 3. We shall bang the drum 8 times, count with me. 1, 2, 3, 4, 5, 6, 7, 8. Now let's bang the drum 3 times. 1, 2, 3. **Which number had more drum beats? Which number had fewer? What does this mean?**

1	2	3	4	5
6	7	8	9	10
11	12	13	14	15
16	17	18	19	20
21	22	23	24	25
26	27	28	29	30

Photocopiable resource 1.4 – Number Cards to 30.

Stations

- Tell the children they are going to sell seeds in their shop. Label seed packets (envelopes) with numbers and then place seeds inside. Make sure some of the packets have more than, some have fewer than and some are the same as stated. For example, the label on the packet says we have 5 sunflower seeds, open the packet and see if this is correct. Put 10 sunflowers seeds in. **Does this packet have 5 sunflower seeds? What do we need to do to make this packet have 5? Do we need to add some more? Do we need to take some out?** Make the packet have the correct amount. Repeat with the next packet. The children could count or use tens frames to compare.
- **Farmyard** – Tell the children that the cows want to go in the barn as it is nearly milking time. I am going to put 8 cows in this barn. Watch as I place (count) them into the barn. 1, 2, 3, 4, 5, 6, 7, 8. There are 3 cows in this barn. Watch as I place (count) them into the barn. 1, 2, 3. **Which barn has the fewest? Which barn has the most? Does this barn have the most cows? How do you know? How can you prove that you are right?** Take the cows out of the first barn and place them on a tens frame, do the same for the cows in barn 2. **Were you right?** Model the statement, 'This barn had more cows as 8 is more than 3' or 'This barn has fewer cows than this barn because 3 is a smaller number than 8.'
- **True or False?** – Make a set of true and false cards with labels, 'Set A' and 'Set B' and 'Set A has … items than Set B' and 'Set B has … items than Set A' (see Photocopiable resource 1.5). Place items in both sets to compare, e.g. 5 ducks under the Set A label and 3 ducks and 2 chicks under the Set B label. Say, 'Set A has more animals than Set B.' Ask: **True or false? What else can you say?** Use counting to check. Encourage them to look at the similarities and differences too. Use different sized items too!

Set A has		items than Set B
Set B has		items than Set A
Set A has		items than Set B
Set B has		items than Set A

Photocopiable resource 1.5 – True and False Statement Cards.

Unit 1.3 Estimating and comparing collections by using benchmarks

Whole class

- **Linking Cubes as Benchmark** – Using a loose collection of linking cubes as a benchmark, e.g. show them a pile of 5 and say, 'Here are 5 linking cubes.' Ask: **Can you reach into the box of cubes and pull out 5 in one go? How many did you pull out? Count them? Were you close? How do you know?** Repeat with different benchmarks.
- **Mystery Box** – Each child has a box with a mystery number of pound coins inside, which they count and keep the amount a secret. Show them a filled Five Frame and ask them to move around the classroom to the music. When the music stops, find their nearest partner and ask them to state whether they have more than, fewer than or the same as five coins. **Does your partner agree?** Swap boxes. Move around to the music again and repeat. **Who has the most pound coins? Who has the fewest pound coins?** Say, 'I have 5.' Ask: **Does anyone have the same as me?** Prove it. **Does anyone have more than me?** Prove it. Get the children to lay the coins out on an empty five frame on their table and place any extra pound coins onto the table. This can also be done on a tens frame, or ten dot strip.

Stations

- **Photographs** – Take photographs of familiar collections, e.g. conkers (see Photocopiable resource 1.6), after a visit to the woods. Bring a collection of conkers back to class. Say, 'Look at the amount of conkers on this picture.' Ask: **How many conkers are there?** Encourage them to count them. 'Collect more conkers than are in this picture.' 'Collect fewer conkers than are in this picture.' 'Collect the same amount of conkers that are in this picture.' 'Collect a lot more/far fewer).' **How will you know if you have been successful?** Encourage them to count them or to put conkers on top of the conker pictures and know what this means if they have more/fewer conkers than the images.
- **Painting** – Tell the children they have to paint a picture of objects with more than 8 objects on the right and fewer than 8 objects on the left of the printed number. **How do you know you are right?** Children may use a counting strategy or place counters on the painted objects and use a subitising pattern to compare.

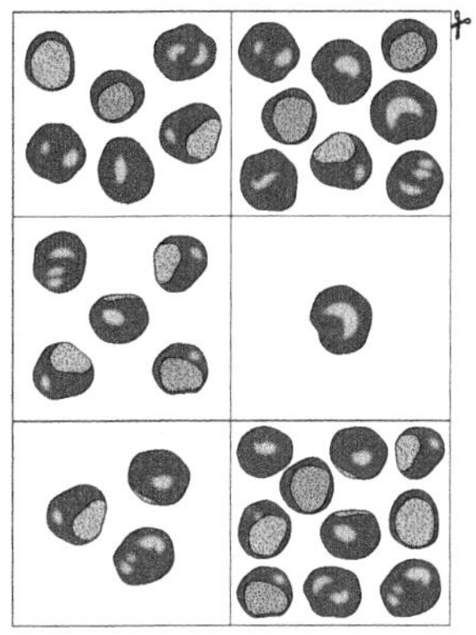

Photocopiable resource 1.6 – Conker Collections.

	8	
fewer than		more than

Unit 1.4 Making equal collections

Whole class

- **Equal Groups in P.E.** – Make a group of 4 children in P.E. Ask the rest of the children to make the same sized group. Depending on the number, some groups may have to have more or fewer than the others. Ask them to point to a group that has the same number of children in as this one. Point to a group that has more children in it. **How do we know?** Tell the children to check how many children are in their group by choosing someone to be the number one and then count around the group. Model the statement 'You were correct, you have more children in your group because 5 is a bigger number than 4.'
- **Collections of 10 and Under** – Place a pile of counters on each table. Take a ten spaced egg box with a lid and place a collection of counters in within the range of 0–10. Walk around the room singing:

  Heigh Ho, Heigh Ho,

  With nought to ten we go,

  Do you have more?

  Or do you have fewer?

  Heigh Ho, Heigh Ho!

Stop next to the nearest person and open your lids. **Do you have more or fewer than your partner? Hands up if you have more counters. Hands up if you have fewer counters. Hands up anyone who has the same amount. How can you make your boxes have the same amount of counters?** 'I need 2 of yours. You can have 3 of mine. Do we have the same? When you have the same amount or 'nearly the same' close your lid and sit down'. When everyone has finished, you can start walking around the room again and singing the song. Repeat questions and task to make the boxes equal.

Stations

- **More than/fewer than boxes** – Use laminated cards showing three boxes with a ladybird in each box, 'fewer than' box on left and 'more than' box on right (see Photocopiable resource 1.7). Ask children to draw spots on the middle one. Draw more spots on the one on the right and fewer in the box to the left. **How do you know you are correct?** Count the spots in each box and record the amount. Use transparent counters to cover the three numbers for the amount of spots on a number line. **What do you notice?** Point out that the amount with the fewest is lower down the number line and that the number with the most spots is further up the number line.

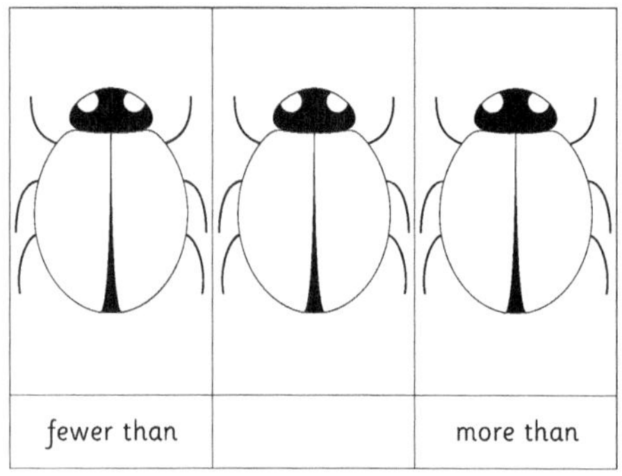

Photocopiable resource 1.7 – Ladybird Spots.

- **The Equals Trees –** Take two identical pictures, e.g. 2 trees (see Photocopiable resource 1.8) and roll a 1–20 dice to find the number of birds in one of the trees. Represent the number of birds using counters. Tell them to place a peg on that number on a large numeral track up to 20. Roll the dice again to find the number of birds in the other tree. Represent the number of birds using counters and place a peg on that number on the numeral track. **Are there more or fewer birds in the first tree? How can we check?** If needs be, mention that 18 is higher up the track than 15. This means that 18 is a bigger number than 15. **How can we make the number of birds equal in both trees?** They may add some more to one tree or take some away. Model the actions taken. 'You took three counters away to make both trees have the same amount,' or 'You added more counters to make them equal.' **How many did you add?**

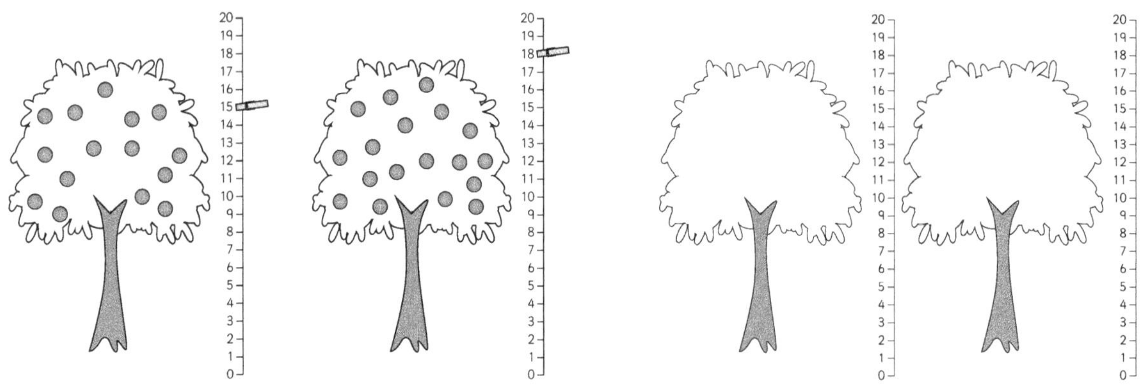

Photocopiable resource 1.8 – Equals Trees.

Unit 1.5 Estimating and comparing measures

Whole class

- **A Necklace for a Doll** – Place a beaded necklace on a doll and provide a collection of beads which are the same size but different colours/patterns. Tell the children that they have 5 minutes to make a necklace which is longer than the one on the doll. Ask the children to hold up their necklace by one end of the lace. **Which is the longest/shortest necklace? Which necklace has the most beads? Which has the fewest beads?** Encourage the children to lay their necklace on the table and line them up against the one that was on the doll (this must be untied and laid out flat). This activity could be repeated with beads which are not the same size. Explain how difficult it is to compare items of varying sizes. **How can we organise the beads to make it easier to compare?** Take them off the lace and pair them up with each other, one bead at a time. **Who has used the most/fewest beads?**
- **Bricks** – Make a pile of the same-sized foam bricks on the floor. **Can you make two towers using all the bricks placed one on top of the other? Which tower has more bricks? Which tower has fewer bricks? Are both the towers the same height? How do you know?** Make sure the towers are close enough to be compared accurately. Model the relevant statement:
    - Both towers are the same height which means they have the same number of bricks.
    - This tower is taller than the other one. This has more bricks than this one.
    - This tower is shorter than this tower. So it must have fewer bricks.

Stations

Vegetable Shop – Set up a table of different vegetables and some balance scales. Place cards with pictures of each vegetable for sale face down on the counter and provide a large dice. Tell the children that one of them is going to be the shopkeeper and one can be the customer. Say, 'If you are the customer, you must roll the dice and choose a vegetable card. If you are the shopkeeper, you must place the number of vegetables in the balance scales.' For example, the shopkeeper places four carrots in the balance scales. 'Now you must choose another vegetable card and roll the dice again. If you are the shopkeeper, you must put that number of vegetables in the other pan.' For example, three sweet potatoes. **Which is heavier? Which is lighter? Four carrots or three sweet potatoes?**

Photocopiable resource 1.9 – Vegetable Shop.

- **Role Play** – Set up a table for Goldilocks and the Three Bears; provide three bowls of the same size, a range of different spoon sizes (from teaspoon to ladle), five wise tens frames (1–10) face down on the table and some porridge oats (or real porridge!). Tell the children that Goldilocks didn't think it was fair that the bowls were different sizes. 'Choose a tens frame and pick up a spoon. If you have chosen a '4' you need to put four spoonfuls into one of the bowls. Now choose a different spoon and put four spoonfuls into the next bowl. Choose a different spoon again and place four spoonfuls into the last bowl.' **Which bowl has the most porridge? Which bowl has the least? Who will you give the most porridge to? Why?**

Unit 1.6 Reasonable estimates

Whole class

- **Story** – Many stories have 'unreasonable' outcomes. Ensure there are many opportunities to discuss the likelihood of these events happening. Take the Aboriginal story of Tiddalik, the frog who drank all the water in the lake? **Could you drink the same amount of water? How much water could you drink if you were very thirsty**? Model statements: 'I cannot drink as much water as Tiddalik.' 'I drink much less water than Tiddalik the frog.'
- **Mystery Box** – Make a 'mystery box' out of an old box and place statements inside; make sure there are some outrageous, some reasonable and some 'possible' statements. For example, 'Jared can run round the playground before I can count to ten.' 'The head teacher is nearly as tall as Mrs Christie.' 'Calum is heavier than a banana.' **Do you agree or disagree? How could we check?** Allow time for some of the statements to be investigated. **Do you still agree/disagree? Would anyone like to change their mind?**

Stations

- **True or False?** – Set up a true or false station with a voting box (shoebox with slit in lid), a collection of laminated 'T' cards and 'F' cards (see Photocopiable resource 1.10) and two bowls. Place 18 pompoms in one bowl and 16 pompoms in another. Write out this statement and place it on the table, 'Set A has fewer items than Set B.' **Is this true or false?** 'Write your name on the 'T' card if you think this is true or the 'F' card if you think this is false. Post your card in the voting box.' The box should be emptied and the votes counted. **How many people said it was true? How many people said this was false? Did more people think it was true or did more people think it was false? Who was right? How can we check?** Items could be spaced out instead of in a bowl to check if children are looking at the numerosity of the set. Different-sized items can be used too, e.g. Set A could have five cows and Set B could have two sheep and three pigs. The statement could read, 'Set A has more animals than Set B.' Repeat the vote, the subsequent count and the discussion.

T	T	T	T
F	F	F	F

Photocopiable resource 1.10 – True or False Letter Cards.

What to observe

- Children can state which collection has 'more' or 'fewer' when there is a large visual difference.
- Children can use a pairing strategy for comparing two sets.
- Children can use a lining up strategy for comparing two sets by looking at the difference in size.
- Do children successfully use one-to-one correspondence to count a small collection?
- Children may be able to compare two numbers (within 1–20 range), e.g. 14 is more than 8, because 14 is a bigger number.
- Children may be able to predict how many there will be if you add or take one away (in 0–10 range) by using subitising rather than counting.
- Children are beginning to use familiar 'benchmarks' of 5 or 10 to compare collections.

- Children are beginning to use concrete materials to compare pictorial representations, where they cannot move the items.
- Children are beginning to understand that if items are removed from a collection, the amount left will be fewer than before.
- Children can make collections which can show 'more than', 'fewer than' and 'the same as'.
- Children can compare a collection by counting but may have to re-count if they are asked to start at a different place as they may not trust this count will be the same.

2 Number and Number Processes: Number Recognition and Place Value

Number Recognition and Place Value A

Section overview

Units of learning	Review units	Prior learning	Benchmarks/Es&Os
Unit 2.1 **Principles of counting (1 and 2)** Beginning to understand that to say 'how many' they must touch an item once and say the counting numbers in the right order.		• Knows some numbers and uses them to count even if unreliable.	**MNU 0-02a** • Explains that zero means there is none of a particular quantity and is represented by the numeral 0. • Recalls the number sequence forwards within the range 0–30, from any given number. • Recalls the number sequence backwards from 20. • Identifies and recognises numbers from 0 to 20. • Orders all numbers forwards and backwards within the range 0–20. • Identifies the number before, the number after and missing numbers in a sequence within 20. • Uses one-to-one correspondence to count a given number of objects to 20. • Identifies 'how many?' in regular dot patterns, for example, arrays, five frames, ten frames, dice and irregular dot patterns, without having to count (subitising).
Unit 2.2 **Principles of counting (3)** Beginning to know that the last number said means the number of things counted.		• Uses some numbers in the context of counting even if unreliable.	
Unit 2.3 **Principles of counting (4)** Beginning to understand that I can count anything.		• Hearing variety of collections being counted which are familiar to the child, e.g. pieces of apple on plate.	
Unit 2.4 **Principles of counting (5)** Learning to count things in any order.		• Having experiences where small counts are done and then recounted in a different order.	
Unit 2.5 **Forwards number sequences to 10** Able to say the forwards number sequence up to 10.		• Knows some numbers but may not put them in the correct order.	
Unit 2.6 **Backward number sequences from 10** Can say the backwards number sequence up to 10.		• Knows some numbers but may not put them in the correct order.	
Unit 2.7 **Forward number sequences to 20** Can say the forwards number sequence up to 20.		• Knows some numbers but may not put them in the correct order.	

Units of learning	Review units	Prior learning	Benchmarks/Es&Os
Unit 2.8 **What's the number after?** Can say the number that comes after in the range of 1 to 10.		• Beginning to understand the term 'after' in familiar contexts. E.g. you can have your ice cream after your sandwich.	• Groups items recognising that the appearance of the group has no effect on the overall total (conservation of number). • Uses ordinal numbers in real life contexts, for example, 'I am third in the line'. • Uses the language of before, after and in-between. • When counting objects, understands that the number name of the last object counted is the name given to the total number of objects in the group. I have explored numbers, understanding that they represent quantities, and I can use them to count, create sequences and describe order.
Unit 2.9 **What's the number before?** Can say the number that comes before in the range of 10 to 1.		• Beginning to understand the term 'before' in familiar contexts. E.g. you need to do your teeth before we go out.	
Unit 2.10 **Reading numbers** Can recognise and identify some of the numbers up to 10.		• May be able to recognise that numbers are not letters.	
Unit 2.11 **Representing numbers** Can make marks to represent some of the numbers up to 10.		• May use marks to represent numbers.	
Unit 2.12 **Sequencing numbers** Can place the numbers 1 to 5 in the correct order.		• May be able to recognise that numbers are not letters. May know some of the numbers from 1 to 5 range.	
Unit 2.13 **Ordering numbers** May be able to order numbers within a 1 to 5 range.		• May be able to recognise that numbers are not letters. May know some of the numbers from 1 to 5 range.	
Unit 2.14 **Missing numbers** May be able to spot missing numbers within a 1 to 5 range.		• Beginning to understand the terms 'in-between' and 'missing' in familiar contexts. E.g. Where is your shoe? It has gone missing.	
Unit 2.15 **What is zero?** Beginning to realise that zero means nothing.		• Knows familiar terms such as 'all gone' or 'no more'.	
Unit 2.16 **Ordinal numbers to 3** Can say if something is in first, second and possibly third position.		• May know what 'you can go first' means.	
Unit 2.17 **Tracking sounds on fingers** Can keep track of movement and sounds on fingers.		• May use fingers to count small numbers.	
Unit 2.18 **Finger patterns** Can 'grow' and may be able to 'throw' finger patterns up to 5.		• May use fingers to represent one or two.	

Units of learning	Review units	Prior learning	Benchmarks/Es&Os
Unit 2.19 **Regular dot patterns to 5** Can recognise most regular dot patterns up to 5 without counting them.		• May know some of the regular patterns to 5.	
Unit 2.20 **Irregular dot patterns to 5** Can recognise most random dot patterns up to 5 without counting them.		• May recognise some irregular dot patterns to 5.	

Conceptual context

Children are fascinated by numbers from a very young age and will include familiar numbers or small quantities in their conversations as soon as they learn to talk, e.g. their age or the number of biscuits on their plate. It is very important that the different uses of number are highlighted when occasions arise as well as in 'planned' number activities, e.g. show and say the numbers being pressed on your phone's handset or read the number plates out loud as they park their ride-ons in the garden area.

It is essential that children explore the connections between spoken number names (one, two, etc.), formal written symbols (1, 2, 3, etc.) and the quantities that they represent. They must be introduced to these concepts by working with concrete materials that can be moved and linking these with the counting numbers. They should be given time to talk about their discoveries.

When simply reciting a forwards counting sequence, they are not actually counting but are learning there is a specific and repeatable order when saying number words. Nursery age children may know parts of the sequence in order but may fill in the gaps with random numbers (e.g. 1, 2, 3, 4, 5, 8, 10, 100). They first learn how to count forwards to 5, then 10 and then to 20. They should also learn how to count backwards alongside learning the forwards numbers.

Repeatedly singing number rhymes that include reciting forwards and backwards number sequences is an excellent way for children to recall the number names and the order. If accompanied with actions such as using finger or dice patterns, the value of the numbers can also be checked by subitising or counting.

The nursery environment should be rich with numbers as labels and to represent quantities. Children should be given opportunities to play and handle large tactile numbers and be encouraged to form numerals using larger movements, e.g. in sand or paintings. They also need to be familiar with the shapes of the numerals using different mediums such as wooden and plastic numbers as well as written numerals so they can both identify and recognise them; these are two very different skills.

- Identifying – Point to a number and ask them to say it. 'What number is this?'
- Recognising – Ask them to find a particular number. 'Can you find the number 7?'

Children often recognise and understand the numerals before they have the skills to write them. Written number words and numerals should be modelled by the adults, and the children should be encouraged to make marks in their own way and to practise correct numeral formation. Resources should be freely available for impromptu recordings, i.e. wooden numbers, pencils, clipboards with squared paper and empty till rolls alongside tools for recording.

Children need experiences of numbers being used in a wide variety of contexts to help them differentiate between the different representations of number. Children should learn that numbers are not solely used to represent a quantity; they can also describe an order or be used for labelling an object.

- Cardinal aspect of number – i.e. how many things there are, the 'six-ness' of a number.
- Ordinal aspect of number – i.e. the order of numbers on a clock face, and position of things, e.g. use of 'first', 'second' and 'third'.
- Nominal aspect of number - i.e. the number 25 bus or their house number.

When children start to use numbers to describe an amount of things, then they can be introduced to the word zero as meaning a set with nothing in it. They also love to include zero as the last number in the backwards counting sequence. It is not advisable however to include the zero on a numeral track as this can cause confusion when looking at ordinal numbers and the position of numbers, i.e. 0 is not the first number when using counting numbers.

There are five important counting principles that children need to internalise before they really trust that counting works.

- Stable Order Principle – Children use a list of number words in the same repeatable order every time they count.
- One-to-One Principle – To be accurate at counting, children must successfully co-ordinate reciting the number words in the correct order while physically pointing or touching the objects once and once only. Be careful that children who choose to use subitising to count small collections are not discouraged by insisting they count each item individually because this ability to recognise small numbers instantly will help them to partition numbers and trust their count later on.
- Cardinal Principle – For children to develop cardinality, they need to understand that the last number said represents 'how many' are in the set. Initially concrete materials that can be moved should be used and strategies for successful counting should be modelled.

  To ensure the final object counted represents the number in the collection, children need to:

  - Touch and count using a one-to-one correspondence.
  - Ensure the stable order of number words has been used.
  - Move objects that have been counted to a designated 'counted area' to ensure objects are only counted once and none have been missed out.
- Abstraction Principle – If given ample opportunities to count 'anything and everything', children will begin to understand what counting is for. They will begin to see that the number name 'three' and the symbol '3' and three counting bears represent the same quantity. These experiences will help them to understand that any group of objects can be counted regardless of whether they are the same shape, size or colour. This can also include things that they cannot see, such as beats of a drum, jumps down the steps or friends hiding in a bush.
- Order Irrelevance Principle – Understanding that the order in which objects are counted is irrelevant to the final count can be one of the trickiest principles to master. Children need lots of opportunities to count collections and then re-count them in a different order, as long as each object has been counted once and once only. For example, counting a small collection when lined up from one end and then counting again from the other end.

Subitising is when you can instantly recognise the number of objects within a group without having to count them one by one. It is an important skill for developing number sense as it helps children to image numbers and start to see how they are made up. If counting is to be discouraged, then flashing objects and dot cards for a few seconds can encourage instant recognition. Perceptual subitising is when a child can instantly state how many are in the group (typically only up to 5 objects). It is also important that children can recognise irregular as well as regular patterns. We want children to be able to recognise 5 dots for example, whatever the arrangements of the dots.

Conceptual subitising is when children can see the smaller numbers hidden inside a number, e.g. on a dice pattern of 5, if asked what numbers can they see hidden inside this number they may answer 4 and 1. Children need to be given opportunities to engage in subitising activities that include a range of representations, including:

- Auditory subitising – recognising sounds such as drum beats or pennies dropped into a jar.
- Visual subitising – counting movements, e.g. skipping, or objects on an out-of-reach shelf.
- Finger patterns – patterns to 5 including doubles to 4. Children will first 'grow' their numbers by putting up fingers one at a time but will move to 'throwing' them when more confident of the number of fingers needed for a particular number.
- Irregular dot patterns on cards or using double-sided counters that will fall randomly.
- Regular dot patterns to 5 – dice, dominoes, patterns on playing cards and five frames.
- Tallies.
- Plastic number shapes with holes.
- Bead strings which have beads arranged in groups of 5.

Resources

- Transparent counters
- Interlocking cubes
- Dice – spotted, numeral
- Containers, e.g. tins, boxes, bowls, cups, shoe boxes, treasure chests
- Collections of toy cars, socks, t-shirts, Lego, musical instruments, chicks, counting teddies, butterflies, farm animals and pom-poms
- Plastic tweezers
- Mini blackboards
- Large ball
- String
- Sand
- Play dough
- Fly swats
- Washing line
- Pegs
- Cardboard crowns
- Sand timers (1, 2 and 5 minutes)
- Variety of 3D numerals
- Numeral tracks to 10
- Numeral cards – various sizes
- Dot cards to 5 – regular, irregular
- Five frames
- Empty number tracks
- Hide and seek numeral tracks
- Finger pattern cards to 5
- Tally cards to 5
- Digital cameras
- iPads
- Headphones

Digital

- Counting tool
- Tree tool
- Tree ants
- Hoopy zoo
- Fingers and toes
- Wobbly tower
- Draw some numbers with me

Record book

Pages 6–16

Key vocabulary

- **Count**
- **Counting**
- **Forward**
- **Backward**
- **Before**
- **After**
- **Missing**
- **Number words** – zero, one, two, three, four, five, six, seven, eight, nine, ten
- **Amount**
- **Many**
- **Much**
- **Dice**
- **Domino**
- **Pattern**
- **Fingers**
- **Match**
- **Order**
- **Sequence**
- **Item**
- **Collection**
- **Five frame**
- **Empty**
- **Domino**
- **Tally**
- **'To take in turns'**
- **First**
- **Second**
- **Third**
- **Numeral track**

Misconceptions

Potential misconception	Addressing the misconception
Units 2.1, 2.2, 2.3, 2.4, 2.5, 2.6, 2.7, 2.8, 2.9 and 2.15 Some children find it tricky to remember the stable order for numbers up to 10. Listen carefully for: Words omitted – one, two, four, five Words in the wrong order – 1 2 4 3 5 Repeated words – 1 2 3 4 5 5 6.	• Allow plenty of opportunities to sing counting songs, play games where the number words can be chanted. Ensure the songs learned include counting forwards to 20 and well as backwards from 10. • Reduce the string of numbers to three numbers and then slowly add the next numbers in the sequence.
Units 2.1, 2.2, 2.3, 2.4, 2.5, 2.6, 2.7, 2.8, 2.9 and 2.15 When counting objects watch out for the children who: • Get the recitation and the pointing completely out of sync. • Point to one of the objects without saying a number word at the same time and then continuing correctly. • Point to the spaces between objects and assigning number names to these spaces. Fail to organise the count and miss out some objects or count the same object twice. • Co-ordinate the recitation with pointing but miss out one of the objects or count an object twice. • Count on syllables for number, i.e. 2 count for seven matches objects to syllables. • May not understand that when counting items each item can only be counted once.	• Use practical resources and demonstrate moving items from an uncounted place to a counted place, to ensure that they are not counted more than once. • Move each item as it is counted. • Encourage the child to describe the completed count, e.g. '1, 2, 3, 4, 5. There are 5 cubes'. • Model 'bad' counting, take some items and count each one more than once. Ask the children what is wrong with your counting.

Units 2.1, 2.2, 2.3, 2.4, 2.5, 2.6, 2.7, 2.8, 2.9 and 2.15
If a collection to be counted is in an unorganised pile, children who choose to count by simply looking or only pointing at an item rather than physically moving them may make more errors.
	• Fewer errors are made by 'physical' counters. Model the strategy by:
	  ○ Placing items to be counted in a designated 'uncounted' space.
	  ○ Point and then move an item to a designated 'counted' space, while saying the appropriate counting word.
	• Say 'We have counted all the items; there are no more to be counted.'

Units 2.1, 2.2, 2.3, 2.4, 2.5, 2.6, 2.7, 2.8, 2.9 and 2.15 When counting a collection some children may not understand that the final count represents the amount of items. If you ask them for four, they may give you the fourth item. They also may simply repeat the final number said without realising this means how many are in the collection.	• Lots of opportunities to collect and count small collections using a wide range of objects. Model the final count. • Lots of opportunities to check amounts rather than told 'to count'. E.g. the jar says there are 4 items inside, are there 4 items? • Evidence in environment and displays of value associated with a number. • Practise forward number sequence up to 10. • One-to-one correspondence practice – they touch, say and move each item in turn and only once! • Change your tone for the last number said to support this as the final total.

Potential misconception	Addressing the misconception
Units 2.17, 2.18, 2.19 and 2.20 Watch carefully for the children who can use subitising to count a small collection under five, and not one-to-one counting, e.g. some children will grab a collection of counters say 2 then 3 and say they have 5.	• This strategy should not be discouraged but it should be explained to children that for collections larger than five it may not be consistently accurate.
Units 2.1, 2.2, 2.3, 2.4, 2.5, 2.6, 2.7, 2.8, 2.9 and 2.15 Some children will find it very tricky to start counting a collection without starting from 2. E.g. Give them 5 bears and ask them to start their count at 3, they may say there are 7 bears.	• Model counting a small collection of different items and repeat the count starting at a different item each time. E.g. a collection of 5 plastic animals, first start with the cow. How many animals are there? If we start our count with the pig, how many animals will there be? • Place the items in different positions and start at a new place each time. • Line the items up, start the count from both ends and the middle. Does it change the final amount?
Units 2.5, 2.6, 2.7, 2.8 and 2.9 Sometimes children will blend the numbers together as they do not realise they are separate numbers. For example, they may say 'onetwothree' as one word.	• Provide visuals with numerals on and point to each item while saying the number words.
Units 2.5, 2.6, 2.7, 2.8 and 2.9 Some children cannot break the forward number sequence and are unable to start from a different number other than one.	• Give lots of practical opportunities to count objects while sometimes stopping the count momentarily and then continuing. • Ask them to identify the number on a numeral track and then ask them what the next number is, and the next. • Give them experiences of estimating where numbers are sitting on a hidden number line. Where is the number 7? Are you right? Reveal the numerals. Show them one or two numbers before 7 so they have a 'run up' to that number. '5, 6 … What's the next number?'
Units 2.5, 2.6, 2.7, 2.8 and 2.9 Crossing over the decade can be tricky for some children, they omit some numbers, e.g. 9, 10, 12 when counting forwards or 13, 12, 10 when counting backwards.	• Lots of opportunities to say the sequence and see the sequence on both vertical and horizontal numeral tracks.
Unit 2.7 Some children get confused when saying the teen numbers. They may confuse the 'teen' and 'ty' sounds and say: • 30, 40 and 50 for 13, 14 and 15 • twen teen for 20 • ten teen for 10 • five teen for 15.	• Model the words in the correct sequence; make sure the 'teen' sound can be clearly heard. • Ask the children to notice what happens to the position of their mouths as they say the word 'teen'.
Units 2.5, 2.6, 2.7, 2.8, 2.9, 2.10, 2.11, 2.12, 2.13, 2.14 and 2.16 They might confuse the number word before for the number word after.	• Pointing to a hidden space on a numeral track to the right for number word after and to the left for number word before.

Potential misconception	Addressing the misconception
Units 2.5, 2.6, 2.7, 2.8, 2.9, 2.10, 2.11, 2.12, 2.13, 2.14 and 2.16 Children who find giving the next 'number word after or before difficult may have to drop back to one and count up to the number before being able to give the next number after or before.	• Give them two or three numbers before, for example: 'What's the number before 6?' Say '8, 7, 6 ...?' 'What's the number after 7?' Say, '5, 6, 7 ...?'
Units 2.10, 2.11, 2.12, 2.13 and 2.14 Difficulties can arise when identifying numerals; the numerals 6 and 9, 2 and 5 and 3 and 8 are often reversed.	• Give the children lots of opportunities to touch large concrete numbers and encourage them to explore the shapes and describe them. • Change the order the numerals are presented so they are not always shown in order.
Units 2.10 and 2.11 Difficulties forming the correct numerals.	• Practise making the numbers by using large motor movements in the air, paint and sand for example. • Give lots of opportunities to practise forming numerals by providing materials such as clip boards, pencils and numeral cards to accompany appropriate activities. • Chant mnemonics to accompany recording of numerals that are being modelled to the children.
Units 2.1, 2.2, 2.3, 2.4, 2.5, 2.6, 2.7, 2.8, 2.9, 2.10, 2.11, 2.12, 2.13, 2.14 and 2.15 Some children may not distinguish very much between numerals and letters	• Frequent activities with the numerals and naming them.
Units 2.1, 2.2, 2.3, 2.4, 2.5, 2.6, 2.7, 2.8, 2.9 and 2.15 Be aware that some children may allocate the word 'zero' to the first object counted.	• Do not use number tracks/lines with zero as the first number. • Opportunities to count collections as well as zero collections.
Units 2.17 and 2.18 When tracking a movement or sound on fingers it can prove tricky for some children to assign one finger to one sound or movement.	• Reduce the number of sounds or movements to be counted to 2 or 3. • Keep the count regular (not too quick or slow!). • Practise the forward number sequence to 5.
Units 2.17 and 2.18 Some children may raise a random number of fingers without looking at their hands first to check.	• Practise the forward number sequence to 5. • Practise one-to-one counting with small collections. • Provide lots of opportunities to collect, name and record small collections.
Units 2.19 and 2.20 Look out for the children who are still counting the dot patterns in ones rather than subitising. They may recognise some of the patterns within 5 but not all of them.	• Reduce the number of dots to 3 until they can recognise the patterns without counting them, then include 4 dot patterns and then 5.
Units 2.19 and 2.20 Some children who can subitise regular patterns dot patterns to 5 resort to counting in ones for irregular patterns.	• Make sure children have plenty of experiences using random patterns as well as regular ones.

Teaching ideas

Unit 2.1 Principles of counting (1 and 2)

Whole class

- **Songs with Visuals** – Songs and nursery rhymes are a fabulous way to introduce children to counting. Songs like 'One tomato, two tomatoes' with finger puppets or cards can be used. Say 'Put your cards up one at a time as we sing the song.' **How many tomatoes have we got up now? What number have we just said?** 'Let's count them from the beginning of this line.'
- **S Cube Drop** – Place a tray and a row of cups on the floor and count them out as you place them down. (Start with just one cup, then two, etc.) **Who would like to count how many cups we have?** Drop one interlocking cube into each cup, again saying the number word sequence one at a time. **How many cubes do you think there are altogether?** Empty the cubes onto the tray and group them all at one end. Count them all by moving them one by one to the other end of the tray while saying the number words in sequence. **How many cubes did we count? Did you expect to have this number of cubes? How did you know?** Click the cubes together (not for 1) and place behind the numeral. (Repeat with two cups, then three, etc.)

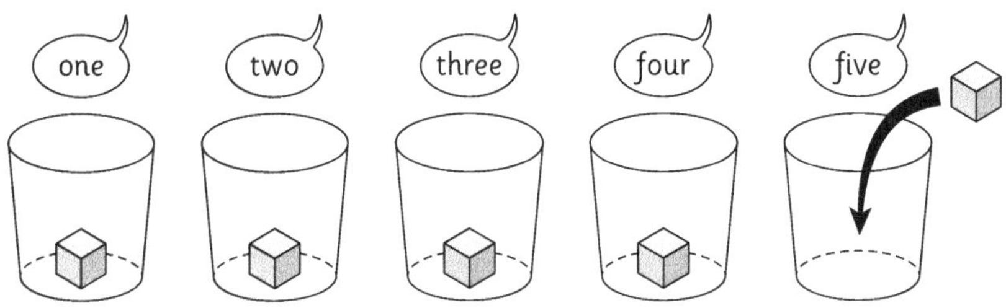

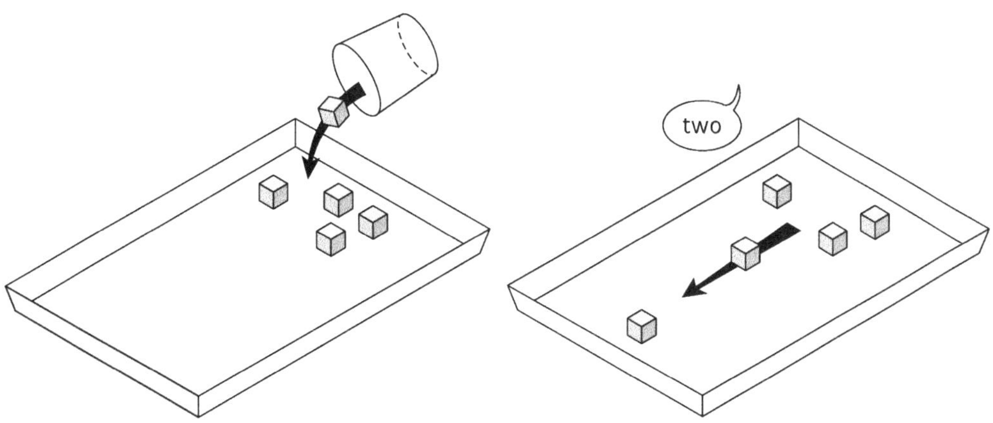

Stations

- **Sticky Dots** – Place large numerals with furry dots on to represent their value, a bowl of pom-poms and a pair of tweezers at the station. Ask the children to pick up one pom-pom at a time using the tweezers and place it on one of the dots. **How many dots are there on this number? How many pom-poms are there? What does this number say?**

- **Fill the Cup** – Have a collection of interlocking cubes and a cup for each player. Say, 'You are going to take it in turns to roll the dice and take out that many cubes from the box. Say the number sequence out loud as you pull one cube out of the box at a time. Your partner needs to check that you have collected the same amount of cubes as is shown on the dice. They need to say the number words out loud too. Then you can put the cubes into your cup.' **Who do you think will fill their cup up first? Did you agree with your partner's count?** 'Now it's your partner's turn.' Repeat until someone's cup is full.

Unit 2.2 Principles of counting (3)

Whole class

- **Rhymes with Visuals: Chicks in the Coop** – Take an old shoebox with a lid and place several chicks inside. Say the rhyme and invite children to 'peep' inside the box.

  Here is the chicken coop but where are the chicks?

  Hiding away where no one looks.

  When I peep in, I can see them all. (**Can you see some chicks? How many can you see?**)

  I think there are **7?**

  Let's give them a call!

  Chickadees, chickadees

  Out you come. (Bring out the chicks one at a time while saying the forwards counting sequence: '1, 2, 3, 4, 5, 6').

  Were you right? Were there seven chicks? How many chicks were inside the coop? Repeat with a different amount of chicks.

- **Grumpy Queen** – Make a crown for the Queen and attach a numeral card. Choose a child to wear the crown. Give the other children a small basket. Say, 'We need to make the grumpy Queen happy by giving her a basket that holds the same number of things as the number on her crown.' Give the children one minute to collect their items and then ask them to present them to the Queen. **How do you know you have the correct amount? Can you prove it? Is the Queen happy with your collection? Do the collections have to have the same thing in them or can it be a mixed collection?**

Stations

- **Goldilocks and the Three Bears** – Set up the story corner with props from a familiar story, e.g. Goldilocks and the Three Bears. Provide more than enough blankets, bowls, spoons and beds for each of the bears. Place some subitising cards or numeral cards out too. **How many bowls have you put out? How many spoons did you need?** Encourage the children to count out the number of items. **What if Little Red Riding Hood comes for breakfast? How many cups will you need now?**

- **Numbered Towers** – Build towers to match the numbers on the card. Then turn the card over and see if your tower matches the one on the card (see Photocopiable resource 2.1). **How do you know you have the right amount? Can you count each cube one at a time? Give me 5 cubes. If you place your tower on top of the picture, does it match?**

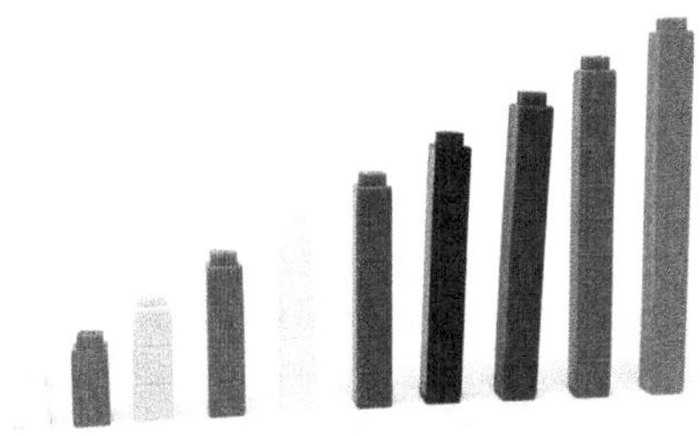

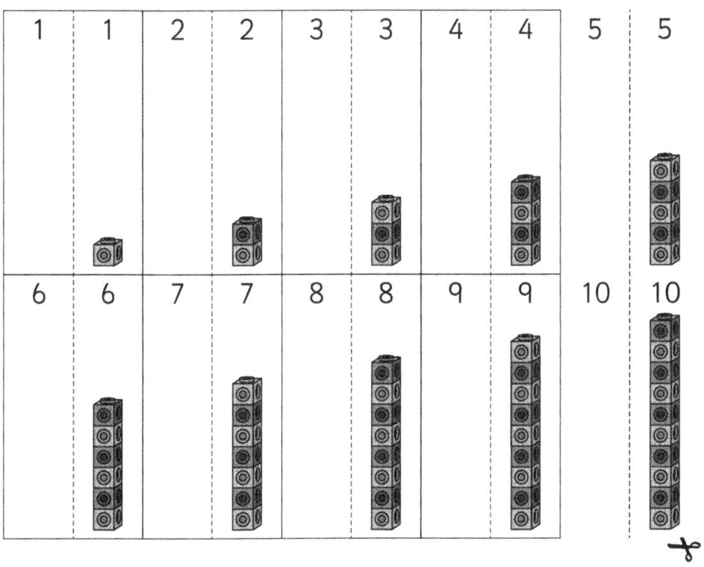

Photocopiable resource 2.1 – Number Towers.

- **Race to Build Towers**– Have a basket of two different colours of Lego bricks for two players. Tell the children they are in a race to see who can build the most towers. They must take it in turns to roll the dice and collect that amount of Lego and then place it on the board (see Photocopiable resource 2.2). If there are no spaces left for a particular sized tower, they miss a go. 'Show me a tower of 5.' **How do you know it has 5? Which is the smallest tower? How many Lego bricks does that have?**

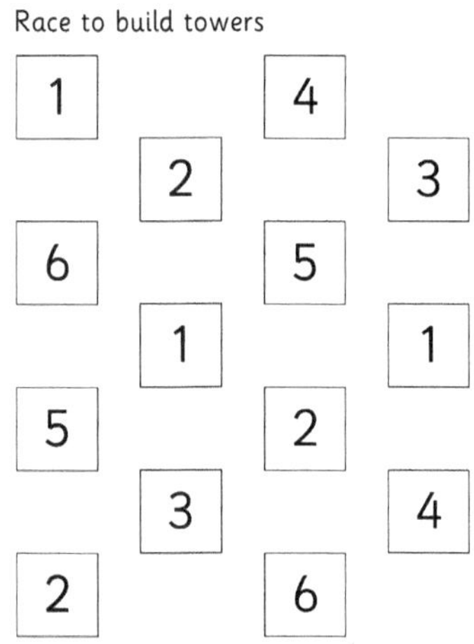

Photocopiable resource 2.2 – Building Towers.

Unit 2.3 Principles of counting (4)

Whole class

- **Number Hunt** – Go for a number hunt around the school and the local area. Let the children take photographs for a large floor book. **Why is there a number 9 on that door? Do you think 9 people live in that house? Look at this car registration plate, how many numbers can you identify? Can you point to the number 3? There are birthday cards in the window, what does the 5 mean? He has the number 15 on his t-shirt; is this how old he is?** The book can be added to another trip out or the camera can be sent home to collect more photographs.
- **Stories** – Stories that are rich in number talk, e.g. *Alfie's Numbers* by Shirley Hughes, can be used to explore the idea that anything can be counted. 'There are five people getting ready for the party, let's count them.' **Have you ever been to a party with five people? How many balloons can we see? Will everyone get a balloon?**
- **Teacher Says** – This is just like Simon Says, where a number has been said or shown. For example, if the number 6 is chosen: 'Teacher says jump six times.' 'Teacher says clap six times.' Tell the children they have to shout the numbers out as they do the action. Include counting in their head too. If the teacher does not tell them to do it, they must stand still or be out, e.g. 'Tap your head six times!'

Stations

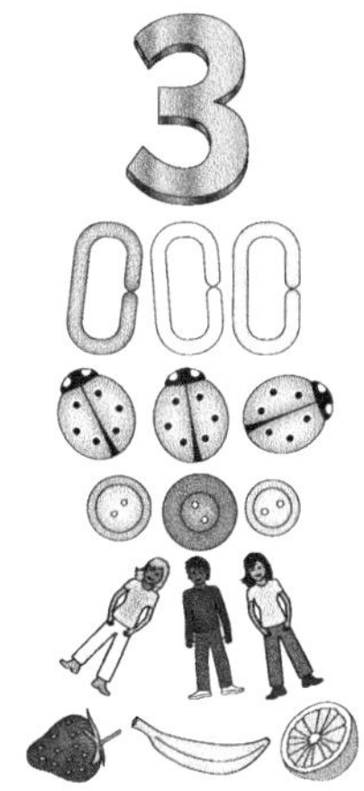

- **Collections** – Put bowls of different collections out on the table and a set of wooden numerals. Place a wooden numeral on the table, e.g. the number 3. Then place three of each collection under the number. Ask the children to do the same with a different number. **How many bugs have you counted? How many fruits have you got? How many buttons have you got?**
- **Nature Table** – Go for a walk in the woods and collect items in small bags. Place all the items on the nature table. Ask: **I wonder how many sticks we have collected? I wonder how many flowers we have got?** 'Let's count them.' Encourage the children to line them up next to each other as they count them. Write numbers down on cards that represent the collection. 'We have four fir cones.' **Do we have four of anything else?** 'We have six flowers.' **Can you check by counting them?**

Unit 2.4 Principles of counting (5)

Whole class

- **Counting Teddies** – Place some counting teddies in a bowl, grab a handful of them and line them up. Starting at one end, point and touch each teddy bear, saying the counting words from one. When you have touched the final teddy bear, say how many there are altogether, e.g. '1, 2, 3, 4, 5, ***6***. There are six of them altogether'. Make sure the emphasis is put on the last number said. Ask a child to repeat the count but this time starting at the other end. Point this out to the children. 'Let's start counting at the other end.' **Can we start the count in the middle? Will we still get the same amount? What will happen if we start counting with this red one?**
- **Lining Up to Go Outside** – Line the children up in one long line and tap them on the shoulder as you walk down the line saying one number in the forward counting sequence at a time. State how many children there are altogether, e.g. 'There are 15 children in the line.' Now ask them to split into two lines, e.g. 'Those of you wearing a scarf line up here. Those of you not wearing a scarf, stay in this line.' **How many children will there be now?** 'A moment ago, I counted 15 children.' **Are there more or fewer children now that we have two lines?** Repeat the count as above. Say, 'There are still 15 children altogether.' Do this often and change the number of lines.

Stations

- **Hanging up Washing** – Prepare a washing line and a basket of the same sized items in three different colours, e.g. socks or t-shirts. Tell the children to hang the washing out. **How many t-shirts have you hung up?** 'Let's start at this end and count them.' **How many will there be if I start at this end? Can you count them again? Can you count the red ones first and then the blue and green ones? How many have you got altogether? What would happen if you hung all the red ones together, then all the blue ones and then the green ones? How many would you have altogether?**
- **Counting Butterflies** – Place a box of butterflies (three colours) on the table, some numeral cards (up to 10) and four laminated flowers. Tell the children the butterflies are hungry and need some nectar from the flowers. Can they turn one of the number cards over to find out how many butterflies they need to put out? **How do you know you have eight butterflies? How do you know you have counted them all? What happens if you start counting on this flower? How many red butterflies have you got? Does it matter what order you start counting them?**

Unit 2.5 Forwards number sequences to 10

Whole class

- **Count, Count, Count** – Start off with a very small number of forward counting words and tell the children they are going to copy you. Say, '1, 2, 3, now your turn.' Then say, '2, 3, 4, now your turn.' Then say, '1, 2, 3, 4, now your turn.' Continue with these shorter sequences until the children can confidently say the forward number sequence up to ten.
- **Number Songs** – There are numerous forwards counting songs for children to learn, such as 'Five currant buns', 'One potato, two potato', 'One, two, buckle my shoe' and 'One, two, three four, five, once I caught a fish alive'. Tell the children they are learning how to say the forward counting numbers up to 10 and will be using props to help them keep count. Give 10 children one rubber fish each and ask them to place it down on the floor when you point to them. Sing, 'One, two, three four, five' (while simultaneously pointing to the five children in turn to get them to place their fish on the floor), 'once I caught a fish alive.' Then again for 'six, seven, eight, nine, ten, then I let it go again'. The song should then be practised for fluency while you point to the fish.

Stations

- **Counting the Bounce** – Provide a large ball and a clipboard with photographs of the children on. Tell the children there is a competition to see how many times you can bounce the ball to each other in two minutes. Say, 'For the bounce to count you must take it in turns to say the number out loud in the correct order. For example, Carra bounces the ball to Isobel and shouts 'one' and Isobel catches it and bounces it back to Carra, while saying 'two'.' They record the number of bounces in two minutes and someone else can try to beat their record.
- **Giant Steps** – Start at one end of the hall and tell the children they are giants and they are going to see how many giant steps it will take to reach the other end of the hall. Say, 'On each step you make, you must say the forward number sequence out loud. There is only one problem though. Each time you hear a bell you must freeze for 10 seconds and say the counting forward numbers out loud. Then you can continue to make giant steps across the hall once you hear the second bell.'

Unit 2.6 Backward number sequences from 10

Whole class

There are many stories that have the backwards counting sequence, such as *The Shopping Basket* by John Burningham. Read the story aloud and allow the children to copy the sequence with the numbers and the items and the numbers on their own. For example, 'On his way to the store to get six eggs, five bananas, four apples, three oranges, two doughnuts, and a bag of chips.' 'Let's repeat that list. You say the numbers and I'll say the items: **six** apples, **five** bananas, **four** apples, …' Show the appropriate finger pattern as you are saying the numbers.

- **Counting Children** – Get ten children to stand at the front of the class. Count them forwards to check there are ten, place a stringed number card around their necks. Say, 'We are going to practise counting backwards from 10.' Stand behind the children and ask, **How many children do we have here?** 'Yes, ten!' Tap one on the head and ask them to sit down. **How many children do we have now?** 'Yes, nine. So we had ten, now we have nine. You say it – ten, nine.' Repeat. 'So we had ten, nine, now we have eight. You say it, ten, nine, eight.' Repeat until there are no children left.

Stations

- **Drawing Pictures** – Drawing pictures of songs or books that have counting sequences in. For example, 'Can you draw the five frogs on the log?' Ask them, **Can you say this sequence backwards?** Say a small string first, 'Three, two, one. Now it's your turn; four, three, two.' Build the sequence up to five numbers; give a run up to the sequence if they are struggling, e.g. you say, 'Five, four' and the child continues 'Three, two, one'.
- **Loud and Soft Numbers** – Say the backwards counting sequence using alternate loud and soft voices. Start from 5 down to 1, then 7 to 1, then 10 to 1. Change the starting number. Don't always count down to one.

Unit 2.7 Forward number sequences to 20

Whole class

- **Body Actions** – Tell the children they are going to put movements to the forward counting numbers up to 20 to help them remember them. Tell them they are going to say the numbers out loud. For example, 'One, two, three, four, …' Co-ordinate one clap, stamp feet, tap head with each number or with alternate numbers, clap hands on odd numbers and tap knees on even numbers.
- **Circle Counting** – Tell the children they are going to count to 20 but different people will be asked to count. Tell them if they get tapped on the head, they must continue the count. Stop at 20. The next person in the circle starts again from 1. Listen very carefully to any confused numbers and then model how to say the teen sound with a lot of emphasis on the 'teen' word. 'Now we are going to count to 20, but this time we are going to start at a different number.'

Stations

- **Woodland Steps** – Tell the children they are going to see how many steps it will take them to get from A to B by pretending to be different animals. (Make sure it will be somewhere in the teen numbers). For example, 'Let's start from this log.' **How many steps will it take to get to the tree if we are a bear?** 'Ready, nice big strides, one two, …' **What if we are mice? How many steps would it be to go from the tree to the campfire?** The children could walk with a partner who has to listen to their counting. '**That was 17 steps, how many more to 20? What if we get to 20, what's next?**
- **Hide and Count** – One child (the Pointer) stands in the middle of the circle while the other children stand in a circle facing outwards. Tell the children they are going to take 20 steps forward while saying the forward number sequence to 20 out loud. The 'Pointer' shouts 'Turn around, turn around, and face back home.' The children turn around and face the centre. The Pointer chooses one (or more) children who must try to get back to the centre while counting 20 steps forwards. If it is someone who may struggle to say the numbers in order, the class could help support.

Unit 2.8 What's the number after?

Whole class

- **The Language of After** – Have a variety of visuals to support the familiar activities of nursery children, such as brushing teeth and getting dressed. Show these to the children to help them to decide. As a class discuss the concept of 'after' by discussing questions such as: **Did you take your outdoor shoes off 'after' you put your indoor shoes on? Did you put your socks on 'after' you put your shoes on? What did you do straight 'after' your dinner? Do you eat your lunch 'after' going out to play?** 'Let's look at the pictures again and put them on this line.' **Which one happens after the other one?**

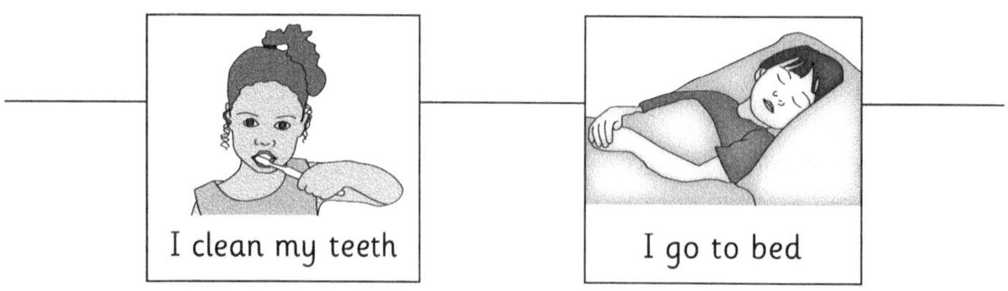

Stations

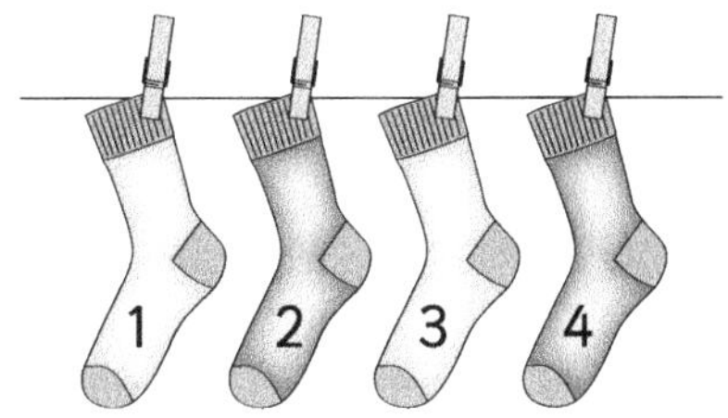

- **Washing Line** – Hang a washing line up outside and ask the children to peg out the socks with the number 1 on it. **What's the number after 1?** 'The number after 1 is 2.' **Can you find the sock with the number 2 on?** Peg that out. **What's the number after 2?** Give a run up if needed, '1, 2, . . .' 'Peg out the number 3.' Repeat and include numbers 6 to 9 if they know the number words after 1 to 4.
- **Parking Cars** – Put numeral stickers (1–10) on ten toy cars and place a numeral track big enough to park the cars in each individual square. **Can you line the cars up so they match the number on the track? Which number is after the number 4? Where would you park the number 3? What number car sits in the space after the number 8?**

Unit 2.9 What's the number before?

Whole class

- **The Language of Before** – Have a variety of visuals to support the familiar day-to-day activities of nursery children, such as eating breakfast and playing outside. Show these to the children to help them to decide. As a class, discuss the concept of 'before' by discussing questions such as, **Would you eat ice cream 'before' you go to school? What do you do just 'before' you get into bed?** 'Let's look at the pictures again and put them on this line.' **Which one happens before the other one?**

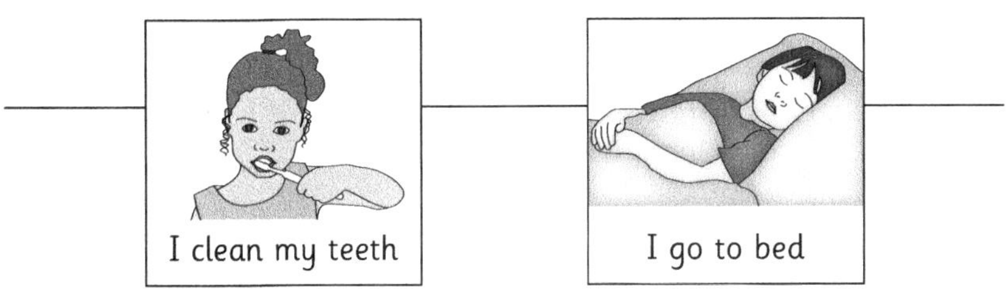

Stations

- **Picture Cards** – Place some photo cards of the day-to-day routines in the nursery (limit them to four or five). Say, 'I need your help to put these pictures in order so the girls and boys know what to do to get ready for their snack time.' **Would the photo of the girl drying her hands go before them putting soap onto their hands? Which one comes before she starts eating the apple?**

- **Drawing Pictures** – Ask the children to draw pictures of something that they like to do before, during and after they come to nursery. Make a display of the pictures and discuss the similarities and differences between each child's favourite things to do. The display could change regularly to include themes, such as 'Things I like to eat ...' or 'People I like to see ... before, during and after I come to nursery.'

Unit 2.10 Reading numbers

Whole class

- **Hidden Numbers** – Make a wall out of Lego and ask the children if they can spot which number it is as it gets revealed a little bit at a time. **Look at the shape of it? Does it have a round edge or is it straight at the top?** 'Draw in the air what you can see, look at the next bit.' **Which number do you think it is?**

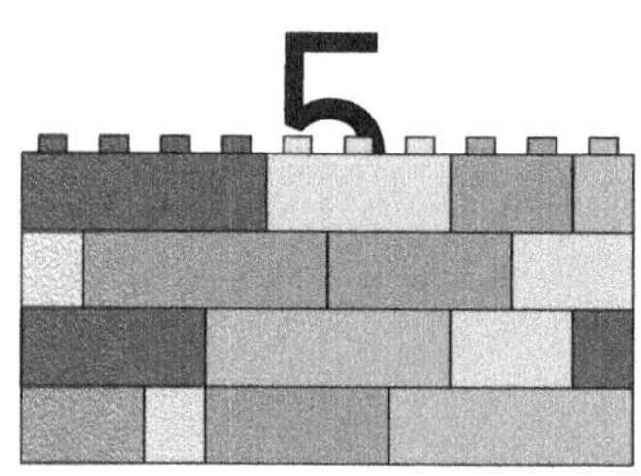

- **Numeral Swap** – Sit the children in a circle and give each child a wooden numeral. Say, 'You need to listen carefully. If I call your number out, you have to stand up and swap places with someone else that has the same number as you. Swap places if you have the number 2, ...' **How do you know you are correct? Did you look at the person's number that sat in your old seat?**
- **Numbered Treasure Chests** – Put four treasure chests out on the carpet, each with a different number on from the 1–10 range. Hide a piece of treasure in one of them when the children have their eyes closed. **Where is the secret treasure hidden?** Tell the children they must individually say which number they think the treasure is hiding under and then go and sit by that number, e.g. 'I think the treasure is in chest number 7.' Once everyone has had a go, one child from each group can open the lid. The numbers can be altered on the boxes.

Stations

- **Fly Snap** – Provide two fly swatters and a packet of large playing cards with the numbers 1–10 in. Deal the cards out equally and place them face down. Each child reveals their top card at the same time and if they match, they can swat them with the fly swatter and shout snap. To get the cards, they must name them. If they get the name of the number wrong, the other person collects the pair of cards. **How do you know if you are right? Does your friend agree?** 'Check on the number line and say the numbers from one to ten to check that you are correct.'
- **Secret Caves** – Provide a collection of small cardboard boxes each with a number from 1 to 10 on the top. Tell the children they need to choose three of the boxes and hide a plastic cow under one of them without anyone seeing. The other person has to use the number to guess where the cow is hiding, e.g. 'I think the cow is hiding under the number 6'. Lift up the number 6 to check.
- **All the Number Twos** – Have a variety of number cards in different fonts, colours and patterns to ensure the children can 'recognise' the number 2 in different ways. Ask them to go on a 'number 2 hunt'. Say, 'There are lots of number cards hidden around the room. Go and find as many as you can but do not bring any numbers back that are not 2'.

Unit 2.11 Representing numbers

Whole class

- **Sand Drawing –** Give each child a small tray with sand that covers the bottom. 'We are going to practise writing numbers from one to ten.' Guide them through the starting place, direction and shape they are making using a rhyme, e.g. 'First his hat, and then his neck, big round tummy, that's five, oh heck! First write it in the air. Next write it on your partner's back, finally write it in your sand box.' The children can then gently shake the box ready for the next number.

Stations

- **Dough Numbers** – Provide different coloured dough and some numeral mats. Roll the dough into snakes and form the number on the card. Take different coloured dough and roll out small balls that represent the number, e.g. 'You've made the number three'. **How many balls will you need to stick on?**
- **Number Grab** – Prepare some laminated hand outlines, whiteboard pens and some boxes of interlocking cubes. Ask the children to grab a pile of cubes and then count them. 'Write the amount in the palm of the hand.' **How do you know you are correct? '**Check your partner's amount and the number they have written.' **Do you agree?**

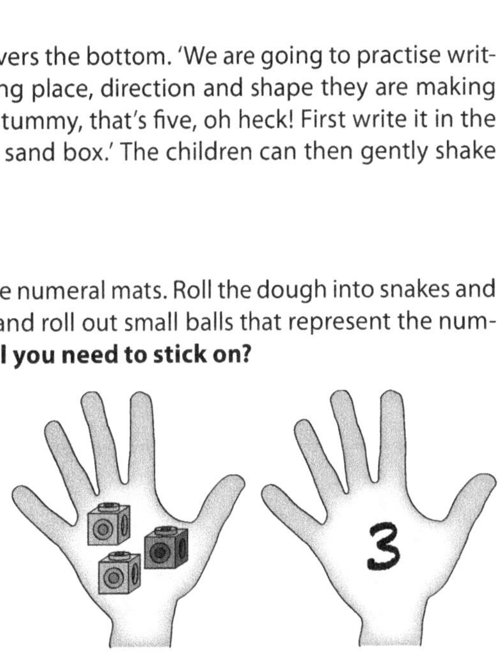

Unit 2.12 Sequencing numbers

Whole class

- **Using the Five Frames and Number Cards** – Put five carpet dots (or large card dots) on the floor with the first one labelled '1'. Flash some large five frames to the children. Ask: **Where would you place the five frames if they have to be in order from the smallest number to the largest?** For example, flash a 3 and ask a child to put it down on the correct dot. **How do you know this is the correct place?** 'Let's count from the dot at the end, 1, 2, 3.' **Is this where you put your three?** Next sequence them from largest to smallest.
- **Numbers on a Number Track** – Lay an empty numeral track down on the floor and have five numeral cards randomly placed face up. Tell the children you need help putting the cards in order from smallest to largest (largest to smallest). Point to one of the numerals, e.g. 4. **Where should this go? How do you know? Would counting this first space from 1 help?** Continue until the track is filled. **Who can read all the numbers on this track?**

Stations

- **Tidying Up** – Place stopping numbers for the items to be organised. Have labels on items in the nursery that will help children to tidy away and to know how many missing and where they should be placed. For example, 'Could you help dry up the plastic cups and put them away. Make sure you hang them up in the right order'. Write numbers 1 to 5 in different colours on the bottom of the cups and have a row of hooks in groups of five for each colour. **How do you know you've put them in the right order?**
- **Number Puzzles in Order** – Place wooden number puzzles out, tell them you would like them to put the puzzle back together. **How do you know if you've put the numbers in the correct order? Could you say the numbers in a forwards/backwards order?**

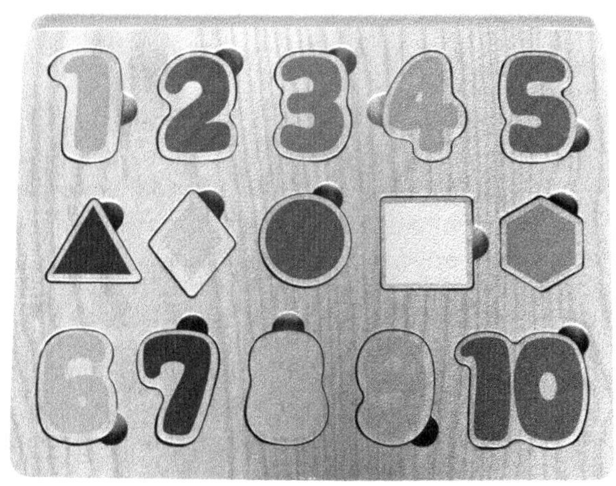

Unit 2.13 Ordering numbers

Whole class

- **Mischievous Ted** – Tell the children that 'Mischievous Ted' has been up to his old tricks and this time he has muddled up the number line and you need his help. Don't tell them yet that he has also hidden some numbers around the room. **Can you put the numbers in order from the smallest number to the largest? Can we still do this if some are missing?**
- **Pens in Pots with Labels On** – Put pencils in labelled pots that represent the numerical value. Leave two of the pots out. **Can you put these pots in the right order? Can you put them from largest to smallest?** Do the children notice there are some missing from the sequence?

Stations

- **Building Towers to Represent the Number** – Provide a dice, numeral cards from 1 to 6 and a box of Duplo bricks (or similar stacking blocks). Tell the children they have to make three towers. You need to roll the dice and make the tower with the number on the dice. If you roll the same number, you can throw again. Place the three towers in order from shortest to tallest and place a numeral card in front of each tower. **What do you notice? How do you know you have put them in the correct order?**

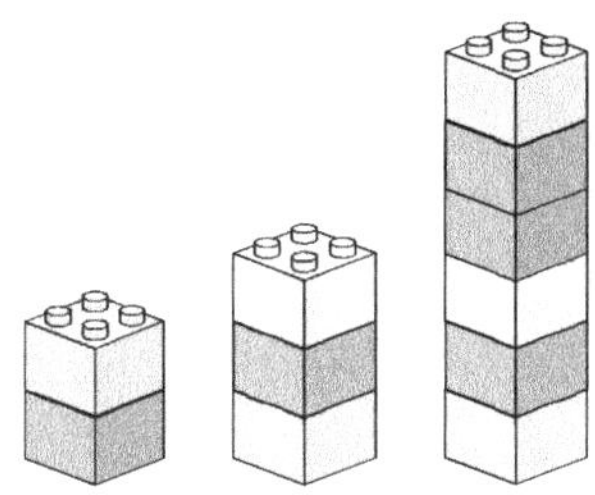

- **Numbers and Dot Cards in Order** – Make some flash cards with the numerals 0 to 5 at one end and the respective dice pattern on the other (see Photocopiable resource 2.3). Turn them face down. Tell them 'You must choose three dominoes and place them in order, starting with the largest number and ending with the smallest number.' **How do you know you've put them in order properly? Can you put them in another order?**

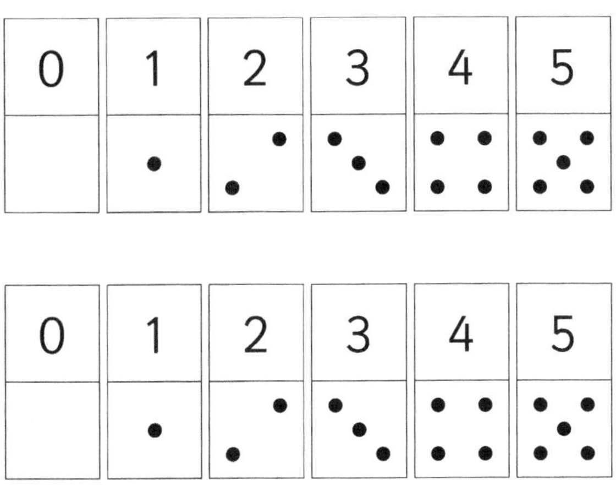

Photocopiable resource 2.3 – Number and Dot Flash Cards.

Unit 2.14 Missing numbers

Whole class

- **What's Under the Cloth?** – Lay five items on a tray with an accompanying number card. Each item should have the appropriate number of parts to represent the supporting numeral card, e.g. number card 1 for one dinosaur, number card 2 for two plastic figures holding hands, number card 3 for three interlocking cubes joined together, number card 4 for four beads on a string and number card 5 for five dots on a dice pattern card. Tell the children, 'I'm going to give you 20 seconds to look closely at the items and then I'm going to cover them up with a cloth. When I tell you to close your eyes, I will remove one of the items.' **What did I take away? How do you know? Have you noticed anything about the items and the number next to them?** To make this more difficult, move the items around once one has been removed.
- **Numeral Flip and See Track** – Place a number strip under the numeral track lids, allow the children to see all the numbers apart from one. **Which number is missing? How do you know?** Make sure the numbers are not always sitting in the same position, otherwise the children will always expect the number 3 to be in the third box along. Two lids could be closed and the first one could be flashed for 2 seconds. **What number is this one? If this one says 2, what's the next number along?**

Stations

- **Floor Walking** – Draw a large number track on the floor from 10 to 1, cover up a number in the 1 to 5 range. Say, 'I want you to walk along the number track and say the numbers as you stand on them. Stop when the number has been covered up.' **Can you tell me which number has been covered up? How do you know? Are you counting backwards or forwards?**
- **Missing Numbers** – Have a number line out at the station and a box of plastic numbers. Say, 'I have hidden a number under this box, can you find out which number is missing without looking underneath?' **How did you find the missing number? Did you have to look at all the other numbers? Did you use the number line to help you?**

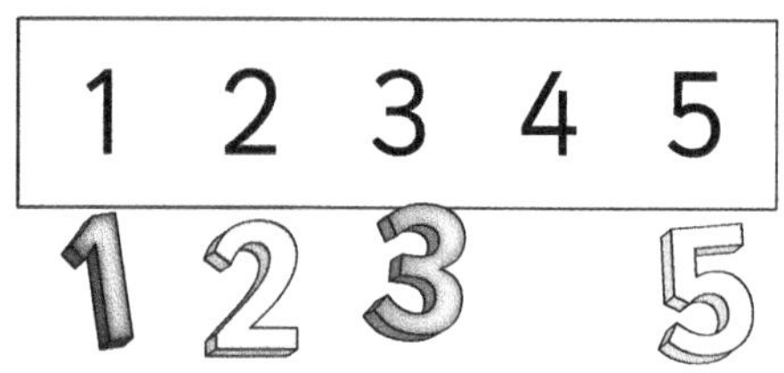

Unit 2.15 What is zero?

Whole class

- **Dandelion Heads** – Carefully give each child a dandelion head full of seeds, tell them they are going to see who can blow all the seeds off so the stem has none left. **What has happened to all the tiny seeds? What other words could we use instead of 'all gone'?** Tell them zero means nothing left or none left. Show them the digit 0 and explain that this means nothing. Have a collection of flower stems with a varied number of petals still on and at least one with no petals. Ask them to place numeral cards next to each flower to show how many petals it has. **Why does this flower have the zero card?**

- **Songs** – Counting back songs are a good way to discuss what zero means. Songs like 'Five Currant Buns' are ideal. **What happens when the last one is bought? How many buns will there be left? How can we get some more? What does zero buns mean?**

Stations

- **Action Cards with Plastic Numbers in the Bag** – Have some Animal action cards in one feely bag, e.g. scratch like a monkey, sting like a bee (see Photocopiable resource 2.4) and numerals 0 to 5 in another feely bag. Say, 'You need to pick a card and a numeral and tell the class what to do. For example, hop like a bunny four times, clap like a seal zero times. Keep the numerals out of the bag until there are none left and then you can put them back in and start again.'
- **Snack Time** – During snack time set up situations that involve you having a conversation about there being nothing (zero) left on the plate. For example, no rice cakes on one plate and 4 on another one. 'This plate is now empty; there are no rice cakes left on this plate.' **Who can tell me what number of rice cakes are left on this plate?** 'There are zero rice cakes left on this plate.'

Photocopiable resource 2.4 – Animal Actions.

Unit 2.16 Ordinal numbers to 3

Whole class

- **Familiar Routines** – Familiar routines like preparing for home time or having a snack are a great way to introduce the children to ordinal numbers. For example, when getting ready to go to the woods say, 'First we need to take our indoor shoes off and put our wellies on.' **What is the next thing to do?** 'Yes, second we need to put our coats on.' **What is the third thing we need to do?** 'Yes, third we need to put on our hat and gloves.'
- **Crossing or Standing in the Line** – Introduce the language of ordinal numbers when it occurs naturally, e.g. when children are lining up for home time. **Who is the first person in the line?** 'Carolyn is first in the line.' **Who is second? What position are you in Abbie?** 'First, second, you must be in third position Lisa. Let's swap everyone around.' **Lisa, are you still in third position? Is this place always second place?**

Stations

- **Building a Snowman** – After an outdoor activity, cards can be made from any photographs taken that show a clear order to the event. Ask the children to look carefully at the cards and find the first card, then the second, etc. For example, before the children go outside to build snowmen, the order can be discussed. **Could we put the hat on before we've rolled the head? What do we need to do first? What's the last thing to do? Can you put these cards in order from first to last so we can see how to make a snowman?**

- **Three Guesses** – Have three identical cups and hide an object under each one of them. Say to the children, 'I'm going to lift each of the cups for a few seconds so you can see what is underneath each one. Then I'm going to move them around, so watch carefully to see if you can follow where the objects are.' Move the cups around in front of the children without lifting the lids off the table. **Who wants to take the first guess and say what object they think is hiding underneath this cup? Who would like to have the second guess?**

Unit 2.17 Tracking sounds on fingers

Whole class

- **Pendulum Count** – Attach a heavy weight to the end of a 50 cm piece of string to make a pendulum. Hold the pendulum and say 'I would like you to keep track of the number of times the pendulum swings past my feet. Keep a track on your fingers and stop counting once it has stopped.' **Was it easy to count the swings? What made it harder?**
- **Pennies in a Tin** – Tell the children it's easier to keep track of sounds that have a single beat on your fingers. Say, 'Watch me as I drop a penny into the tin. Each time you hear it drop to the bottom I want you to raise one finger and count forwards.' **Is it easy to hear the sound? What happens if the penny jumps up and makes another noise; will you count that one too? How many pennies are in the jar?**

Stations

- **Instrument Bash** – Place six different instruments on the carpet (use instruments for which playing a single beat is easy to hear, e.g. drum, recorder, triangle). Have one dice and six number cards from 1 to 6. Number the instruments from 1 to 6. Six children pick a number card and collect the appropriate instrument. One child then rolls the dice so no one can see the number. The rest of the class then close their eyes. 'You have to make the number of sounds that are on your dice.' The children have to count the beats on their fingers. **How many beats did you hear? Did anyone guess correctly? Was it easy to hear the beats?** Repeat with six new children.
- **Body Tap** – Tell the children to get a partner and sit behind them ready to draw on their back. Say, 'you are going to tap some dots on your partner's back but no more than five taps. Your partner has to keep a count of the taps on their fingers and then tell you how many taps you did.' **Did they guess correctly? Did you remember how many taps you did? Did you remember to not make more than 5 taps?**

Unit 2.18 Finger patterns

Whole class

We are going to use our fingers to make a number. **Can you put up one finger like me? How many fingers have you got up?** Now put your finger down. **Can you put up two fingers like me?** Put one finger up at a time. **How many fingers have you got up?** Now put your fingers down. Repeat with numbers 3 to 5. **Can you make them on your other hand?** Repeat sequence as before. Now choose the numbers from 1 to 5 randomly and watch how they 'grow' their fingers. Ask, can you show me 4 fingers? Make finger patterns on your own hand and ask **How many fingers is this? Can you copy my pattern? Can you show me 4 fingers? How many fingers is this?**

- **Outdoor Nature Trail** – Make a finger pattern spinner (see Photocopiable resource 2.5) and place a string on it so that it can hang around a child's neck. Each child has their own hoop on the ground. Tell the children to spin their own spinner and read how many fingers there are. They then have to run and collect that many sticks to put in their hoop. Once they've done this, they spin the spinner again and repeat. **How many sticks can you collect in 1 minute? Did your spinner land on every finger pattern?**

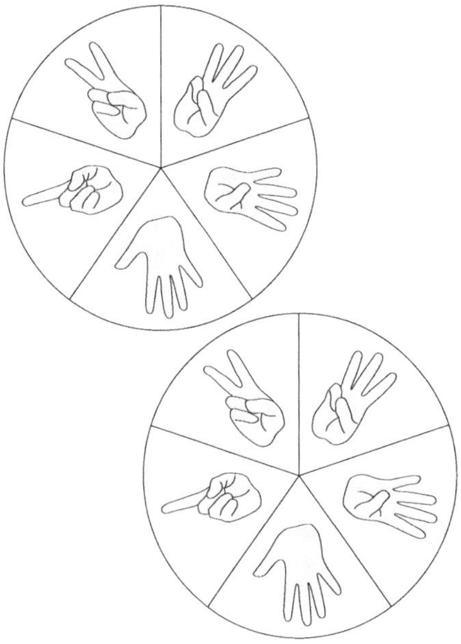

Photocopiable resource 2.5 – Finger Patterns.

Stations

- **Dinosaur Spikes and Finger Flash Challenge** – Place finger pattern cards from 1 to 5 face down on the carpet. Laminate some pictures of dinosaurs that have no horns or back spikes (see Photocopiable resource 2.6). Give children some whiteboard pens and tell them to choose a finger pattern card and draw that number of horns or spikes onto their dinosaur. Now choose another card and do the same on a different dinosaur. Continue until all four dinosaurs have spikes or horns. **Can you make the finger pattern for the number of spikes on each of your dinosaurs? Can you make the finger patterns for your friend's dinosaurs?**

Photocopiable resource 2.6 – Dinosaur Challenge.

- **Finger Dominoes** – Make a set of finger pattern dominoes up to five and five fingers (see Photocopiable resource 2.7). Turn them face down on the carpet. This is a game for 2–4 players. Share the dominoes out. The child with the 5–5 domino places it down on the carpet. The person next to them puts down a domino if they have a matching pattern. If they do not, they miss a go. Play continues until no one else can go.

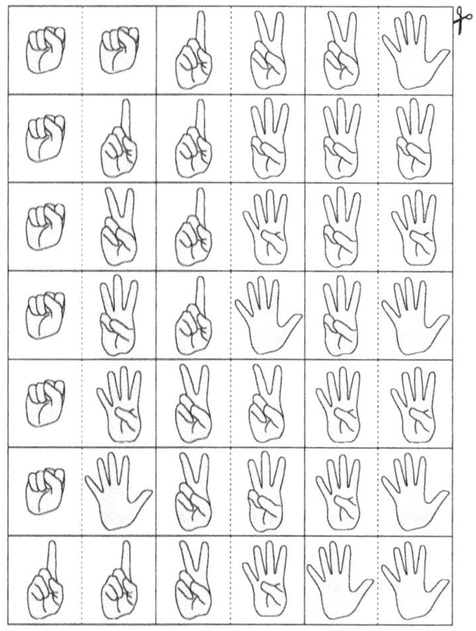

Photocopiable resource 2.7 – Finger Dominoes.

Unit 2.19 Regular dot patterns to 5

Whole class

- **Different Representations of 5** – Show the children a fives frame and explore the number of spaces. Explain that if it is full, there will be five dots. Introduce them to what it looks like with one dot, two dots, etc. (see Photocopiable resource 2.8). **How many dots? How many spaces? What can you see**? Give the children an empty fives frame and five counters of the same colour. Show the frames and ask the children to make what they saw on their own frames. Once confident making the patterns, flash the patterns for 2–3 seconds and ask them to say what they see.

Explore dice patterns and domino cards on another occasion. 'Look at how the dots are spaced out. Look at the space between the dots.' **Can you see the four dots in the corners? Can you see the one dot in the middle of the square?**

- **Look and Snap** – Place six numeral cards face up in front of each child. Briefly flash a dot pattern (dice, domino or fives frame) in the 0–5 range. Say, 'When you see the dot pattern, you have to quickly place your hand on the number of dots you see.' **Did you choose the right card? Did everyone choose the same number?**

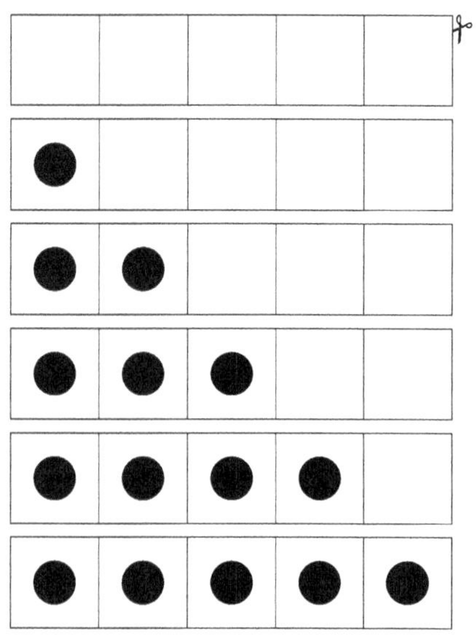

Photocopiable resource 2.8 – Fives Frames.

Stations

- **Chalked Dots** – Provide some mini blackboards, chalk and a collection of regular dot cards from 0 to 5 (see Photocopiable resource 2.9). Tell the children to hide a dot card in the magic box (cardboard pizza box). One person is the magician; they have to flash the pattern by briefly raising the lid. The other children have to draw on their blackboard the number of dots they saw. They then get to see the pattern and check if they were correct. The magician then makes that pattern disappear and a new one arrives. Repeat. **How many did you see? Did you draw enough dots?**
- **Dress for the Seasons** – This game is like the Beetle Game. Each child has an outline of a body, a laminated hat, coat, boots and gloves (see Photocopiable resource 2.10). Each item of clothing is labelled with a number, e.g. 1 for the body, 2 for the coat, etc. Tell the children to take it in turns to roll the dice. They must roll a 1 to start. Once they've got their body, they can collect the rest of the items. If they already have enough of an item, they miss a go. The first child to dress the body for winter is the winner. This can be done for different seasons by changing the outfit.

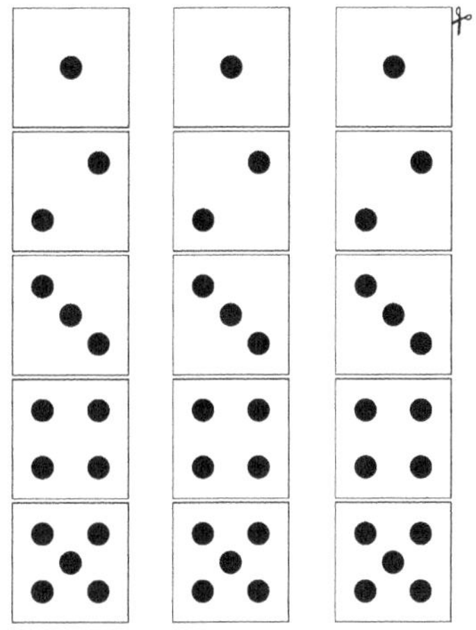

Photocopiable resource 2.9 – Regular Dot Patterns to 5.

Photocopiable resource 2.10 – Getting Dressed.

Unit 2.20 Irregular dot patterns to 5

Whole class

- **Copy Cats** – Prepare 15 dots cards with a variety of representations of the numbers from one to five (one set of regular and 2 sets of irregular cards) and five numeral cards from 1 to 5 per child (see Photocopiable resource 2.11). Tell them you are going to show them a card; call out 'Ready, steady, look!' Show them the pattern for 2–3 seconds and then hide it. Ask them to place their hands on one of the numbers cards from 1 to 5. Tell them to make the pattern in the air to prove it. **Did you guess the correct number? What did you see? Which patterns were easier to see than others?**

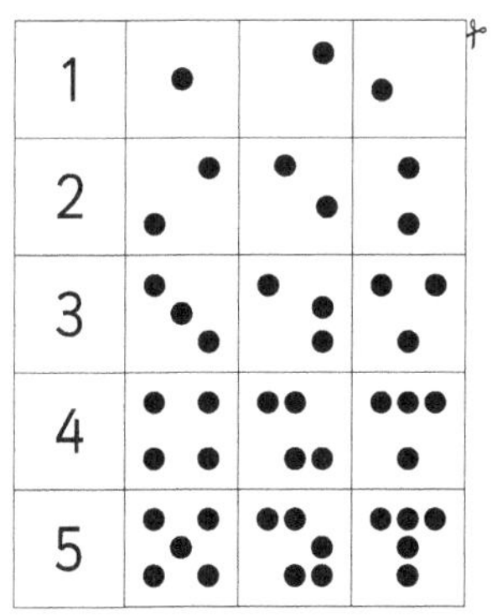

Photocopiable resource 2.11 – Mixed Dot Patterns to 5.

- **Flash and Circle** – Flash an irregular pattern of dots for 2–3 seconds and then ask the children to draw around that many dots on the paper (see Photocopiable resource 2.12). **How many did you see? Do you have that many dots left on your paper?**

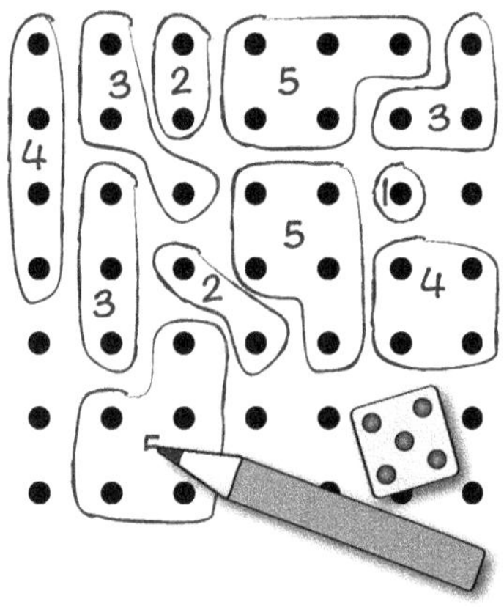

Photocopiable resource 2.12 – Dotty Paper.

Stations

- **Irregular Snap** – Provide 20 cards with four sets of cards from 1 to 5 (see Photocopiable resource 2.11 – Mixed Dot Patterns to 5). Tell the children they are going to play 'Snap' and match two cards with the same number of dots. Say, 'One of you should share the cards out equally. Then take it in turns to turn your cards over after each other. If you turn over a card with the same number of dots as the previous card, the first person to slap their hand down on top of the cards, wins the entire pile of cards.'
- **Blindfold Feely Dots** – Provide a blindfold, some felt dot cards with irregular patterns to 5 (see Photocopiable resource 2.13) and some counters. Tell the child to put the blindfold on and pick up a feely dot card. Then turn the card over and make a copy of the pattern using counters. **Did you use the same amount of counters? Did you use the same pattern? What did you see?**

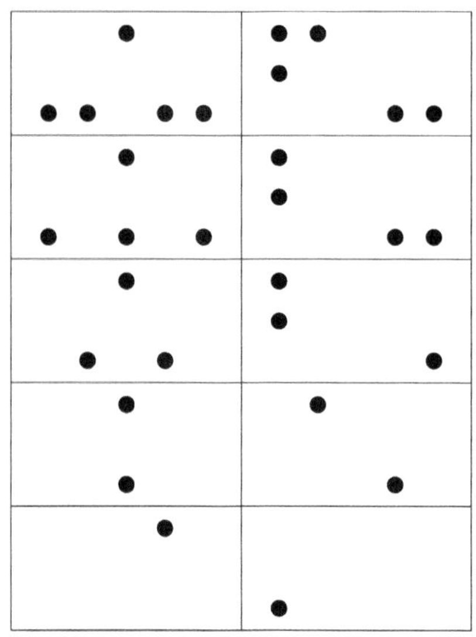

Stick felt circles on top of the dots to make them a tactile resource.

Photocopiable resource 2.13 – Irregular Dots to 5.

What to observe

- Children are beginning to touch and move items when counting collections.
- Children are beginning to use a one-to-one correspondence alongside the forward number word sequence to at least 5.
- Children may understand that the final number word said while counting a collection (within 1–5) is the amount of items.
- Children know that anything can be counted, including concrete materials, movements and sounds, e.g. steps and claps.
- Children may count things in different orders and recognise that it is the same amount but not fully trust the count, so will re-count to check.
- Children can reliably say the forward and backward number word sequence to and from 10.
- Children can say most of the forward number word sequence to 20 in the correct order.
- Children can give some of the number words before and after in the range 1–10.
- Children may recognise numerals of personal significance, e.g. age, bus, house number.
- Children can identify and recognise most of the numerals from 1 to 10.
- Children can record some of the numerals from 1 to 10.
- Children can put the numbers 1–5 in sequence.
- Children can order the numbers in the range 1–5 and identify the missing numbers.
- May use the word zero at end of counting back from 10 but may not realise this means 'nothing'.
- Children understand the first, second and third indicate the position in a sequence, e.g. I put the red dinosaur second.
- Children can 'throw' finger patterns to 5 without putting them up one at a time.
- Children can recognise sounds and movements (1–5 range) without counting.
- Children use subitising to say how many are in a small collection (1–5), without counting them one by one.
- Children can recognise small collections (1–5), that have been placed in an irregular pattern by subitising.

Number Recognition and Place Value B

Section overview

Units of learning	Review units	Prior learning	Benchmarks/ Es&Os
Unit 2.1 **Principles of counting (1 and 2)** Understanding to touch an item once and say the counting numbers in the right order.	**Number Recognition and Place Value A** Units 2.1	• Can count forward to 20 and backwards from 10. • Beginning to touch and move items to count them but may still be unreliable.	**MNU 0-02a** • Explains that zero means there is none of a particular quantity and is represented by the numeral 0. • Recalls the number sequence forwards within the range 0–30, from any given number. • Recalls the number sequence backwards from 20. • Identifies and recognises numbers from 0 to 20. • Orders all numbers forwards and backwards within the range 0–20. • Identifies the number before, the number after and missing numbers in a sequence within 20. • Uses one-to-one correspondence to count a given number of objects to 20.
Unit 2.2 **Principles of counting (3)** Understanding that the last number said means the number of things counted	**Number Recognition and Place Value A** Units 2.2	• Can count forward to 20 and backwards from 10. • Can count small collections under five reliably.	
Unit 2.3 **Principles of counting (4)** Understanding that anything can be counted, including sounds, movement and objects.	**Number Recognition and Place Value A** Units 2.3	• Experience counting variety of collections, sounds and actions.	
Unit 2.4 **Principles of counting (5)** Learning to count things in any order.	**Number Recognition and Place Value A** Units 2.4	• Experience counting variety of collections, sounds and actions.	
Unit 2.5 **Backward number sequences** Saying the backwards number sequence up to 20.	**Number Recognition and Place Value A** Units 2.6	• Can reliably say backwards number sequence from 10.	
Unit 2.6 **Forward number sequences** Saying the forwards number sequence up to 30.	**Number Recognition and Place Value A** Units 2.5 and 2.7	• Can reliably say forwards number sequence to 20.	
Unit 2.7 **What's the number after?** Saying the number that comes after in the range of 1 to 20.	**Number Recognition and Place Value A** Unit 2.8	• Can say the number word after in a 1-9 range.	
Unit 2.8 **What's the number before?** Saying the number that comes before in the range of 20 to 2.	**Number Recognition and Place Value A** Unit 2.9	• Can say the number word before in a 10 to 2 range.	
Unit 2.9 **Reading numbers** Recognising and identify the numbers up to 20.	**Number Recognition and Place Value A** Unit 2.10	• Can recognise and identify the numbers up to 10.	
Unit 2.10 **Writing numbers** Writing the numbers from 1 to 20.	**Number Recognition and Place Value A** Unit 2.11	• Can make marks for numbers up to 10 and can write some of these numbers reliably.	

Units of learning	Review units	Prior learning	Benchmarks/ Es&Os
Unit 2.11 **Sequencing numbers** Placing numbers within the 1 to 20 range in sequence.	**Number Recognition and Place Value A** Unit 2.12	• Can sequence numbers in the 1 to 5 range, forwards and backwards.	• Identifies 'how many?' in regular dot patterns, for example, arrays, five frames, ten frames, dice and irregular dot patterns, without having to count (subitising). • Groups items recognising that the appearance of the group has no effect on the overall total (conservation of number). • Uses ordinal numbers in real life contexts, for example, 'I am third in the line'. • Uses the language of before, after and in-between. • When counting objects, understands that the number name of the last object counted is the name given to the total number of objects in the group. I have explored numbers, understanding that they represent quantities, and I can use them to count, create sequences and describe order.
Unit 2.12 **Ordering numbers** Placing numbers within the 1 to 20 range in order.	**Number Recognition and Place Value A** Unit 2.13	• Can order numbers in the 1 to 5 range, forwards and backwards.	
Unit 2.13 **Missing numbers** Spotting any missing numbers in a sequence in the 1-20 range and use the language of 'in-between'.	**Number Recognition and Place Value A** Unit 2.14	• Knows what the terms 'missing' and 'in-between' mean.	
Unit 2.14 **What is zero?** Beginning to realise that zero means nothing.	**Number Recognition and Place Value A** Unit 2.15	• Use the word zero when counting backwards	
Unit 2.15 **Ordinal numbers** Describing the position of objects using words from first to tenth.	**Number Recognition and Place Value A** Units 2.16	• Knows what first to third position means and can use it in familiar positions.	
Unit 2.16 **Finger patterns** Being able to 'throw' finger patterns to 10.	**Number Recognition and Place Value A** Units 2.17 and 2.18	• Can 'throw' fingers to 5 on one hand and can 'grow' some numbers to 10.	
Unit 2.17 **Groupings to 10** Recognising how many dots there are without counting.	**Number Recognition and Place Value A** Units 2.19 and 2.20	• Can recognise dot patterns to 5 without counting.	
Unit 2.18 **Fact families to 5** Knowing which two numbers when added together will make 5.	**Number Recognition and Place Value A** Units 2.19 and 2.20 **Number Recognition and Place Value B** Units 2.16 and 2.17	• Understands that numbers can be made in two groups, e.g. can make 5 fingers on two hands.	
Unit 2.19 **Fact families within 5** Knowing which two numbers when added together will be within 5.	**Number Recognition and Place Value A** Units 2.19 and 2.20 **Number Recognition and Place Value B** 2.16 and 2.17	• Understands that numbers can be made in two groups, e.g. can make 3 fingers on two hands.	

Units of learning	Review units	Prior learning	Benchmarks/ Es&Os
Unit 2.20 **Five plus facts** Knowing all my five plus facts up to 10.	**Number Recognition and Place Value A** Units 2.19 and 2.20 **Number Recognition and Place Value B** 2.16, 2.17, 2.18 and 2.19	• Understands that numbers can be made in two groups.	
Unit 2.21 **Skip counting sequences** Beginning to know some of the skip count sequences in 2s, 5s and 10s.	**Number Recognition and Place Value A** Units 2.5 and 2.6	• Can count reliably in ones to 30.	

Conceptual concept

Children should learn how to reliably say the spoken forwards counting sequence up to 30 and the backwards sequence from 20. They should also learn how to count backwards alongside the forwards numbers. The numbers up to 13 are learned by memorisation as there is no obvious pattern, however from 14 to 19 they can hear the pattern from four to nine being repeated (apart from fifteen instead of 'fiveteen').

An excellent way for children to understand the value of numbers is to use visuals such as puppets or finger and dot patterns alongside number rhymes and stories. Once the songs are familiar to the children they could be sung without saying the number words and instead use an action to keep track of the count, e.g. a nod of the head. This internalisation supports children to move from 'counting out loud' to 'counting in their heads'.

Children need to see how numbers are used to record quantities; let them see you counting money and labelling the bag or recording the lunch choices, e.g. 14 for roast beef, 3 for salmon nibbles and 9 for packed lunch.

They also need to be familiar with the shapes of the numerals using different mediums such as wooden and plastic numbers as well as written so they can both identify and recognise them; these are two very different skills.

- Identifying – Point to a number and ask them to say it. 'What number is this?'
- Recognising – Ask them to find a particular number. 'Can you find the number seven?'

Written number words and numerals should be modelled by the adults and children should be encouraged to practise correct numeral formation. Resources should be freely available for impromptu recordings, i.e. wooden numbers, pencils, clipboards with squared paper and empty till rolls alongside tools for recording. Extra practice of number formation can be done in handwriting lessons.

Children should learn that numbers are not solely used to represent a quantity they can also describe an order or be used for labelling an object.

- Cardinal aspect of number – i.e. how many things there are, the 'fourteen-ness' of a number.
- Ordinal aspect of number – i.e. the order of numbers on a clock face, and position of things, e.g. use of 'fifth, sixth and tenth'.
- Nominal aspect of number – i.e. the number 75 bus or their Granny's age.

By P1 they should also understand the relationship between numbers and know that 7 is more than 3, which will allow them to sequence and order numbers, for example, from largest to smallest.

Number tracks are an important resource to help support children's understanding of order and sequence but so as not to confuse children it is best if zero is not included. The third position on a number track can be found when you count up to 3 and this can help support the association between ordinal and cardinal numbers.

Children should start to use the word zero as meaning *a set with nothing in it*. At P1 children should be able to identify and recognise the 0 numeral and understand that this is associated with 'none' in a set.

There are five important counting principles that children need to internalise before they really trust that counting works.

- Stable Order Principle – Children use a list of number words in the same repeatable order every time they count.
- One-to-one Principle – To be accurate at counting children must successfully co-ordinate reciting the number words in the correct order while physically pointing or touching the objects once and once only. Be careful that children who choose to use subitising to count small collections are not discouraged by insisting they count each item individually as this ability to recognise small numbers instantly will help them to partition numbers and trust their count later on.
- Cardinal Principle – For children to develop cardinality, they need to understand that the last number said represents 'how many' are in the set. Initially, concrete materials that can be moved should be used and strategies for successful counting should be modelled.

  To ensure the final object counted represents the number in the collection, children need to:

  - Touch and count using a one-to-one correspondence.
  - Ensure the stable order of number words has been used.
  - Move objects that have been counted to a designated 'counted area' to ensure objects are only counted once and none have been missed out.
- Abstraction Principle – If given ample opportunities to count 'anything and everything' they will begin to know what counting is for? They will understand that the number name 'twelve' and the symbol '12' and twelve cubes represent the same quantity. They will also understand that any group of objects can be counted regardless of whether they are the same shape, size or colour. This can also include things that they cannot see, such as beats of a drum, jumps down the steps or friends hiding in a bush. By the end of P1 they will hopefully be able to connect the concept of numbers with a quantity without having to have the concrete materials in front of them. This will help them to count by imaging the numbers.
- Order Irrelevance Principle – Understanding that the order in which objects are counted is irrelevant to the final count can be one of the trickiest principles to master. Children need lots of opportunities to count collections and then re-count them in a different order, as long as each object has been counted once and once only. For example, counting the class as they line up for playtime and then checking the count starting from the other end of the line. Once a child fully trusts that the order in which objects are counted does not affect how many there are they will then be able to use more sophisticated counting strategies such as counting on rather than having to count everything, starting at one.

Subitising is when you can instantly recognise the number of objects within a group without having to count them one by one. If counting is to be discouraged, then flashing objects and dot cards for a few seconds can encourage instant recognition. It is also important that children can recognise irregular as well as regular patterns. Conceptual subitising is when children can see the smaller numbers hidden inside a number, e.g. on a five wise tens frame, the number 7 can be seen as 5 on top row and 2 on the bottom. Conceptual subitising allows them to see the part- part- whole relationships of numbers. It also allows them to see all the different combinations which can make the same number. This in turn this will help them to learn their instant addition and subtraction facts.

Children need to be given opportunities to engage in subitising activities which include a range of representations, including:

- Auditory subitising – recognising sounds such as drum beats or pennies dropped into a jar.
- Visual subitising – counting movements, e.g. skipping or objects on an out-of-reach shelf.

- Finger patterns – patterns to 10, including doubles. Children will first 'grow' their numbers by putting up fingers one at a time but will move to 'throwing' them when more confident of the number of fingers needed for a particular number.
- Irregular dot patterns on cards or using double sided counters that will fall randomly.
- Regular dot patterns to 10 – dice, dominoes, patterns on playing cards and five frames.
- Tallies to 10.
- Tens frames – five wise and pair wise.
- Plastic number shapes with holes, for example, Numicon plates.
- Rekenreks which clearly show the 'five plus' facts up to 10.
- Bead Strings which have beads arranged in groups of five.

Subitising is an excellent way to encourage children to trust that groups of the same numbers can be counted more efficiently by skip counting and that the quantity is the same as if counted in ones. At this stage they may start to learn the number word sequence of skip counting but this should be accompanied by lots of practical experiences so they begin to trust the count, e.g. placing eggs into an egg box two at a time will give the same amount if checked in ones.

Resources

- Counters – transparent, double sided
- Dice – spotted, numeral, coloured
- Pebbles– numbered
- Containers – e.g. tins, boxes, bowls, cups, shoe boxes, baskets
- Collections – toy cars, plastic fruit and vegetables, interlocking cubes, Lego, magnetic fish, trains, 2D shapes, farm animals, £1 coins
- Plastic tweezers
- Paintbrushes
- Beads and laces
- Robot
- Puppet
- Blindfolds
- Large carpet spots
- Balloons
- Paper plates
- Masking tape
- Cardboard tube – large
- Wellie boot
- Washing line
- Pegs
- Sand timers – 1, 2 and 5 minutes
- 3D numerals – variety
- Numeral tracks/lines – to 30
- Numeral cards – various sizes
- Dot cards to 5 and dominoes to 12 – regular
- Dot cards to 10 – irregular
- Tens frames – magnetic, five-wise and pair wise
- Numeral tracks – hide and see
- Finger pattern cards to 10
- Tally cards to 10
- Rekenreks
- Bead strings to 30
- Talking tins
- Digital cameras
- iPads
- Headphones

Digital

- Tree Tool
- Ants on the Bus
- Counting Tool

Key vocabulary

- **Count**
- **Counting**
- **Forward**
- **Backward**
- **More than**
- **Less than**
- **Before**
- **After**
- **Next to**
- **Hidden**
- **Number words** – zero, one, two, three, four, five, six, seven, eight, nine, ten, eleven, twelve, thirteen, fourteen, fifteen, sixteen, seventeen, eighteen, nineteen, twenty
- **Is the same as**
- **Is equal to**
- **Equal**
- **Combines with**
- **Joins**
- **Take away**
- **How many more**
- **Altogether**
- **Makes**
- **How much**
- **Amount**
- **Total**
- **Sum**
- **Add**
- **Subtract**
- **Ten-frame**
- **Empty**
- **Full**
- **Top row**
- **Bottom row**
- **Represent**
- **Dice**
- **Domino**
- **Pattern**
- **Fingers**
- **Match**
- **Number word**
- **Order**
- **Sequence**
- **Item**
- **Collection**
- **Group**
- **Ordinal number words** – first, second, third, fourth, fifth, sixth, seventh, eighth, ninth, tenth
- **Five wise** – numbers shown as e.g. dots on a tens frame arranged in lines counting to 5
- **Pair wise** – numbers shown as e.g. dots on a tens frame arranged as pairs
- **Five plus** – numbers shown as e.g. dots on a tens frame arranged as 5 and patterns, triangle, square, circle,
- **Rekenrek** – an abacus of 20 beads arranged to show 5 and 10 patterns
- **Bead string**
- **Skip counting** – counting by 2s, 3s, 4s, etc.
- **Numeral track**
- **Number line**

Misconceptions

Potential misconceptions	Addressing the misconception
Units 2.1, 2.2, 2.3, 2.4, 2.5 and 2.6 When counting objects watch out for the children who: • Miscount because they do not organise their collections into a pre-counted and counted pile and then count items twice or miss items out. • Count 2 counts for teen numbers, e.g. seven – teen. Using a visual strategy (looking at items) or pointing strategy and miscount the collection.	• Use practical resources and demonstrate moving items from an uncounted place to a counted area, to ensure that they are not counted more than once. • Move each item as it is counted if it is not organised in a line. • Encourage the child to describe the completed count, e.g. …16, 17, 18. There are 18 cubes. • Model 'bad' counting, count some items more than once, miss some out. Ask the children what is wrong with your counting.
Units 2.1, 2.2, 2.3, 2.4, 2.5 and 2.6 When counting a collection some children may not understand that the final count represents the amount of items. If you ask them for 12, they may give you the twelfth item. They also may simply repeat the final number said without realising this means how many are in the collection.	• Lots of opportunities to collect and count small collections using a wide range of objects. Model the final count. • Lots of opportunities to check amounts rather than told 'to count', e.g. the jar says there are 15 items inside, are there 15 items? • Evidence in environment and displays of value associated with a number. • Practise forward number sequence up to 30. • One-to-one correspondence practice – do they touch, say and move each item in turn and only once? • Change your tone for the last number said to support this as the final total.
Units 2.1, 2.2, 2.3, 2.4, 2.5, 2.6, 2.7 and 2.8 Some children will find it very tricky to start counting a collection without starting from 2. E.g. Give them 6 different cubes and ask them to start their count at 3, they may say there are 8 cubes.	• Model counting a small collection of different items and repeat the count starting at a different item each time. E.g. a collection of different coloured counters. Ask them to start their count with a red counter. Change the starting colour. • Place the items in different positions and start at a new place each time. • Line the items up, start the count from both ends and the middle. Does it change the final amount?
Units 2.1, 2.2, 2.3, 2.4, 2.5, 2.6, 2.7 and 2.8 Some children cannot break the forward number sequence and are unable to start from a different number other than one.	• Give lots of practical opportunities to count objects while sometimes stopping the count momentarily and then continuing. • Ask them to identify the number on a numeral track and then ask them what the next number is, and the next. • Give them experiences of estimating where numbers are sitting on a hidden number line. Where is the number 23? Are you right? Reveal the numerals. Show them one or two numbers before 23 so they have a 'run up' to that number. What's the next number?
Units 2.5, 2.6, 2.7 and 2.8 Crossing over the decade can be tricky for some children, they omit some numbers, e.g. 19, 21, 22 when counting forwards or 20, 18, 17 when counting backwards.	• Lots of opportunities to say the sequence and see the sequence on numeral tracks both vertical and horizontal.
Units 2.5, 2.6, 2.7 and 2.8 They might confuse the number word after for the number word before.	• Pointing to a hidden space on a numeral track to the right for number word after and to the left for number word before.

Potential misconceptions	Addressing the misconception
Units 2.1, 2.2, 2.3, 2.4, 2.5, 2.6, 2.7, 2.8, 2.11, 2.12, 2.13 and 2.14 Many children do not realise that the next number in a sequence is one more and the number word before a number is one less.	• Building staircases using cubes or number rods can help develop this understanding. • Opportunities to help children recognise that if they add one more, they will get the next number in the sequence and if they take one away they will get the previous number.
Units 2.1, 2.2, 2.3, 2.4, 2.5, 2.6, 2.7 and 2.8 Children who find giving the next 'number word after or before' may have to drop back to one and count up to the number before being able to give the next number after or before.	• Give them two or three numbers before, e.g. what's the number before 16, say '18, 17, 16, ...?' What's the number after 23? Say, '24, 25, 26 ...?'
Units 2.9, 2.10, 2.11, 2.12 and 2.13 Difficulties can arise when identifying numerals in the twenties because the decade number comes first unlike the teen numbers which say the ones column first. E.g. 26 can be confused with 16.	• Give the children opportunities to organise numbers in sequence and to match numbers on numeral tracks.
Units 2.9, 2.10, 2.11, 2.12 and 2.13 Some children have difficulties recording the teen numbers because they cannot hear the digit sound clearly before the teen word or it has no resemblance to the digit at all: • Eleven and twelve do not sound like 1 or 2 • 'thir'teen and 'fif'teen – they cannot hear the 3 or 5 part before the word teen. • We say the four before the ten watch out for reversals 14 and 41	• Plenty of writing and reading of numeral practice to accompany songs, games and activities.
Units 2.1, 2.2, 2.3, 2.4 and 2.14 Be aware that some children may allocate the word 'zero' to the first object counted.	• Do not use number tracks/lines with zero as the first number. • Opportunities to count collections as well as zero collections.
Unit 2.16 When tracking a movement or sound on fingers it can prove tricky for some children to assign one finger to one sound or movement.	• Reduce the number of sounds or movements to be counted to 2 or 3. • Keep the count regular (not too quick or slow!). • Practise forward number sequence to 5.
Unit 2.16 Watch out for the children who are still growing their fingers instead of 'throwing' them as this means they are still using a count by one strategy.	• Practise making the numbers and playing games to make them more familiar with finger patterns up to ten.
Units 2.16 and 2.17 Children may use a count from one or a count on strategy to see dot patterns within ten.	• Lots of opportunities to explore patterns on 'five wise' and 'pair wise' tens frames, dice patterns and finger patterns. • Give a longer time for the patterns to be seen and then shorten this to enable them to recognise the pattern within a 2 second flash.
Units 2.1, 2.2, 2.3, 2.4, 2.7, 2.16, 2.17, 2.18 and 2.19 Children may use a count from one or a count on strategy to say their 'facts to 5 or within 5'.	• Provide opportunities for the children to see the part part whole nature of a number under 5. • Use concrete materials, e.g. double side counters for the children to make and see the patterns, 2 and 3 makes 5.

Potential misconceptions	Addressing the misconception
Units 2.1, 2.2, 2.3, 2.4, 2.7, 2.16, 2.17 and 2.19 Children may use a count from one or a count on strategy to say their 'five plus facts'.	• Present the five pattern and flash the other pattern to encourage children to 'see' it in their mind. Tap, point or clap the number hidden. • Keep both the seen and screened dots in a regular pattern until this knowledge is secure.
Units 2.1, 2.2, 2.3, 2.4 and 2.21 Many children do not realise that skip counting is also a way to tell how many there are of something.	• Skip counting must be introduced with materials and not just by reciting numbers or jumping along a number line. • Equal groups of items should be used and moved as the next number in the sequence is said.

Teaching ideas

Unit 2.1 Principles of counting (1 and 2)

Whole class

- **Photo Shoot** – Prepare a large floor book with separate pages for the numbers 0 to 10. Choose one number to focus on at a time, tell the children that they are going to take photographs of collections of three objects for example. Ask them to look around to see if they can see three things that are the same and to bring them to the photo booth (a table and a screen). Get them to count the objects as they place them on the table. Take the photograph and then the next child finds three new items. In between children's turns, quietly take photos of collections that are 'not 3'. Once everyone has had a turn, look at the photographs with the children. **Here are some teddy bears, are there three of them? How do we know? How can we check?** Print and paste the photographs into the appropriately numbered page. Have a different focus number on another occasion and repeat. **Why does the zero page have nothing on it? What happens to the size of the collections as we go through the book? Why does this happen?**
- **Stories with Puppets** – Stories with accompanying puppets or visuals can help to support children's counting. Familiar stories such as 'Billy Goats' Gruff' and 'Goldilocks and the Three Bears' can be acted out, with the puppets standing in a line and being tapped as they are counted. **How many bears are there at the table? How many people are in the house now? How many goats have crossed the bridge? How do you know?**

Stations

- **Number Flowers** – Prepare paper petals, stems and flower centres. Place large dots on the edge of the centres and ask the children to stick a petal on each dot. Ask them to count the petals they have stuck down one by one. **How many petals have you stuck on? How do you know? Show me how you counted them?** Observe if they have counted one petal and said one number word at a time. Model a clockwise direction to ensure all petals have been included in the count.

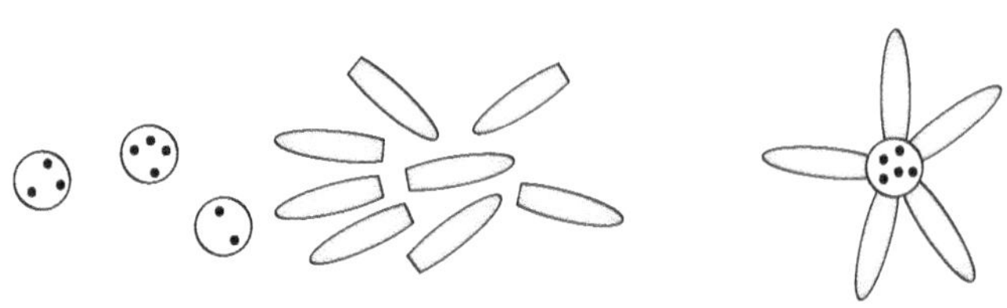

- **Talking Tin** – 'You are going to record yourself counting out the jewels using a Talking Tin. Roll the dice and start the recording tin. Use the tweezers to pick up that number of jewels and say the counting words in the right order as you put them on the tray. Listen to your counting. Now count them again to check you have the correct number of jewels. Touch each jewel as you count them.'

- **Fruit and Veg Stall** – Place a selection of fruit and vegetables out and have some pictorial shopping lists prepared. Ask the children to choose a list before they go shopping. Encourage them to count the pictures and ask for the correct amount. **How many bananas do you have on your list?** Model asking for this amount; say, 'You have two bananas on your list', point and count them, 'One, two. Please can I have two bananas?' **How do you know you have the right amount?** 'Put your fruit on top of the picture to check.'

Unit 2.2 Principles of counting (3)

Whole class

- **Numbered Chair Hunt –** Place the chairs in a circle and place a number card on each chair (in the range 10–20). Have collections of items in small boxes that correspond to the numbers on the seats. Say, 'I want you to choose a box and count how many items are hiding inside. First you must find a space big enough to empty the contents onto the carpet. Then pick up the items one by one. Say the forward counting names as you put them back in the box. When I clap my hands, I want you to find the chair with the number of items you have counted. Sit on the chair and put the number card on top of your box.' **Does anyone think they have the same number as you? If only one of you can be correct, what could you do? Do you know what the number looks like? How could you get help?**
- **Count to 20** – Mark out two large tens frames on the carpet so that a large carpet spot or laminated circle can fit into each square. Tell them you are the 'Master Tens Maker'. Choose someone to roll the dice and tap the same amount of children on the head. 'You must shout out the number sequence from one as you tap them on the head.' Instruct the children who got tapped on the head to stand up and ask another child to give them each a large carpet spot, counting them out loud as they are handed out. 'Come and fill up the tens frame. Filling the top line up first with no gaps.' **How many spots do we have? How many more spots can we fit in? How do we know? Point and touch each spot as the children say the sequence of counting names from one.** Roll the dice again and repeat until both tens frames are filled.

Stations

- **Coloured Towers** – Prepare a box of Lego, a 5-minute sand timer, a 1–6 spotted dice and a coloured dice.

Ask, 'What size of rocket can you build in five minutes? You need to roll both dice. The first one will tell you the colour of bricks and the second dice will tell you how many to collect. Repeat until the sand timer runs out.' **How many red bricks have you used? How many blue? How many bricks have you used altogether?**

- **At the Bank** – Use four money boxes (tins) with labels in the teen numbers. Have a jar of £1 coins. 'Please post the correct number of pound coins into each money box. Take the boxes to the bank to check that you have put the correct amount of money in.' **Did you put enough coins in? Do you need to add some more? How could you check?** 'Use two tens frames to organise your coins when checking.'
- **Shape Bank** – Place a box of 2D shapes of different sizes and colours from an attribute shape box on the table. Stick large 2D shapes on the table using masking tape. Place a plastic number in each 2D shape. For example, for a masking taped triangle with the number 6 in it, tell the children 'You need to collect six triangles and place them in the triangle space.' **Can you place them in a dice pattern (or a tens frame) to prove you have the correct amount? How many shapes have you got altogether?** Make sure this is not beyond the children's counting range. 'I give you one more triangle.' **How many will you have now?** 'You give me one of the squares.' **How many squares will you have left?**

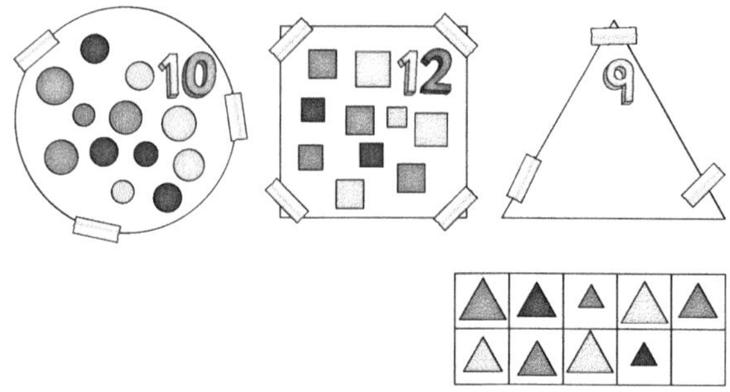

Unit 2.3 Principles of counting (4)

Whole Class

- **Bossy Boots** – Place some plastic numbers in an old wellie boot and ask a child to pull out a number. Put the boot to your ear and pretend it is being very bossy. Order the child to do something associated with the number. For example, the bossy boot says you have pulled out the number 4 so it wants four paintbrushes! Choose another child to help count as the boot is no good at counting. They must count the paintbrushes to prove there are four. **How will you know that you have counted them all? How do you know you haven't counted the same one twice? Are there four paintbrushes?** Once the boot is satisfied, a new number can be pulled out. The commands should include making sounds (e.g. beating a drum), or movements (e.g. hops) or finding something with a particular number on.
- **P.E.** – Children need to be positioned at stations which are set up around the gym hall, with a different aerobic activity at each one. Tell the children to find a partner. A number is chosen and each exercise must be completed that many times. One child will do the exercise while their partner counts the moves, then they swap over. For example, if the number 12 is chosen, those at Station One will do 12 star jumps, those at Station Two will do 12 toe touches, etc. **How will you know that you have done the right amount? Show me how you counted? Did you count them in your head or out loud? Did you both count the moves? How did you know your partner counted correctly?** The children then move in a clockwise direction to the next station and a new number is chosen.

Stations

- **Five Cube Combinations** – Tell the children they are going to explore different ways to stick five cubes together. 'Before you put it down, check it against the other shapes.' **Are there any that look like yours? What is different about yours? What is the same? Have they all got five cubes?** Repeat with other numbers.

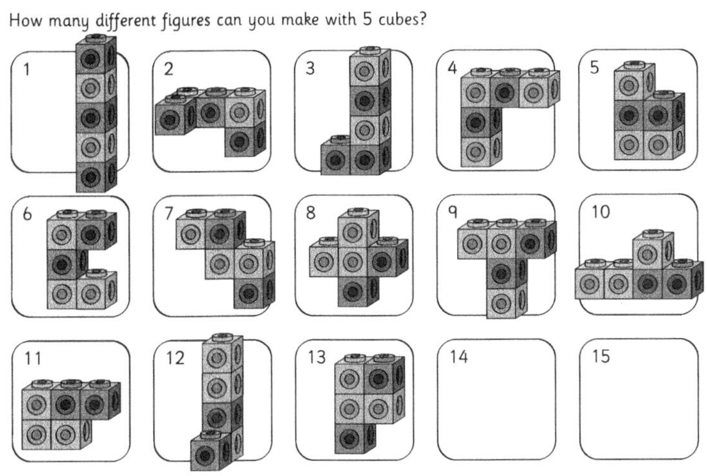

- **Combination Plates** – Cut paper plates into quarters and on each quarter draw a different representation of the same number, use numerals, tally marks, simple item (eg, a bee), dots. Tell the children to turn the plates face down and take it in turns to find a matching pair. If successful they keep the pair. Play until no cards are left. The one with the most pairs is the winner.

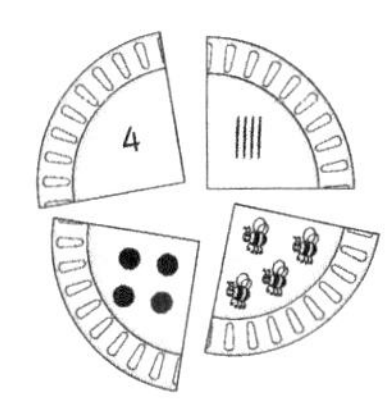

Unit 2.4 Principles of counting (5)

Whole class

- **Class Bar Charts** – Have a daily visual register on two or three tens frames. Ask the children to count how many children are in school today and how many of them are absent. **How many are in school today? How many are away?** 'Let's check by starting the count here, where the children are absent and then count the children that are in school.' **How many children are there in the class when everyone is in school?** Introduce a simple 'Question of the Week' and show the results on a bar chart. For example, two variables: **Who had cereal for breakfast? Who did not?** or more than two variables: lunch choices – school dinners, packed lunch and home lunches. Ask the children to find the square card with their photograph and place it in one of the sorting trays. Then show them how to organise this into a bar chart (horizontal and vertical). **How many children had cereal today? How many people did not have cereal? How many children are in class today? How do we know? How many children are not here today?** Start the count with one variable and then continue onto the next one. Then check by counting variable 2 first and then variable 1. **What do you notice when we change the way we count them?**

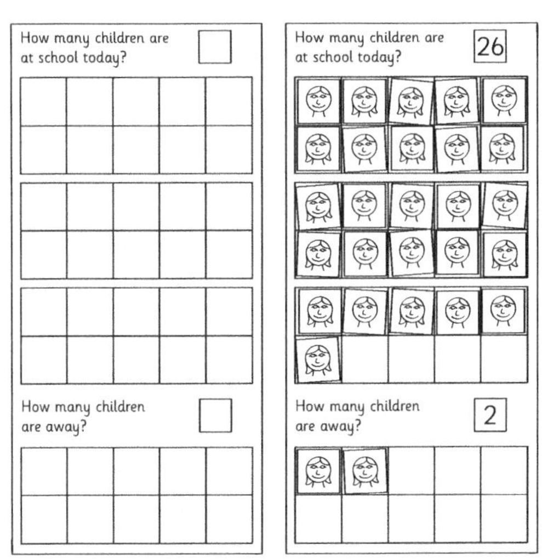

- **Counting Round the Circle** – Organise the class into a circle. 'Let's check how many children are here today, the register says 23, let's see if we agree.' Ask someone to start counting from one and say the number loud and clear, the next person continues the forward count. **Did we count up to 23? What does this number mean? If you include myself, what will the number be? What is one more than 23?** 'Let's check our counting, let's start at this end.' **What if we started here?** 'I can see there are four children here, so you must be number 5.' **Can we start at number 5 and then continue to count in ones? Do we still get 23 children in the circle?**

Stations

- **Bead Bracelets** – Place beads and laces at the station with some number labels. Tell the children to collect a number card and to make a bracelet using that many beads. Then peg the number onto the end of the bracelet. **How do you know you have put the correct amount on? Do you think the amount will change if you start counting from the other end of the bracelet? Can you start counting from the third bead? Do you still have the same amount?**

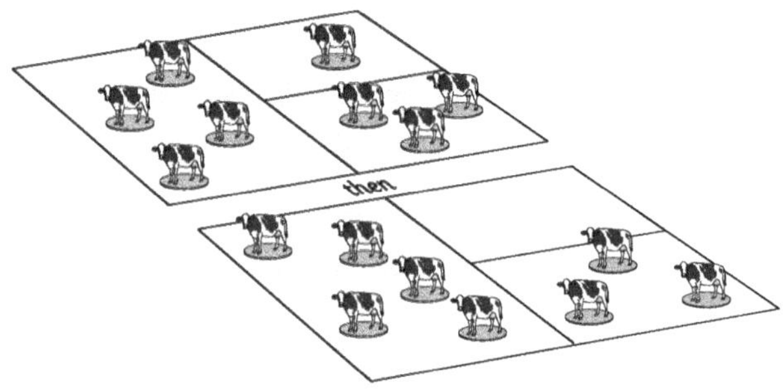

- **Farmyard** – Look at the three fields (drawn on paper). **How many different ways can you organise these eight cows?** 'Draw as many different ways as you can.' 'You have three cows in this field, four in this one and one in this one.' **How many cows are there altogether?** 'What if this cow gets lonely and wants to join his friends in this field. You now have three cows in this field and five in this field.' **How many cows have you got altogether?** Observe if the child knows there are still eight cows or has to re-count to be certain. The child that starts the count again does not fully understand that the purpose of counting is to determine 'how many' there are. Even though they could see that no cows left the farm – they were just repositioned – they have not fully trusted their original count.

- **P.E.** – Get into groups of four. Choose someone to be number 1. Count round the group. **What number do you say last?** 'Choose someone else to be number 1, count round the group again. Don't miss anyone out.' **How many are in your group?** Repeat. **What do you notice?**

Unit 2.5 Backward number sequences

Whole class

- **Robot Voice: Target Reached** – Ask the children to stand in a circle. Have a robot and some numeral cards from 0 to 10. Choose one child and give them the robot, say 'Choose a number to be the target number and then put the robot with this number in the middle of the circle.' Turn to the class and say 'We are going to take it in turns to say the backwards number sequence out loud starting from 15.' When the target number is reached, the child whose turn it is, has to use a robot voice to say 'Target reached'. For example, the target is 8. '15, 14, 13, 12, 11, 10, 9, target reached.' Then that child sits down in the centre with the robot and chooses a new target number. The sequence starts again at 15. Once competent from 15, then they can practise starting with any number up to 20.

- **Sliding to Zero –** Put two tens frames on the board filled with counters stuck on with tac. Tell the children that they are going to count down from 20. First they are going to check that there are 20 counters on the tens frames. **Do we need to count them all to see if there are 20? What else could we use?** 'Let's count them all to check.' **How many do we have?** Take one off and post it into a box with a small slit cut into the lid. Say, 'We had 20 and now we have 19. Your turn – 20, 19.' Take another counter off and post it into the box. Say, 'We had 20, 19 and now we have 18. Your turn – 20, 19, 18.' Continue until the whole sequence has been repeated up to 10. They should know the sequence from ten to zero. Say, 'Look at the number of counters on this tens frame. We can count backwards from 10. Say the numbers when I post a counter; 10, 9, 8, ...'

Stations

- **Songs** – Film or record the children singing familiar backward counting songs but change the lyrics to start from 20. Songs like 'Ten Green Bottles' or 'Ten Little Monkeys' will work well. Have the songs available for the children to listen to on headphones on the iPads. Ask them to sing the songs to you at circle time.

Unit 2.6 Forward number sequences

Whole class

- **Guess Who Said It!** – Someone sits blindfolded in the middle of the circle and decides on a starting number between 10 and 19. The other children take it in turns to say the forward number sequence up to 30. The blindfolded child has to guess who said 30. If they guess correctly, they swap places. If not, they keep the blindfold on and the sequence starts again with a new starting number.
- **Correct the Puppet** – Introduce the children to the puppet, who needs help with his counting. Allow the puppet to count nicely up to 10 but to go wrong by saying ty instead of teen numbers, or missing out 20 and 21. (See the misconceptions for difficulties children have counting to 30.) For example, **Can you help me count to 23? Is this right: 1... 18, 19, 21, 22, 23? What did I do wrong? How should I say it?** 'I thought I could say teen or ty.' **Why can't I say twen-teen? What should I have said?**

Stations

- **Bopping Balloons** – How many times can you bob the balloon without it touching the ground? Each time you hit the balloon, say a forward counting number. **Can your count get to 30? Can you count every second hit and count in twos?**
- **Large Motor Skills** – Tell the children they are going to say the forward number word sequences at the same time as kicking a ball across a field, jumping over log steps in the garden or patting their head. **Can you kick a ball and say the forward counting numbers up to 30? Can you start at 12 and count to 27 as you pat your head once for each number?**

Unit 2.7 What's the number after?

Whole class

- **Hide and See Numeral Stand** – As the children are familiar with forward counting numbers up to 10, introduce the numbers up to 15 next and then up to 20. Have all the numbers from 5 to 15 on show in the numeral stand so the children can see the sequence. Tell the children 'We are going to practise saying all the forward counting numbers up to 20. Listen to me first and then you can say them: 5, 6, 7, 8, 9, 10, 11, 12, 13, 14, 15. Now it's your turn.' Practise starting at different numbers and also going up to 20. **What's the number after 12? What's the number after 16? Can you find the number 19? What's the next number?**

Once children are confident giving the number word when still able to see them, it is time to hide some of the numbers under the 'hide and see' lid. Point to a hidden number. **What number is this? How do you know?** 'Let's check by flipping the lid up.' **Were you correct?** If they are unsure, give them a run up to the hidden number. **What's the number after 13? 10, 11, 12, 13...** Next hide all the numbers apart from one of them. **What's the number after this one?**

- **Number After Bingo** – Give the children bingo boards (see Photocopiable resource 2.14) that have a range of numbers from 10 to 20. Hold a numeral card up. Ask: **Do you have the number after?** They can then cover that number up with a counter. The first to get all of the numbers covered up can shout 'After!'

12	19	13	17	15
20	16	11	14	10

11	14	19	12	20
13	18	15	16	47

Photocopiable resource 2.14 – Bingo Boards.

Stations

- **Number Word After Strips** – Cut up a numeral track into cards with three consecutive numbers on, e.g. 13, 14 and 15, and put them face down on the floor with some counters. The children can play with a partner. 'Hide one or two of the numbers on your strip and ask your partner what the next number is? You cannot cover the number on the far right.' **Why can't you hide this number if you are asking what the number after is? Did your partner guess correctly? How can you prove that?**
- **Number After Race** – Tell the children that three to four players can play the game 'Race' but one person must be the 'caller'. There are two sets of cards; the blue numeral cards from 2 to 20 are placed face up on the carpet. If you are the 'caller', you have a pink set of number cards from 1 to 19, which you hold face down in your hand. You take the top card off and show the group. For example, say, **What is the number after 13?** One child will successfully grab the correct blue numeral card and add it to their winning pile. The caller continues until there are no cards left to get. The child who collected the most cards is the winner.

Unit 2.8 What's the number before?

Whole class

- **Hide and See Numeral Stand** – As the children are familiar with counting numbers back from 10, introduce counting numbers down from 15 and then down from 20. Have all the numbers from 15 to 5 on show on the numeral stand so the children can see the sequence. Tell the children, 'We are going to practise saying all the backward counting numbers down from 20. Listen to me first and then you can say them, 15, 14, 13, 12, 11, 10, 9, 8, 7, 6, 5. Now it's your turn.' Practise starting at different numbers and also down from 20. **What's the number before 12? What's the number before 16? Can you find the number 19? What's the number before it?**

Once children are confident giving the number word before while still able to see them, it is time to hide some of the numbers under the 'hide and see' lid. Point to a number that is hidden behind its lid. **What number is this? How do you know?** 'Let's check by flipping the lid up.' **Were you correct?** If they are unsure, give them a run up to the hidden number. **What's the number before 13? 16, 15, 14, 13...** Next hide all the numbers apart from one of them. **What's the number after this one?**

- **I Spy** – Tell the children they are going to have a game of 'I Spy'. Put number cards from 2 to 20 up on the board. 'I Spy with my little eye, the number before 18.' Choose a child to answer. If they are correct, they can come and collect the number card from the board and they get to ask the next 'I Spy' question. Repeat until there are no cards left.

Stations

- **Before and After Charts** – Prepare charts with three columns and randomly fill in the middle column with numbers in the 9 to 19 range. Column one should be titled 'Before' and column three should be titled 'After'. The answers (2 sets of numbers from 8 to 20) should be cut up and placed face down on the table. The children take it in turns to pick up a number and see if it will fit onto their chart. If it doesn't, they miss a go. The first person to fill their chart is the winner. The chart could be extended to five columns, so the children are looking for two numbers before and after the middle number (see Photocopiable resource 2.15). This can also be filled in by pen where the children simply write in the correct answers.

before	before		after	after
		14		
		19		
		11		
		17		
		9		
		12		
		15		
		10		

before		after
	9	
	15	
	14	
	11	
	18	
	10	
	17	
	16	

Photocopiable resource 2.15 – Before and After Charts.

Unit 2.9 Reading numbers

Whole class

- **Run to a Number** – Hang large numeral posters around the gym hall and shout out instructions with actions. Say, 'You have to listen carefully to the number I shout out and how I want you to travel to that number. For example, I want you to hop to the number 2, skip to the number 8.' Once they have all completed one action, ask them to do another action that represents the value of the number. For example, 'Run to the number 7, now clap your hands 7 times.'
- **Number Hunt** – Ask the children to sit in a circle. Hide plastic numbers (10–20) in a box of sand and place the box in the middle of the circle. Ask for a volunteer. Say, 'You need to pick out one of the numbers and then choose someone to ask what the number is. If you agree with them, you can keep the number and sit down in their place. They now pick a number and choose someone to ask what the number is. If you do not agree, you ask the class for help and choose someone to swap places with. **How could you check who is correct? Would a number track help?**

Stations

- **Teen Splat** – Provide sheets with teen numbers written down multiple times and add a colour code for each individual number at the top of the page (see Photocopiable resource 2.16). Put bingo marker pens out so the children can find the numbers and colour them in the designated colours.

11 12 13 14 15 16 17 18 19

16 12 18 16 17
11 19 12
17 13 14 13 11
12 15 16
18 18
14 13 12 14 13
16 11 14 16 15
18 18 17
15 17 19 13 16
19
14 12 11
13 11 13 14
15
12
14 11 19
18 11
17 19 17 15
18 12
19 15 16 19 17 15

Photocopiable resource 2.16 – Teen Splat.

- **Train Tunnel** – Take a long cardboard tube and some toy trains that can travel through the tube. Place a strip of paper at the end of the tube with a number line up to 20. Say, 'One person at a time will let the train slide down the tube and you have to see which number the train travelled to'. **Who got the furthest away? What number was that? What would happen if it went further than 20?**

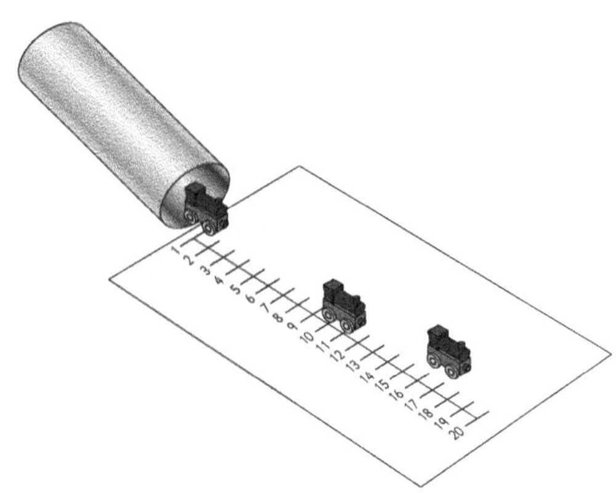

Unit 2.10 Writing numbers

Whole class

- **Nature Walk Tally** – Take clip boards and pencils out on a walk. Prepare a tally sheet and photographs of things the children are likely to see on their walk. Tell the children that each time they see one of the things on the list they need to make a tally mark. Show them how to make the fifth mark. After the walk, tell the children 'now I would like you to count up your tally marks and write the total amount in the final column.' Provide a numeral track to 20 at the bottom of the sheet for support.

item	tally	total
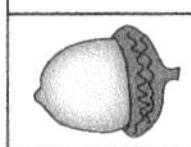	~~IIII~~ ~~IIII~~ IIII	14
	III	3
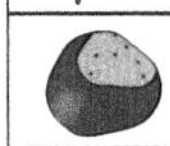	~~IIII~~ II	7
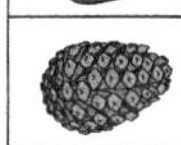	IIII	4

1	2	3	4	5	6	7	8	9	10	11	12	13	14	15	16	17	18	19	20

- **Invisible Back Writing** – Sit the children in groups of six to eight in circles on the carpet. Ask the children to choose a 'starter'. Tell them 'The starter thinks of a number between 10 and 20 and "writes" it on their partner's back using their finger. That person then writes it on the next person's back.' Repeat until the last person in the group has been 'written' on. That person says what the number is and the starter agrees or disagrees.

Stations

- **Water Painting** – On blackboards outside, write the teen numbers in chalk. Provide water in pots and thin paintbrushes. Say, 'I would like you to practise writing your teen numbers. Go over them with water.' **Did you start at the correct place? How do you start writing the number 15?**
- **Roll the Dice** – Have a dice which has teen numbers on it and a chart with the numbers from 11 to 20 on (see Photocopiable resource 2.17). Say 'you have to roll the dice and then write the same number in the correct place on your chart.' The winner is the first person to write one of the numbers six times. The squares could have number outlines in for added support or be blank apart from the first row.

11	12	13	14	15	16	17	18	19	20
11	12	13	14	15	16	17	18	19	20
11	12	13	14	15	16	17	18	19	20
11	12	13	14	15	16	17	18	19	20

11	12	13	14	15	16	17	18	19	20

Photocopiable resource 2.17 – Roll the Dice.

Unit 2.11 Sequencing numbers

Whole class

- **The Postie Has Been** – Tell the children to put their heads on the table and close their eyes. Randomly place 10 envelopes that contain the numbers 1–10 out on the tables. Announce 'The Postie has been!' Say 'If you have an envelope, look inside and pull out the number. Come to the front of the class and sequence the numbers from one to ten (or ten to one).' **Who can read these numbers forwards/backwards?** 'Let's hang the numbers up in order on this washing line. I would like us to start with the biggest number and put them in order from the biggest to the smallest number.' **Who has the biggest number? What's the next number?** 'This time I would like to hang them up in order from the smallest number to the largest number.' **Which number should we start at? What's the next number?**
- **Human Number Line** – Prepare some numeral cards (in the range 1–10). Ask the children to wear them around their necks. Tell them, 'I'm going to call out a number and I would like you to line up behind the person wearing that number.' For example, call out the number 4, so the children with numeral cards 5 to 10 line up behind the number 4. Starting with 4, ask the children to call out their number in sequence. **Why did some of you not get to line up and call out your number? Did we have the line in the right order? Can we call the number sequence backwards? Which number would we need to start at?**

Stations

- **Numbered Pictures in Order** – Cut a laminated picture into 10 strips and number the bottom row from 1 to 10 (see Photocopiable resource 2.18). Ask the children to put the picture back together again using the sequence from 1 to 10. (This should also be done from 10 to 1 for the backwards sequence.) 'Read the sequence out.' **Does this sound correct? How can you tell you have the numbers in the right position?**
- **Dot to Dot** – Provide dot to dot pictures for the children to join the dots in number order to make the picture. **Does the picture look right? Can you read the sequence out forwards and backwards?**

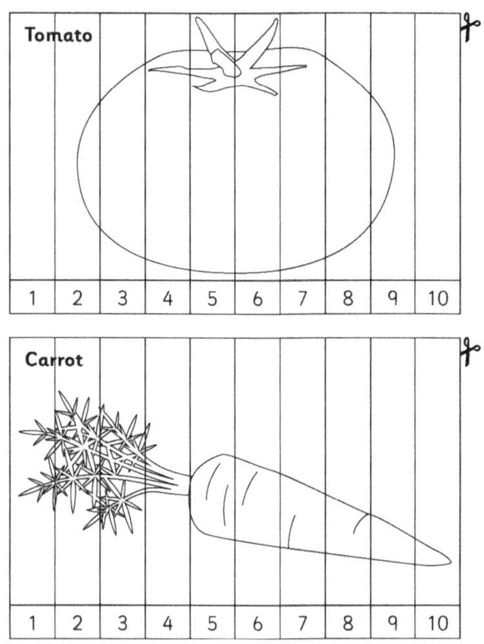

Photocopiable resource 2.18 – Numbered Pictures.

- **Number Snakes** – Use two snake boards (see Photocopiable resource 2.19) and two sets of numeral cards (different colours) in the range 1–10. Mix the numeral cards up and place them face down (in one pile) on the carpet. Say, 'You are going to build two snakes by taking it in turn to pick up a numeral card and place it along the snake's body in the correct sequence from 1 to 10. Match the colours of the cards and the snake boards. You may have to put a card down on your partner's snake.' **What number have you turned over? Where will the blue 4 go?** 'Look carefully at the empty boxes on the snake's body.' **Which one is the fourth box? Is this where the number 4 should go? Now you have a 6, if this is where the 4 goes, where should you put the 6? Which snake will get finished first?** 'You can play again but this time you need to start with the number 10 next to its head.' **Can you say the sequence forwards and backwards?**

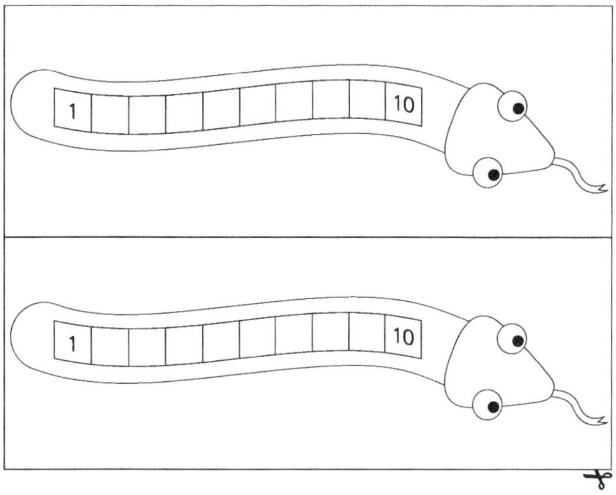

Photocopiable resource 2.19 – Number Snakes.

Unit 2.12 Ordering numbers

Whole class

- **Numeral Necklace** – Choose five children to play this game. Place a pile of numeral cards (1–10) with string attached, face down on the floor. Say, 'Here are the number cards for the sequence 1 to 10. While you are moving around the room to the music; I am going to remove five of them. When the music stops you must pick up one of the numbers and place it around your neck and then arrange yourselves into a line to sequence yourself. **Have you left extra spaces for the missing numbers? How do you know you have the correct order? Which numbers are missing? Who can say the order of the five numbers?**

- **Puppet Making Mistakes** – Children love helping someone if they find something difficult or correcting them if they say something wrong. Introduce a puppet to the children who needs their help. For example, the puppet takes three numeral cards and says she can put the numbers in the correct sequence starting with the smallest number, then the next one and finally the biggest number. Then get the puppet to make a mistake, e.g. I have a 5, 7 and 2. I think the 2 is the smallest number and then it is the 7 and 5 is the biggest number. **Is the puppet correct? How could you help them? Could you use a number track to prove it?**

Stations

- **Giant's Footprints** – Place large giant footprints on the floor (see Photocopiable resource 2.20) but not in the complete sequence. Tell the children they must get from one side of the room to the other stepping on the footprints, but their path must follow a forward counting sequence. Explain that it's okay if not all the numbers are there but you cannot go backwards. For example, they may choose the path that goes 1, 3, 4, 6, 8, 9, 10. **How did you know you were still going forwards? Did you say any of the missing numbers in your head?**

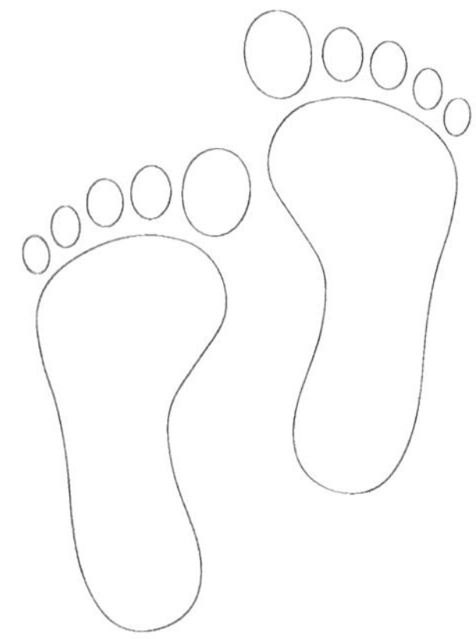

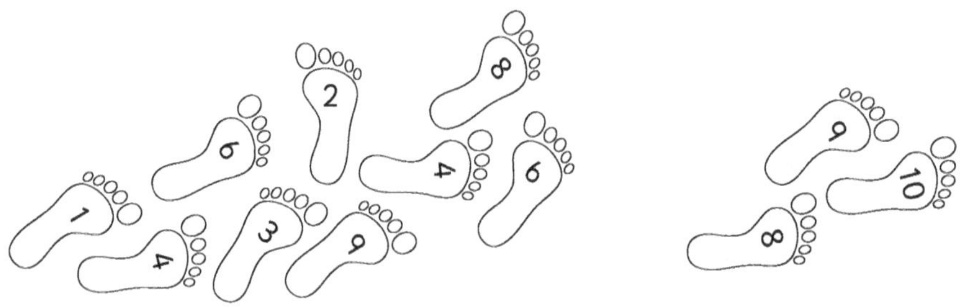

Photocopiable resource 2.20 – Giant's Footprints.

- **Catching Fish** – Place fish (numbered magnetic shapes) from the range 1–10 in a large container with water but leave some numbers from the range out. Say, 'You are going to go fishing but you have to catch the fish in order. There are some numbers missing so you cannot complete the sequence from 1 to 10. Saying this sequence may help you to know which fish to catch next.' **Did you manage to catch the fish in order? How can you prove it? How did you organise the fish once you'd caught them?** This could be done in a backwards order.

Unit 2.13 Missing numbers

Whole class

- **Ten Activities** – Give the children an empty numeral strip to 10 and have 10 activities set out that are labelled from 1–10 (these activities don't have to be completed in one session). Each time they finish an activity, the children stamp the appropriate number box (and/or write the number of the activity in the box). Say 'You do not have to do the activities in order, but you do have to complete them all before you can do any again.' **How will you know you have filled in your track correctly? How did you choose which order to do the activities?**

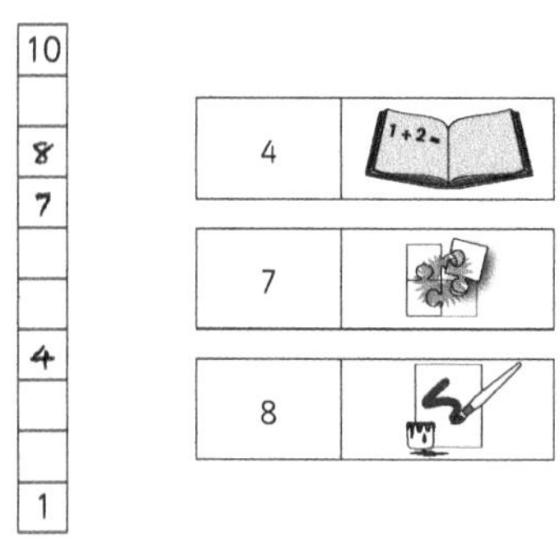

- **Missing Pebbles** – Have a large bag of painted pebbles with at least three collections of the numbers 1–10 but keep out a handful of numbers. Explain to the children that you need some help as you have lost some of the numbers but you are not sure which ones you need to make more of. **How could we organise this? How will we know which numbers are missing? What number is between 4 and 6?**

Stations

- **Numeral Washing Line** – Give the children numeral cards from 10 to 20. Tell the children to hang the numbers up in order on a washing line, starting with the smallest number on the left. They then take it in turns to ask their partner to hide their eyes while they remove one of the numbers. **What number is missing? How do you know? Could you prove it by counting up or down? What number is between 7 and 9?**
- **What's the Missing Number?** – Place some mini sequence strips (see Photocopiable resource 2.21) out on the table and some whiteboard pens. Tell the children to find the missing number on the strip and write it in the correct space. **How did you know which number was missing? Did you say the sequence out loud to check? Do you know what the next three numbers would be, forwards and backwards?**

	17	18		
12	13		15	
14		16		18
10		12		
	16		18	
11				15

Photocopiable resource 2.21 – Missing Numbers.

Unit 2.14 What is zero?

Whole class

- **Drum Beats** – Children each have a musical instrument that can clearly make a single sound, such as a drum or a recorder. Flash a large fives frame pattern at the children and ask a child to copy the number of dots shown using beats on an instrument. Make sure the frame showing zero is shown. Do the same using finger patterns or tens frames to show the number. **What does this one mean? How can you show me this on your instrument?**
- **Labelling Collections Relay Race** – Split the class into four groups. At the far end of the hall, set up a collection of bowls each containing a collection ranging from 0 to 10. Give each group a pack of numeral cards with the numbers 0–10 on them. Ask them to put these face down in a hoop at the head of the line. Tell the children the first person in each group must pick up the first card and run down to the table holding the collections. Look for the bowl that is holding the number of items on the card. Place this number in the bowl. **What if you cannot find a bowl for your number? What does this mean? What number does the empty bowl need?**

Stations

- **Organising Collections** – Have a large box that has been partitioned off into 10 parts (two cutlery drawers) which have been labelled 0–9. Place a large bowl with a collection (e.g. conkers) next to the box. Ask the children to organise the box using the numbers and the conkers. **Are there six conkers in the space labelled 6? What does the 0 mean? What do you notice?**
- **Making Up Stories** – Give the children picture cards with collections of items, such as a basket with four buns, or a nest with three eggs. 'I would like you to make up a story which has the ending … "and that is how there was zero left!"' **What's another word for none left? What number could you use?** Listen to the stories to see if they have understood the concept of zero.

Unit 2.15 Ordinal numbers

Whole class

- **Stories** – Stories like *The First Day of Winter* by Denise Fleming, where a snowman comes alive as the child building it adds pieces during the first 10 days of winter. The days are numbered in order. **What happened on the first day? The tenth day could also be called what day? What do you notice about the way we say the days from the fourth day onwards?**
- **Ordering Cups** – Randomly place 10 cups with the numbers 1–10 on the table in front of the class. Tell the children that you need their help to put them in order from the smallest number to the largest number using the smallest number of moves; however, there is one rule! You can only swap two cups at a time.

For example; the order is 5, 8, 2, 10, 3, 9, 1, 6, 4, 7. **Who wants the first go?** The first child asks to swap cup 1 and cup 5. Now the cup 1 is in the first place. **Are any of the numbers in the right place? Who wants the second go?** The second child swaps cup 8 and cup 2 around. Continue until all the cups are in the correct order. Place a card down in front of the cups to show the order they now sit in represents the position they sit on a number line or track.

Stations

- **List Making** – Give the children a list with five rows, noting the first thing I want to do, the second thing I want to do, etc. Ask the children to either draw or write down the order of things that they would like to do during their soft start in the morning. Read them out the next day. **Nuria, was the jigsaw puzzle the first thing that you did this morning? Who thought that painting was the second thing that they wanted to do? Amman, what did you have down for your fifth choice?**
- **Favourite Foods** – Place 10 photographs of school lunches that children can rank in first to tenth place. Place a large number line on the floor. Say, 'I would like you to look at these photographs of the school lunches and put your favourite lunch in first place next to number one and then your next favourite in second place next to number 2.' **What is the place called if you put it next to the number 3? Number 7? What is last place called? What does it mean if someone has put the macaroni cheese in tenth position?**

- **Racing Cars** – Draw a race track on paper and set up the finishing places, using dot stickers in a line to show children what the respective positions look like, e.g. If the car is in second place, there will be ten dots in a line and they will all be blue apart from the second one which will be red (see Photocopiable resource 2.22). **Which car came first? Which car was second? What place were the last two cars in? What is the last place called? What place did the red car come in?**

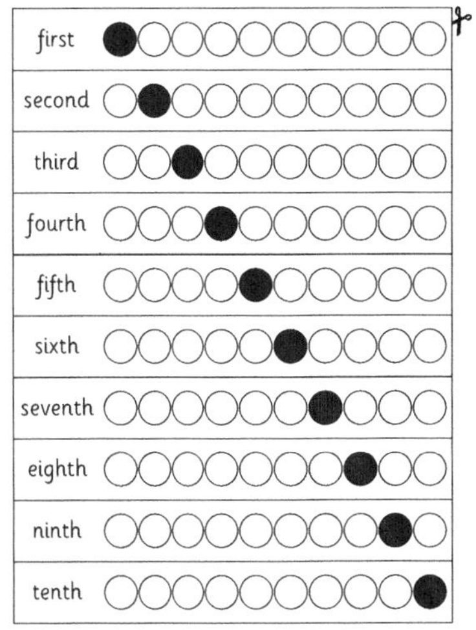

Photocopiable resource 2.22 – First to Tenth Chart.

Unit 2.16 Finger patterns

Whole class

- **Finger Patterns to 10** – Remind the children that they can already make finger patterns to 5, ask them to 'show me 4, show me 2.' **Can you do this without looking at your fingers?** 'Now we are going to make the patterns for 6 to 10.' **Can you show me 6, 7, 8, 9 and 10 on your fingers?** 'Let's look at the different ways you have made 8. Sam you have 4 fingers up on one hand and 4 on the other. Laura you have 5 on one hand and 3 on the other.' Ask the children to get into pairs; one child stands behind the other. The child in front makes patterns behind their backs and their partner has to copy the same pattern (or make another pattern). **Does your partner 'grow' their fingers or can they put the number of fingers up instantly?**

Stations

- **Finger Patterns** – Place two sets of numeral cards from 0 to 10 face down on the floor. Say, 'You have to roll the dice (5–10) and make the number patterns with your fingers. If your partner agrees with your pattern, you get to keep the card. The person with the most cards at the end is the winner.'
- **Finger Pattern Pairs** – Place two sets of cards from 6 to 10 using a set of finger pattern cards (five plus or doubles) (see Photocopiable resource 2.6) and another set of either tally marks (see Photocopiable resource 2.23), numerals or domino patterns. Turn them face down on the carpet and mix them up. Say, 'You can only turn two cards over at a time. You have to try and match up the finger pattern card with a card that has the same amount.'

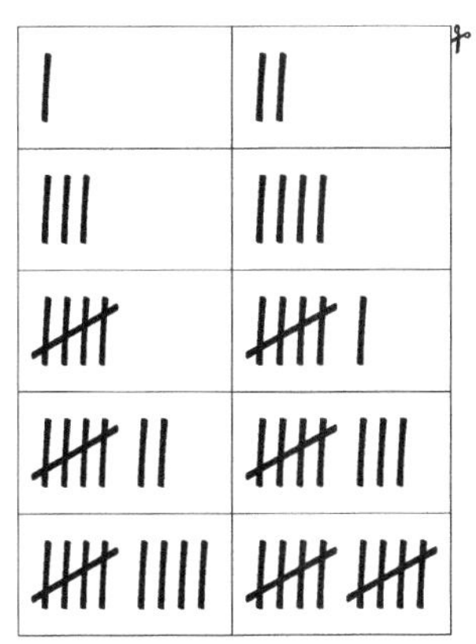

Photocopiable resource 2.23 – Tally Mark Cards.

Unit 2.17 Groupings to 10

Whole class

- **Tens Frames Patterns Five Wise** – Model how to fill a tens frame using the 'five wise' pattern (from left to right, top row, then bottom row). **Can anyone think why it is called a tens frame? How many spaces are there when it is empty? How many dots are there when it is filled?** Have some large tens frames already filled with the five wise patterns and show them to the class. **What do you see? How do you know? How many dots are there? How many spaces do you see? How many are there on the top row when it is full?**

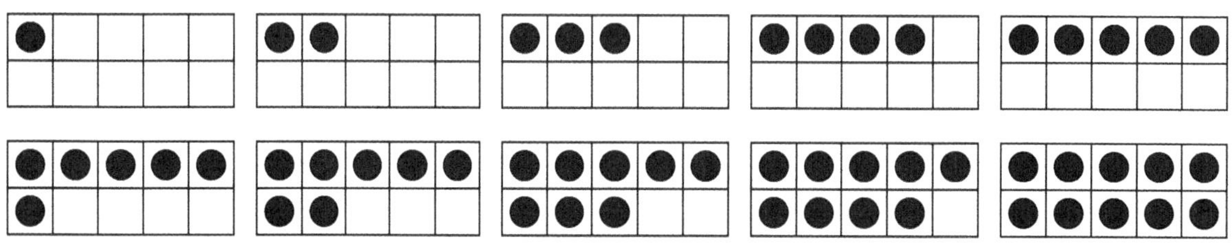

- **Ten Frames Patterns Pair Wise** – Model how to fill a tens frame using the 'pair wise' pattern; put one dot on the top row and then one on the bottom row. **Which numbers are the easiest to see? What does the 4 look like? Look at the 5, what do you see?** Tell them, 'You are going to make some tens frame patterns of your own. Take one number and make it using both the five wise pattern and the pair wise pattern. Tell your friend what you can see.' **How many are on the top row? How many on the bottom row? How many spaces? What's the same about the two patterns? What's different?**

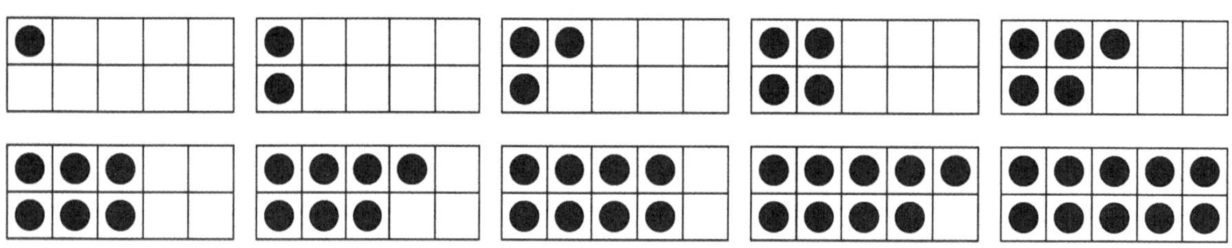

Stations

- **Tens Frame Back to Back** – Have a set of five wise tens frames in your hand and sit back to back with your partner. Describe what you see, then ask your partner to guess the number. Turn around and let them see. If they were correct, they can keep the card. Now swap over. This can also be played with the pair wise cards or with mixed sets.
- **Drawing the Tens Frame** – Have a set of pair wise tens frames in your hand and sit opposite your partner. Flash one of the frames for 2 seconds and then ask them to draw what they saw and label the collection. **Do the tens frame and the pictures look the same? How could we check? What did you see?**

Unit 2.18 Fact families to 5

Whole class

- **Ladybird Composition** – Make a large paper leaf and place 5 red counters (ladybirds) on it. Tell the children that sometimes the ladybirds like to hide underneath the leaf. Screen the leaf and place two ladybirds underneath. **How many ladybirds are underneath the leaf if there are 3 on top? How do they know? What do you add to 3 to make 5?** Flash the three ladybirds if they need to see them. Repeat with different numbers on top and underneath but make sure they always make 5.
- **Stories** – Make up stories involving five children where the children can be seen together and then in two distinct groups. For example, 'This afternoon there are five children going camping. Three of them go in the tent. **How many children will be left outside? Could you picture three of the children in the tent? How did you know this?** 'Let's put three children under this blanket and pretend it is a tent.' **How many children are not in the tent? Were you correct?**

Stations

- **Story of Five Boxes** – Decorate the inside of a shoe box; for example, as a farm environment and place farm animals inside. Ask the children to tell stories where they start with the five animals and then some of them hide on the farm. They must ask their partner some questions to see if they can guess how many are hiding. For example, 'There were five animals on the farm and some hide in the barn. There are three left in the field.' **How many are hiding in the barn? How could you prove it? Could you make finger patterns to prove it?**

We started with five ducklings. There are three ducklings in the pond. How many ducklings are hiding in the barn?

- **Birds' Nests** – Provide two nests (circles on paper) and five eggs (counters). Ask the children to find all the different ways to arrange the five eggs in the two nests. They could then be asked to photograph or draw a picture of their arrangements. 'Let's look at all the different ways the children found to place the five eggs.' **Can you write an addition sum for these pictures? Is 4 + 1 the same as 1 + 4?**

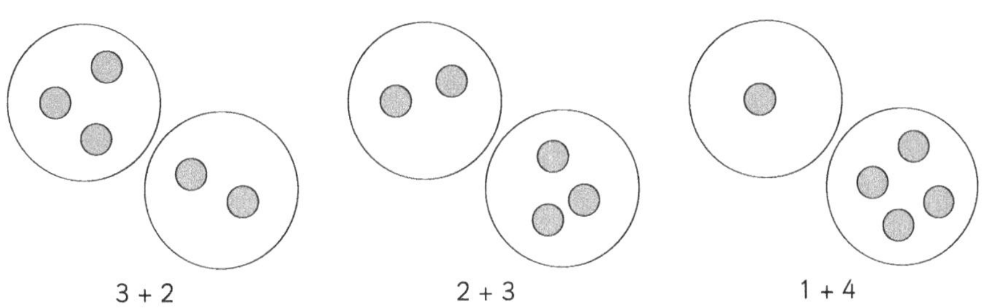

Unit 2.19 Fact families within 5

Whole class

- **Dice Flash** – Provide dice patterns for two, three and four dots (see Photocopiable resource 2.24) that show the addition facts within 5 using two different colours for the pattern. Flash dice patterns for 2–3 seconds and ask the children to draw what they see. **Can you write the addition or subtraction sentence to go with this pattern? Is it easier to write addition or subtraction sentences? What dot patterns did you see? Which are the easiest ones to see?**
- **Double Sided Counter Throw** – Place four double sided counters in a cup. Tell the children they are going to write some addition and subtraction sums to go with the number 4. Throw the counters onto a tray and hide them from the children. Say, 'I have three yellow counters, how many red counters do I have?' **What is the addition sum for this? Could it be written in a different way? How could we write this as a take away sum? How many counters did we start with? How many did we subtract?** Repeat with different numbers of counters.

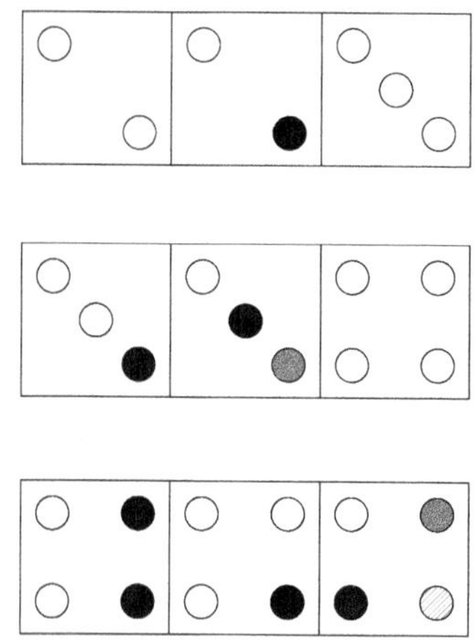

Photocopiable resource 2.24 – Dice Flash.

Stations

- **Part-Part-Whole Boards** – Prepare some part-part-whole boards where the whole number ranges from 2 to 5 (see Photocopiable resource 2.25) and supply a tub of double sided counters. Say, 'You must collect the number of counters that are in the big circle and put them in the counting cup. Then shake the cup onto the carpet.' **How many red do you see? How many yellow?** Put the coloured counters in the two separate smaller circles so the red ones are all together and the yellow ones are all together. **If 4 is the whole, what are the parts?** 'You have three red counters and one yellow counter.' **How would you write this as an addition sum?**
- **Two Handed Finger Patterns** – Place out 12 numeral cards from 2 to 4 face down in one pile (four of each number). Tell the children they will play this game with a partner. They take it in turns to pick up a card without their partner seeing. They then make a finger pattern using two hands (one hand can represent zero) to equal the number on the card. The partner makes a different pattern for the same number. If they cannot, the first player keeps the card; if they can, the partner keeps the card.

Photocopiable resource 2.25 – Part-Part-Whole Board.

Unit 2.20 Five plus facts

Whole class

- **Five Plus Shout** – Provide large domino cards in the range of 5 to 10 with the 5 always filled in on one side (see Photocopiable resource 2.26). First look at the patterns and ask: **What do you see? What have all these dominoes got in common?** Explain that these are called the 'five plus' patterns. Flash the patterns to the children for 2–3 seconds and ask them to make the same pattern on their fingers. **Is 3 and 5 the same pattern as 5 and 3? Does it matter which hand you put up with five fingers? What do you see? Can you 'throw' your fingers?** Watch out for the children who still need to raise their fingers sequentially.

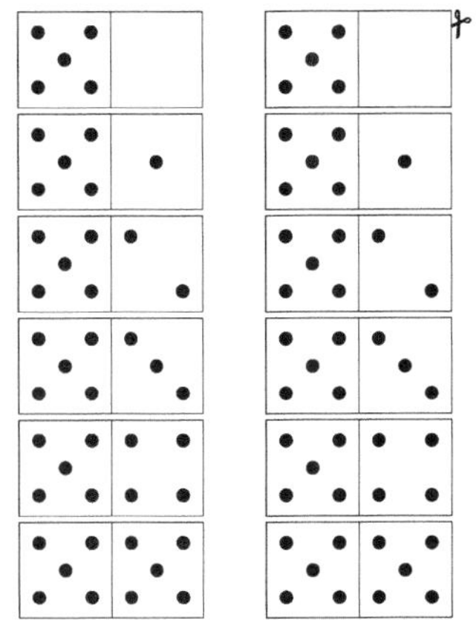

Photocopiable resource 2.26 – Domino Cards 5 to 10.

Stations

- **Throw a Five** – You will need a 1-minute timer and a dice with numerals from 5 to 10. Tell the children to sit facing their partners and say, 'Get ready to make some "five plus" finger patterns.' Decide who is going to throw a five (they can wear a card crown with the dice five pattern on) and who is going to throw the other number. Turn the sand timer over and roll the dice. For example, the first child rolls an 8, they instantly put up five fingers and their partner instantly puts up 3.
- **Bead Strings and Rekenrek** – Tell the children to sit opposite each other behind a screen. Say, 'One of you picks a number card from 5 to 10 and has to make that number on a bead string. You must describe the bead string to your partner without them seeing. They have to make the same pattern on their bead string and also pick up the number card they think they have made.'

Unit 2.21 Skip counting sequences

Whole class

- **Body Music** – Show a dice pattern for 5. Ask the children to make up some class moves to represent the 5 dots. For example, tap their head on the first count, touch their nose on the second, lips on the third, left shoulder on the fourth and finally the right shoulder for the fifth count. Practise this until the children can make these moves at the same time all together. Stress the multiple of 5 count so the children can hear the pattern and hear the numbers 1, 2, 3, 4, **5**, 6, 7, 8, 9, **10**, etc.
- **Double Counting** – Ask the children to find a partner and then to form a circle where the taller person of each pair is standing behind the smaller person. The children are then asked to count from 1, but the outside person whispers the numbers and the inside person shouts the numbers. **What numbers are really easy to hear?** 'I heard 2, 4, 6, 8, …' This is called skip counting in twos. It's a faster way to count people. **What would happen if we counted in ones loudly? Would we end on the same number?** This can also be done for counting in fives, where four children stand behind the inside person.

Stations

- **Dotty Cards** – Provide a box of two dot cards (see Photocopiable resource 2.27) and a dice. Place them on the carpet. Tell the children they are going to collect as many cards as they roll on the dice. Then they are going to count them up in twos by whispering the first dot count and saying the second one louder. For example, they have collected 12 'two dot cards' and they count 1, **2,** 3, **4**, 5, **6**, 7, **8**, 9,**10**, 11, **12**, 13, **14**, 15, **16**, 17, **18,** 19, **20,** 21, **22,** 23, **24. Do you have to count all of them in ones or do you know some of the skip counting sequence? Can you say 2, 4, 6, 8?**

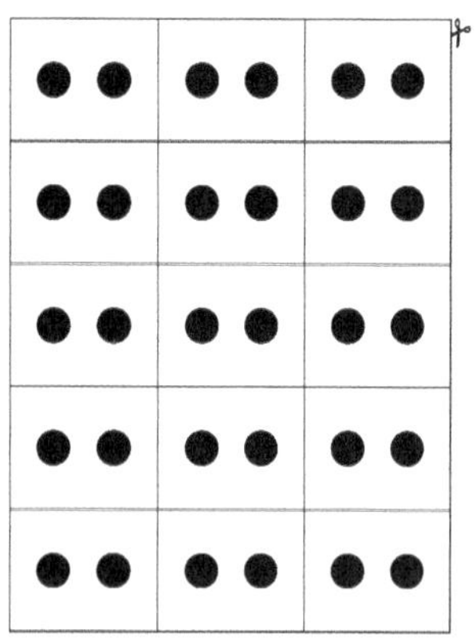

Photocopiable resource 2.27 – Dotty Cards.

- **Necklace Patterns in Twos** – Ask: **Can you copy a necklace pattern that has two of the same beads, then the next two are the same but different from the first two?** Pattern AA BB CC AA BB CC. **How many beads have you used altogether? Can you count them in twos? Can you pull two beads over at a time to count them?**

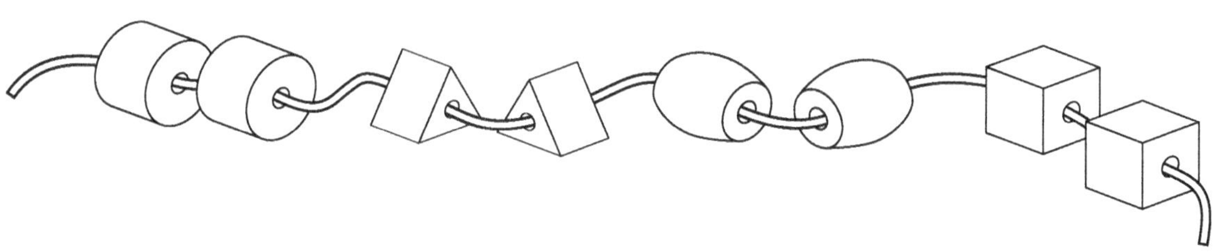

What to observe

- Children understand that to count a collection they must count an item once and once only and say the forward number words in the correct sequence.
- Children understand that the final number said while counting a collection is the amount of items.
- Children understand that anything can be counted, including concrete materials, pictures, movements and sounds.
- Children may count things in different orders and recognise that it is the same amount but not fully trust the count in larger collections.
- Children can reliably say the forward number word sequence to 30.
- Children can say the backward number word sequence to 20 in the correct order.
- Children can give most of the number words before and after in the range of 1 to 20.
- Children can identify and recognise most of the numerals from 1 to 20.
- Children can record most of the numerals from 1 to 20.
- Children can put the numbers 1 to 10 in sequence.
- Children can order the numbers in the range of 1 to 10 and identify the missing numbers.
- Children can put concrete items into an order up to tenth place and know this indicates the position in a sequence, e.g. I put the green counter in sixth place.
- Children know that zero means an empty set.
- Children can 'throw' finger patterns to 10 without putting them up one at a time.
- Children use subitising to say how many are in a small collection up to 10 (using five-wise or pair-wise tens frames and irregular patterns) without counting them one by one.
- Children can recall instantly the two numbers which when added together make 5 and within 5.
- Children can recall instantly their five plus facts to 10.
- Children know some skip count sequences in 2s, 5s and 10s but may not realise they are a quicker way to count.

3 Number and Number Processes: Four Operations

An Introduction to Addition and Subtraction A

Section overview

Units of learning	Review units	Prior learning	Benchmarks/Es&Os
Unit 3.1 **Subitise** Subsitising flashed patterns to 5 (dice, five frames and irregular/random).	**Number Recognition and Place Value A** Unit 2.12	• From their lives children may have a recognition of regular number patterns such as those seen on a dice and dominoes through playing games with their families and friends.	**MNU 0-03a** • Partitions quantities to 10 into two or more parts and recognises that this does not affect the total. • Adds and subtracts mentally to 10. • Uses appropriately the mathematical symbols +, – and =. • Solves simple missing number problems. • Counts on and back in ones to add and subtract. • Doubles numbers to a total of 10 mentally. I use practical materials and can 'count on and back' to help me understand addition and subtraction, recording my ideas and solutions in different ways.
Unit 3.2 **Language of quantity** Learn to explore and talk about quantity using appropriate vocabulary, e.g. more, less, makes, add, take way, altogether.	**An Introduction to Addition and Subtraction A** Unit 3.1	• From their lives the children will have an ever expanding level of vocabulary and a range of personal experiences of addition and subtraction to draw upon. These experiences will arise naturally when the children are sharing toys, getting ready for lunch, playing games and shopping.	
Unit 3.3 **Addition and subtraction using concrete materials** Joining sets using concrete materials in range 0-10 starting count at one.	**Number Recognition and Place Value A** Units 2.1, 2.2, 2.13, 2.14, 2.15, 2.16 and 2.18 **An Introduction to Addition and Subtraction A** Unit 3.2	• Through play and routines, the children may have built up some experience of addition and subtraction through combining amounts and taking some away. An example of this may be five cars in the garage and two drive off. How many are left? Three sheep in a farm and two lambs. How many altogether?	
Unit 3.4 **Principles of counting** Know that the last number said is different to the others since it both 'labels' the last object and tells how many there are in the whole group.	**Number Recognition and Place Value A** Unit 2.2	• The children should be developing counting principles through experiences in play. They should also know the forwards and backwards counting sequences, be developing an understanding of quantity and using the language of number in their play and routines.	
Unit 3.5 **Doubles patterns** 1 + 1, 2 + 2.	**Number Recognition and Place Value A** Unit 2.20	• The children should understand that a double is two of the same and be using 'double' in their play and conversations.	

Conceptual context

The children should learn how to subitise – identify an amount without counting. This will start with subitise patterns to five, before extending. The children should also visualise small numbers as groups in different combinations, such as seven as a group of 4 and 3; 5 and 2; 2, 4 and 1, and so on. By visualising numbers in this way, it helps to lay the foundations for number bonds and aids the children's ability to recall number bonds mentally. Structured resources such as five/tens frames, bar models and Rekenreks help the children to 'see' number and develop an understanding of the quantity and combinations of that number. This in turn helps the children to develop number bonds and have good mental recall of number facts. The children should develop their ability to recall number facts from memory, building up quick and accurate recall of number bonds. This will support more complex learning of number strategies in future learning, such as partitioning in advanced addition and subtraction, rounding and using doubles/near doubles to become confident and competent manipulating numbers.

The children should understand the process of the four operations arising from natural situations based on concrete experiences. Children must have many opportunities to explore number and use language to explain their thinking. The children need to feel, move, manipulate and experience number in concrete forms, then move on to imaging these materials in their mind and subsequently move on to the abstract form of number. Concrete experiences provide the children with a base to refer to when imaging the number and operations in an abstract form.

The children should also have their attention drawn to the link between the four operations. The children should be able to combine and separate collections as well as use language to describe what they are doing. Meta-cognitive skills – being able to think about their thinking – is vitally important in numeracy and every opportunity should be taken to allow the children to discuss their learning and thought processes. Calculations should be based on the manipulating of materials and the acting out of number problems. The children should begin to see patterns and links in number calculations. The part-part-whole model should be used to highlight the link between addition and subtraction. If the children understand the link, then they can choose how best to perform a calculation when facing more challenging calculations in future learning.

Place value is a fundamental concept the children must be secure with. This sits beside addition and subtraction. When moving onto two-digit numbers, the children must be able to see the teens number as ten and ones; for example, 13 as 10 and 3. To manipulate bigger numbers mentally, the children must understand the value of a digit and use this in their mental calculation. The children may choose to partition a two- or three-digit number into tens and ones or hundreds, tens and ones, perform the calculations and then put the component parts back together again. This can only be done if the children are secure with place value. Formal addition and subtraction, as in column sums, should not be introduced until children can add and subtract mentally in many ways using a number of strategies and can explain their thinking. Once formal addition and subtraction is introduced, the correct terminology must be used referring to place value; for example, you carry a ten not one! The children will then understand these calculations and not just follow a formula. Many children will not use these formal calculations as the mental strategies they have already developed are quicker and more accurate. The early years in numeracy provide the foundations for all this future learning and therefore the children must secure their understanding through deep and varied concrete experiences.

Resources

- Dice
- Dominoes
- Playing cards
- Dot patterns – regular and irregular
- Subitising cards
- Counters
- Frames – fives and tens
- Chalk
- Small plastic toys
- Pipe cleaners
- Beads
- Chalk
- Washing line
- Dolly clothes
- Fancy boxes – different sizes
- Buttons
- Teddy bears
- Plates
- Picnic food
- Role play
- Sweetie jar and sweetie templates
- Empty number line
- Spotty animal frames
- Pom-poms
- Giant tweezers
- Flip-flops
- Digit cards 0–10
- Feely bag
- Rekenrek
- Multi-link
- Grab mat
- Rabbit puppet
- Raisins
- Garage
- Cars

Digital

- Hoopy Zoo
- Fingers and Toes
- Wobbly Tower

Record book

Pages 6–16

Key vocabulary

- **More**
- **Less**
- **Add**
- **Take away**
- **Altogether**
- **Subitise** – recognise an amount without counting
- **One more/less**
- **Double** – two of the same object/amount
- **Opposite**
- **Enough**
- **Many**
- **The same as**
- **Most**
- **Fewest**

Misconceptions

Potential misconception	Addressing the misconception
Units 3.1–3.5 As language is still developing, the children may not understand the instruction in the language.	• Model language clearly and visually, then address any misuse of words.
Units 3.3–3.5 The children do not count the objects with one to one correspondence.	• The children need lots of experience in counting objects and they need to know the counting sequence well and show the basic counting principles. Ensure they touch all the objects when counting to guarantee one to one correspondence.
Units 3.3–3.5 The children miscount.	• Encourage the children to move objects if they need to so they can ensure none are missed or counted twice.

Teaching ideas

Unit 3.1 Subitise

Stations

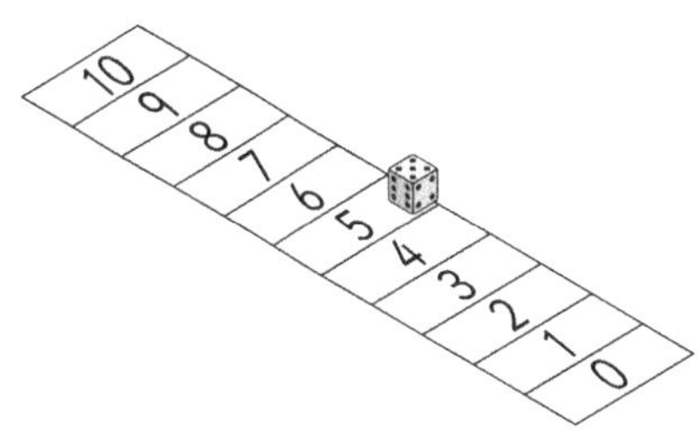

- **Roll and Say** – Roll the dice and say the number aloud. The children should just be able to say the number and not have to count the dots. Dice tool can be used as a variation to a real dice.
- **Roll and Find** – Roll the dice and find the number. A chalk number line could be drawn for the children to find the number and stand on, or as an extra challenge the digit cards 1–6 could be randomly placed on the carpet for the children to locate and stand on. Ask the children to do this quickly so they don't have time to count the dots. Dice tool can be used as a variation to a real dice.
- **Roll and Clap** – Roll the dice and clap the amount. The aim is for the children to recognise the amount, not count it. Clapping can change to jumping, stamping, shoulder taps and so on. Dice tool can be used as a variation to a real dice.
- **Domino** – Place dominoes in a feely bag. Ask the children to pull a domino out. **How many dots do you see? How do you know this?**
- **Five Frames** – Show the children the five frames (see Photocopiable resource 3.1) and discuss. **How many dots? How many spaces? How many more to make five?** Show the children the five frame quickly and ask them to say the amount, not count. Counting can be used to check. If the children need to, they can use counters to fill up the five frame and find how many more are needed to make 5.

Photocopiable resource 3.1 – Five Frames.

- **Ten Frames** - Show the children the tens frames (see Photocopiable resource 3.2) and discuss. **How many dots? How many spaces? How many more to make ten?** Show the children the tens frame quickly and ask them to say the amount, not count. Counting can be used to check. If the children need to, they can use counters to fill up the tens frame and find how many more are needed to make 10. Discuss the patterns of numbers that you can see. Different children may see a different pattern.

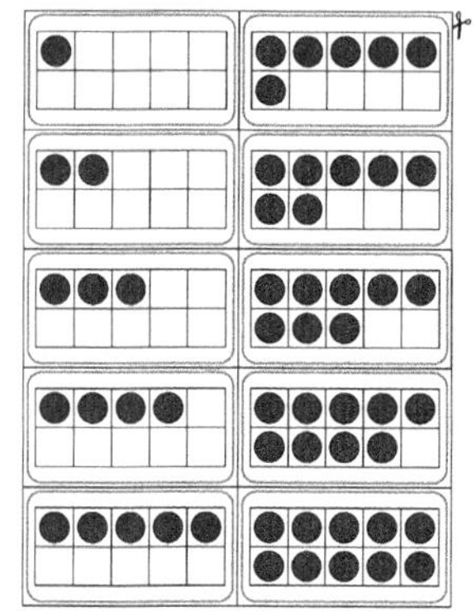

Photocopiable resource 3.2 – Tens Frames.

- **Quick Flash 1** – Quickly flash a regular dot pattern (see Photocopiable resource 3.3). Ask the children how many they saw. Ask them to explain their thinking describing what they saw. Spend time looking at the different number patterns within the amount. Counting can be used to check. Start with amounts to 6 linking to the dice, then extend to 10.

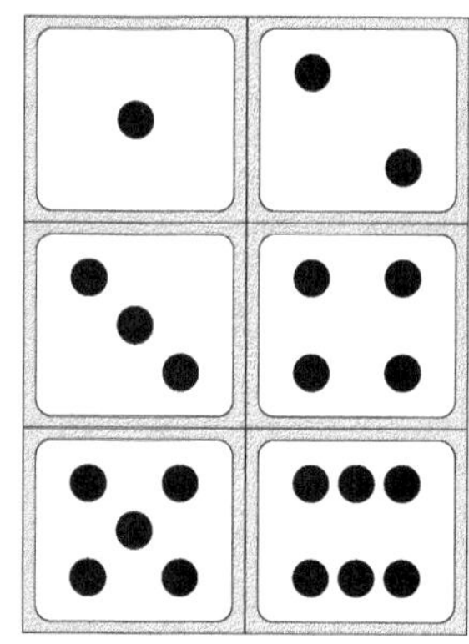

Photocopiable resource 3.3 – Regular Dot Patterns.

- **Quick Flash 2** – Quickly flash an irregular dot pattern (see Photocopiable resource 3.4). Ask the children how many they saw. Ask them to explain their thinking describing what they saw. Spend time looking at the different number patterns within the amount. Counting can be used to check. Start with amounts to 5, then extend to 10, then 20.

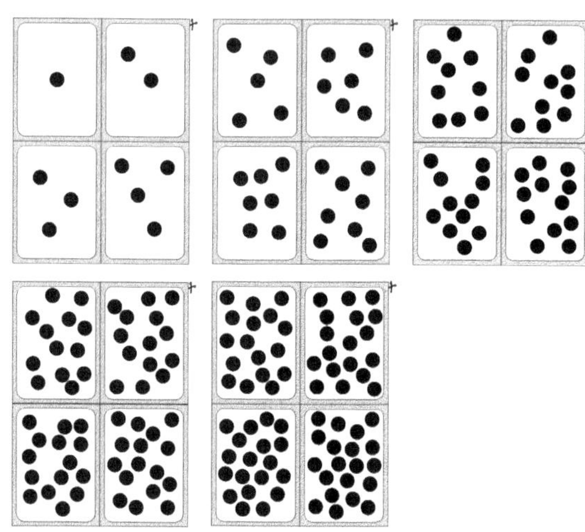

Photocopiable resource 3.4 – Irregular Dot Patterns.

- **Finger Flash** – Ask the children to show you a number, first within 5 then extending to 10. Ask the children to show you using one hand or two. For example, 3 on one hand holding up three fingers and 3 on two hands with two fingers on one hand and one on the other. Ask the children to show you an amount one way and then another. For example, 8 could be five fingers on one hand and three on the other or four fingers on each hand. Discuss what the children show you highlighting all the different ways as well as capitalising on language, for example, pointing out 4 on each hand is a double. Play the game the other way round with you flashing the fingers and the children shouting out the amount. Ask the children to explain how they knew.

- **Bunny Ears** – The children place their hands each side of their head and finger flash the requested amount. The children find this harder as they cannot see their fingers. Discuss all the number patterns you have been shown. Then ask, now show me one more/less and discuss what the children did. Play the game the other way around with you flashing the fingers and the children shouting out the amount. Ask the children to explain how they knew.
- **Hidden Counters** – Hide some counters under your hand. Quickly lift your hand and let the children see the counters before replacing your hand. Go around the group asking the children to say the amount. Ask them how they knew, discussing any number patterns they spotted. Compare their guesses with others in the group. Remove your hand and check to see how many counters there were highlighting any good guesses and their thinking. For example, Isla you thought there were five because you saw three together and two at the side. You were correct because 3 and 2 make 5 (count to prove).
- **Rekenrek** – Using a 20-bead frame with two rows of 10 split into two colours highlighting 5. Ask the children to show you an amount up top (using the top bar only). Try to get the children to do this in one big push. Start with numbers up to 5 then extend to 10. Discuss how the children knew where to push the amount. For example, if 4 was requested, the children may say, 'It is one less than five so I knew to leave one red bead behind'. Discuss. **How many are left?** Again, try to get the children to answer without counting but recognising the amount, using the 5 and 10 reference points. The children can also be asked to push across the same amount but using both bars. Discuss all the options the children show you. For example, if 4 was requested, it could be three up top and one below or one up top and three below. Point out how these are opposites of each other. Or it could be two on the top and bottom, the double. Discuss how all these different options still give 4.

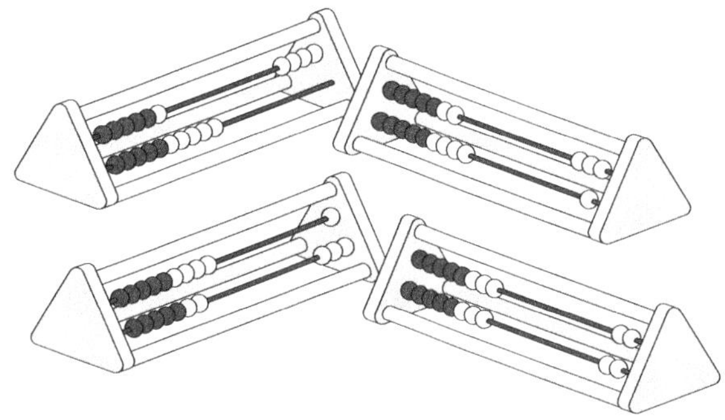

- **Spotty Animals** – Ask the children to roll the dice and place that number of pom-poms on their animal. Using giant tweezers to do this also improves their fine motor skills. The children keep rolling and covering the spots until there are none left. Encouraging the children to race against another child helps them to subitise and not count.

- **Empty Number Line** – The children pick a subitising card and place where they think it should go on an empty number line. Once a few cards have been placed discuss the dot patterns. For example, 'This one is near the end, which is ten, because it has lots of spots. The five is the same as the dice therefore it goes in the middle. This card has one spot, so it is the smallest number at the start of the number line.'
- **Sweeties** – Cut out a sweetie jar template and have a set amount of sweetie cut outs. Get the children to roll the dice and place that number of sweeties in the jar. Play continues until no sweeties are left. The children can play this in two teams to see who wins.
- **Subitise Hunt** – The children look around the setting to find a number. For example, 2 – could be two windows together or two flowers in a jar. Ensure the children are spotting the items and not counting. The objects can be photographed and displayed in a book or on the wall.
- **Race to 10** – Take turns with a partner. Roll the dice and place that number of counters in a ten frame. The winner is the first to complete the ten frame.

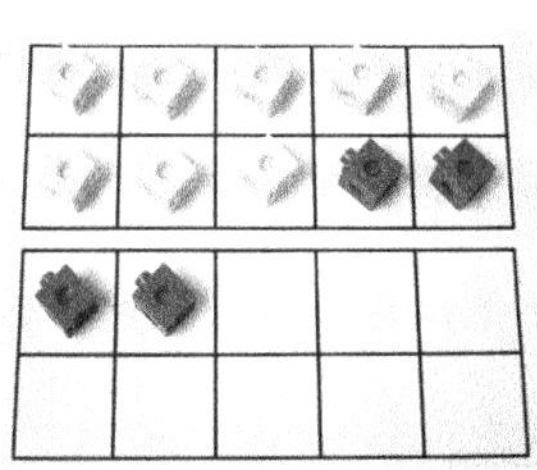

- **Subitise Match** – Have cards with different amounts of ladybirds on leaves or fish in bowls. The children can start with amounts to 5 and extend to 10. There should be multiples of the same amount of ladybirds on leaves or fish in bowls but in different number groups. Have the children match up the cards with the same amounts.

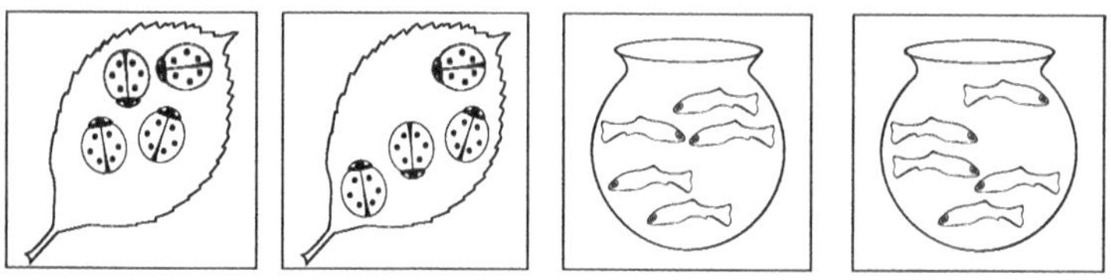

- **Flip Flops** – Ask the children to fold their flip flop (see Photocopiable resource 3.5) to show a number. Discuss the number patterns they made. **Did everyone make it the same way?** Discuss.

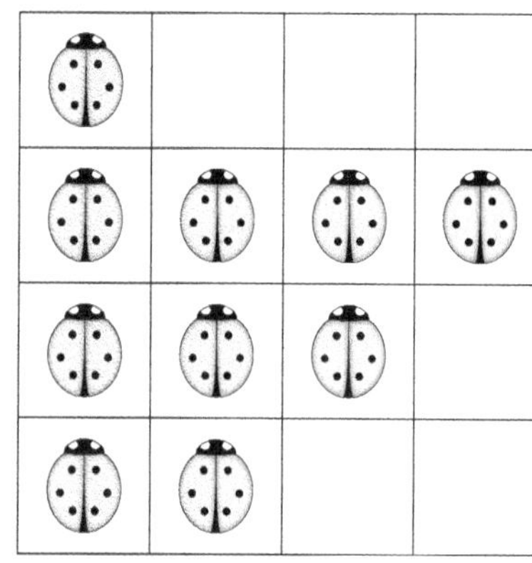

Photocopiable resource 3.5 – Flip Flops.

Unit 3.2 Language of quantity

Whole class

- **Conversation** – Maximise opportunities around the setting to engage the children in conversation about number and number calculations. For example, 'I need one more person to come and play the game.' 'There are four people at snack but only two bananas, how many more do we need?' 'There were five cookies. Finlay ate one so now there is one less. There are four cookies left.'
- **Role Play** – Set up a café where there are great opportunities for setting tables and matching amounts of food to people and so on.
- **Songs and Rhymes** – Sing and act out traditional rhymes such as, 'Five currant buns', 'Ten fat sausages', 'Five little monkeys', 'Ten green bottles', 'One, two buckle my shoe', 'One potato, two potato…', '1, 2, 3, 4, 5, once I caught a fish alive', 'Five little ducks went swimming one day', 'There were ten in the bed', 'Baa baa black sheep', 'Two little dickie birds' and 'Five little speckled frogs'. Laminated pictures or props such as cuddly toys and puppets really help to engage the children and also provide visuals for the changing amounts. 'Hoppy Zoo', 'Fingers and Toes' and 'Wobbly Tower' are all available in the digital pack.

Stations

- **Making Teams** – Ask the children to make teams, maximise opportunities to use one more/one less needed. Match the children up to compare amounts. More in this team, less in this team and so on. For example, **I need two teams of six.** The 12 children get into two teams, of 7 and 5. **Let's stand side by side and compare. If I have two teams of six then my lines will be the same length as they have the same amount of children.** Help the children stand side by side and compare.

  Oh this line is longer, it must have more children, let's count to check. Each line is counted aloud by the children as you touch each child on the head.

  So, there are seven in this line and five in this line. But I wanted six in each line. What should we do? Help the children by pointing out, **Seven is one more than six and five is one less than six.** Show this on a number line. Once the children are rearranged count to check we have 6 in each line and recap learning.

  This can be done with any amount of teams and number in a team. Pick numbers that are going to make the children think. It is also good for pointing out odd and even numbers.

- **Number Line** – The children roll a dice and stand on that number on the chalk number line drawn on the carpet. Then ask the children to jump on one more. **What number are you now?** 'Step back one place.' **What number are you on now?** Discuss what happened when they added one more or took one away.

- **Teddy Bears Picnic** – Set four teddies on a picnic rug. Ask the children to give out the picnic food. Have five apples, four drinks, three sandwiches, and so on. Encourage the children to use language such as, 'We need one more' or 'That's just right' or 'We have one too many'.

- **Tip Out** – Tip out some buttons from a jar. Then tip out another pile of buttons. Compare the piles. **Which has the most/fewest?** Line up the two piles to compare and discuss. **What happens if you add some to a pile? What happens if you take some away from a pile?**

- **Washing Line** – Peg up some dolly clothes. **What happens if we add some more? What happens if we take some away? Are there enough pegs for the items?**

- **Towers** – Ask the children to build a tall tower. When the children are finished, discuss. **Who has the most/fewest bricks? How do you know?**

- **What's in the Box?** – Get a few fancy boxes and fill each with a small number of toys. Let the children feel, shake and explore the boxes. Ask the children, **Which box do you think will have the most/least number of toys? Why do you think this?** Count the toys from the boxes and compare the amounts.

- **Doubles** – Use playing cards or dominoes to show doubles. Discuss what a double is. The children can play snap with the cards when they find a double.

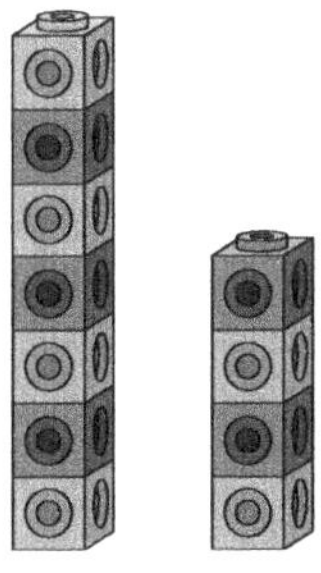

Unit 3.3 Adding and subtracting using concrete materials

Stations for addition

- **Handfuls** – Two children take a handful of counters each. The children put their handfuls together and add up by counting to see the total amount.

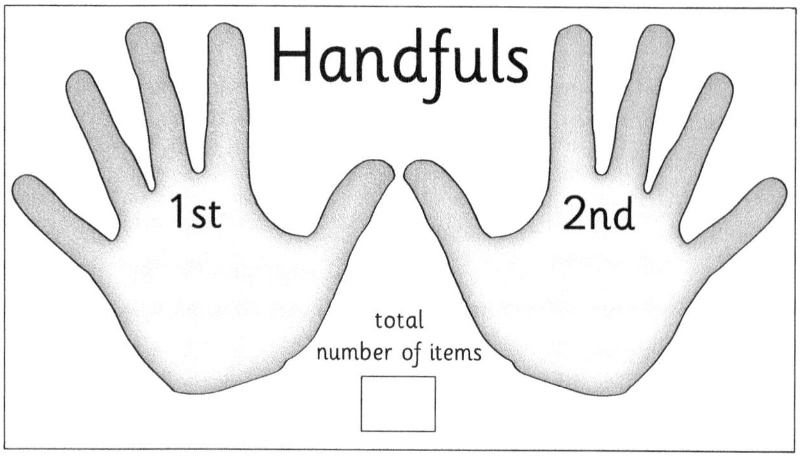

- **Pipe Cleaner** – The children put some coloured beads on one end and another colour of beads on the other end. The children then slide the beads together to count the total number of beads.
- **Friendly Towers** – Two children build a tower and say how many blocks they each used. Then they put the towers together to find the total number of blocks.
- **Split Up** – Take a number of objects. Ask the children to find as many different ways as they can to split the objects into two smaller groups. Ask the children to explain what they have done using mathematical language. For example, if six objects were used, then it could be seen as five and one, three and three, four and two. It can also be used to highlight the commutative aspect of addition if someone else finds one and five or two and four.
- **Garages** – Using two toy garages (these could be made from a cardboard box) and a number of cars, ask the children to find all the different combinations for parking the cars. Discuss their answers.
- **Story Problems 1** – Children can manipulate items or draw pictures to show a number story. For example, 'In the garage there are four red cars and three yellow cars.' **How many cars are there altogether?** 'Nuria has ten beads on her necklace, eight are yellow.' **How many beads are red?** 'There are five children and four biscuits.' **How many more biscuits do we need so we can all have one?**
- **Real Life Situations 1** – Maximise real life situations around the setting. 'Three girls and two boys in line.' **How many children in line?** When counting, ensure the children count aloud with you as you touch each child in turn. Registration in the morning: **How many girls are in my group today and how many boys? How many children are here?** 'There are seven children but only five seats.' **How many more seats do we need?** Take time to highlight the language and operations used. Discuss the thinking behind it.

Stations for subtraction

- **Hungry Rabbit** – Lay out an amount of raisins. **Rabbit ate some.** Use a hand puppet to hide a few raisins. **How many are left?**

- **Hidden counters** – Count an amount of counters laid on the table. Discuss the number patterns that the children can see in the laid out counters. For example, five counters on the table could have three beside each other in a row and two in the next row. The children close their eyes and you hide some. The children open their eyes and say how many were hidden. By thinking of the pattern before they were hidden, this might help the children work out how many were taken. For example, if the row of two were taken and the row of three were left. You should then use language to describe the subtraction sum. 'Five take away two leaves three.'
- **Story Problems 2**– Children can manipulate items or draw pictures to show a number story. For example, 'There were six oranges and Finlay ate two.' **How many oranges are left?** 'There are five birds in a tree, two birds fly away.' **How many birds are left in the tree?**
- **Real Life Situations 2** – Maximise real life situations around the setting. 'There were six apples in the fruit bowl. Nuria had one and so did Amman.' **How many apples are left?** 'There are 12 children in the group. Isla is off today so that's one less than usual.' **So there's how many today?** 'There were four books, I put two away.' **So how many books are left?** Take time to highlight the language and operations used. Discuss the thinking behind it.

Unit 3.4 Principles of counting

Stations

- **Move and Say** – Using two groups of counters, ask the children: **How many counters are there altogether?** Pull the counters towards yourself and repeat the question. **So how many counters were there?** The children should be able to say the total without re-counting. Discuss how they knew the answer. If the children do not realise this, re-count to check. Repeat the activity often so that the children become more secure with conservation of number.
- **Number Line Addition** – Using a giant number line add two amounts by laying objects out along the number line. The children can count and also see that the last number labels the total amount.

- **Jump the Number Line** – Use a giant number line. 'Roll a dice and step on the starting number. Then roll another dice and jump on that number of spaces to find the total.' The children can see the last number they count on is also the total.
- **Equal Groups** – Using everyday objects make groups of the same amount. For example, 'We have three pencils and six glue sticks. Lay them side by side.' **How many more pencils do I need so that I have the same amount of both objects? How did you know?** Discuss the amount and the labels.

- **Number Line Comparison** – Lay different groups of objects along a number line. Discuss the amount and compare them to each other. For example, 'There are seven books and five pencils, so there are more books as seven is a bigger number.'

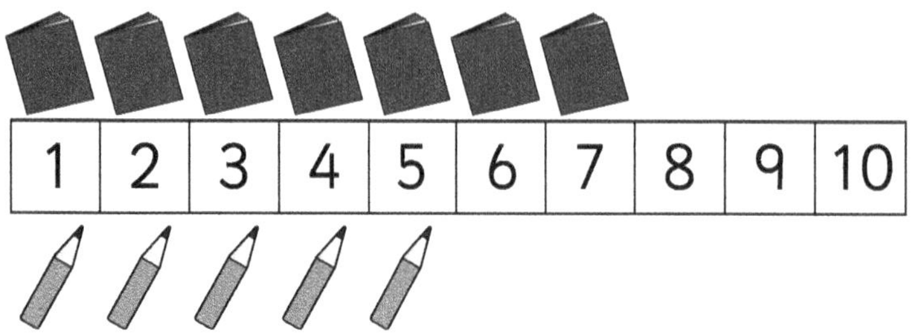

Unit 3.5 Doubles patterns

Stations

- **Rainbow Towers 1** – Choose two different colours of multi-link cubes and make two towers of the same height. Discuss how these are the same and check by counting and measuring.
- **Dominoes** – Challenge the children to pick out all the doubles dominoes from a pack. Count the spots to find the total for each double.
- **Double Finger Flash 1** – Ask the children to show you double four on their fingers. Check the children have the same amount on each hand. Repeat for other numbers.
- **Doubles to Ten Tower Challenge 1** – Using different colours of multi-link, can the children find all the doubles to ten? Skip count in twos up the towers.

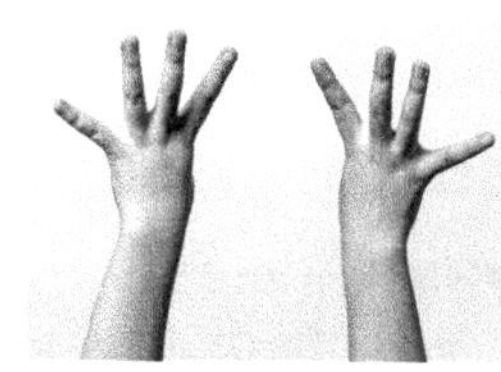

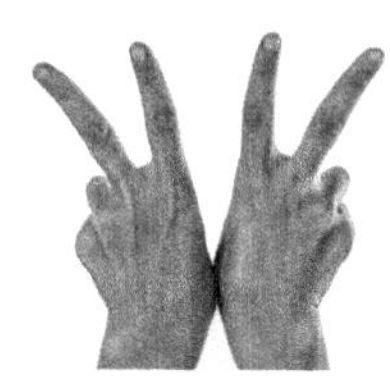

- **Rekenrek Doubles 1** – Ask the children to show you double three by pushing the beads over using top and bottom. **What is the total?** Repeat for all doubles.
- **Ten Frame Doubles** – Using a blank ten frame and two different coloured counters, show and discuss the doubles.
- **Bead String Doubles** – using beads and strings or pipe cleaners, ask the children to show the doubles facts to ten. Start with one add one. Then add another bead on to each side to show two add two and continue until you can see five add five.

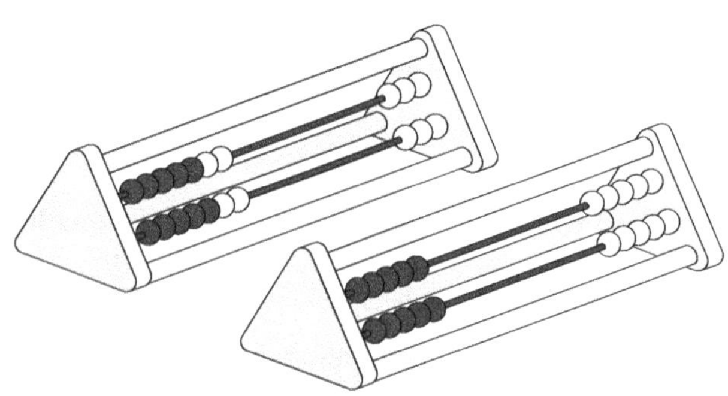

Addition and Subtraction B

Section overview

Units of learning	Review units	Prior learning	Benchmarks/Es&Os
Unit 3.1 **Conceptual subitising**	**An Introduction to Addition and Subtraction A** Unit 3.1	• I can subitise with small amounts. • I can recognise common regular patterns.	**MNU 0-03a** • Partitions quantities to 10 into two or more parts and recognises that this does not affect the total. • Adds and subtracts mentally to 10. • Uses appropriately the mathematical symbols +, – and =. • Solves simple missing number problems. • Counts on and back in ones to add and subtract. • Doubles numbers to a total of 10 mentally. I use practical materials and can 'count on and back' to help me understand addition and subtraction, recording my ideas and solutions in different ways.
Unit 3.2 **Odd and even number patterns**	**Number Recognition and Place Value A** Units 2.1 and 2.2	• I can say the forwards counting sequence. • I can say the backwards counting sequence.	
Unit 3.3 **Doubles number facts within 10**	**An Introduction to Addition and Subtraction A** Unit 3.5	• I can recognise a double pattern.	
Unit 3.4 **Fact families to and within 5**	**An Introduction to Addition and Subtraction A** Units 3.2, 3.3 and 3.4	• I can use the language of addition and subtraction. • I can add and subtract using concrete materials. • I know the last number said labels the amount in the group.	
Unit 3.5 **Mathematical symbols involved in addition and subtraction**	**An Introduction to Addition and Subtraction A** Unit 3.2	• I can use the language of addition and subtraction.	
Unit 3.6 **Adding and subtracting using count all concrete materials and imaging**	**An Introduction to Addition and Subtraction A** Unit 3.2, 3.3 and 3.4 **Addition and Subtraction B** Unit 3.5	• I can use the language of addition and subtraction. • I can add and subtract using concrete materials. • I know the last number said labels the amount in the group. • I can use mathematical symbols involved in addition and subtraction.	
Unit 3.7 **Continuing a sequence (next 2, 3, 4, numbers forwards to 30 and backwards from 20)**	**Number Recognition and Place Value B** Units 2.4 and 2.17	• I can identify the numbers without dropping back. • I understand that when I count forwards the numbers get bigger and backwards the numbers get smaller.	
Unit 3.8 **Counting on and back (from one unscreened numeral to locate a numeral on a screened numeral track)**	**Number Recognition and Place Value B** Unit 2.4	• I can identify the numbers without dropping back.	

Units of learning	Review units	Prior learning	Benchmarks/Es&Os
Unit 3.9 **Counting jumps on a number line**	**Number Recognition and Place Value B** Units 2.1, 2.2 and 2.18	• I know the forwards and backwards counting sequence. • I can use one to one correspondence.	
Unit 3.10 **Addition using concrete materials to count on**	**Number Recognition and Place Value B** Unit 2.15	• I know a group gets bigger when we add more objects.	
Unit 3.11 **Addition using counting on using screened concrete materials**	**Addition and Subtraction B** Unit 3.10	• I can add using concrete materials.	
Unit 3.12 **Addition using counting on by imaging**	**Addition and Subtraction B** Units 3.10 and 3.11	• I can add using concrete materials • I can add using screened materials.	
Unit 3.13 **Subtracting using concrete materials to count back**	**Number Recognition and Place Value B** Unit 2.16	• I know a group gets smaller when we take objects away.	
Unit 3.14 **Subtracting using counting back using screened concrete materials**	**Addition and Subtraction B** Unit 3.13	• I can subtract using concrete materials.	
Unit 3.15 **Subtraction using count back by imaging**	**Addition and Subtraction B** Units 3.13 and 3.14	• I can subtract using concrete materials. • I can subtract using screened materials.	
Unit 3.16 **Fact families for 5 and 10**	**Addition and Subtraction B** Units 3.1 and 3.4	• I can subitise patterns to ten. • I can give fact families to and within 5.	
Unit 3.17 **Partitions to 10**	**Addition and Subtraction B** Unit 3.16	• I can give fact families to 10.	
Unit 3.18 **Adding and subtracting mentally using known facts**	**Addition and Subtraction B** Unit 3.2, 3.3 and 3.16	• I can identify odd and even numbers. • I can give my doubles facts. • I can give my fact families to 5 and 10.	
Unit 3.19 **Solving simple missing number problems**	**Addition and Subtraction B** Units 3.5 and 3.18	• I can use mathematical symbols. • I can add and subtract mentally using known facts.	

Conceptual context

The children should learn how to subitise – identify an amount without counting. This will start with consolidating subitise patterns to 5, before extending to 10. The children should also visualise small numbers as groups in different combinations, such as 7 as a group of 4 and 3, 5 and 2, 2, 4 and 1 and so on. By visualising numbers in this way, it helps to lay the foundations for number bonds and aids the children's ability to recall number facts mentally. Structured resources such as five/tens frames, bar models and Rekenreks help the children to 'see' number and develop an understanding of the quantity and combinations of that number. This in turn helps the children to develop number bonds and have good mental recall of number facts. The children should develop their ability to recall number facts from memory building up quick and accurate recall of number bonds, this will support more complex learning of number strategies in future learning, such as partitioning in advanced addition and subtraction, rounding and using doubles/near doubles to become confident and competent manipulating numbers.

The children should understand the process of the four operations arising from natural situations based on concrete experiences. Children must have many opportunities to explore number and use language to explain their thinking. The children need to feel, move, manipulate and experience number in concrete forms, then move on to imaging these materials in their mind and then finally move on to the abstract form of number. These concrete experiences provide the children with a base to refer to when imaging the number and operations in an abstract form.

The children should be able to combine and separate collections using appropriate language to describe the processes. Symbols should be introduced and used to record calculations. The children must be able to manipulate materials, draw and act out number problems progressing on to imaging and visualising before mentally performing operations. The children should be able to choose appropriate procedures to perform calculations and explain their thinking.

The children should also have their attention drawn to the link between the four operations. The children should be able to combine and separate collections as well as use language to describe what they are doing. Meta-cognitive skills – being able to think about their thinking – is vitally important in numeracy and every opportunity should be taken to allow the children to discuss their learning and thought processes. Calculations should be based on the manipulating of materials and the acting out of number problems. The children should begin to see patterns and links in number calculations. The part-part-whole model should be used to highlight the link between addition and subtraction. If the children understand the link, then they can choose how best to perform a calculation when facing more challenging calculations in future learning.

Place value is a fundamental concept the children must be secure with. This sits beside addition and subtraction. When moving onto two-digit numbers, the children must be able to see the teens number as ten and ones; for example, 13 as 10 and 3. To manipulate bigger numbers mentally, the children must understand the value of a digit and use this in their mental calculation. The children may choose to partition a two- or three-digit number into tens and ones or hundreds, tens and ones, perform the calculations and then put the component parts back together again. This can only be done if the children are secure with place value. Formal addition and subtraction, as in column sums, should not be introduced until children can add and subtract mentally in many ways using a number of strategies and can explain their thinking. Once formal addition and subtraction is introduced, the correct terminology must be used referring to place value; for example, you carry a 10 not 1! The children will then understand these calculations and not just follow a formula. Many children will not use these formal calculations as the mental strategies they have already developed are quicker and more accurate. The early years in numeracy provide the foundations for all this future learning and therefore the children must have a secure understanding through deep and varied concrete experiences.

Resources

- Dice
- Dominoes
- Playing cards
- Dot patterns in regular and irregular patterns
- Counters
- Five frames
- Tens frames
- Chalk
- Pipe cleaners
- Beads
- Chalk
- Fancy boxes in different sizes
- Buttons
- Teddy bears
- Empty number line
- Counting stick
- Digit cards
- 20 track
- 1–10 grid
- 0–30 board
- A4 number cards
- Multi-link
- Cubes
- Rekenrek
- Diennes
- Sand timer
- 100 board
- Interlocking chains
- Grab mat
- Think board
- Strategy board
- Whiteboard and pens
- Plus/minus cards
- Part-part-whole template
- Balance scales
- Paper plates
- Paper cups
- Number line
- Toy frog
- Feely bag
- Pendulum

Digital

Fishy Facts – Recall addition and subtraction facts to 20

Ants on a Bus – Recall addition and subtraction facts to 20

Double the Fun – Solve problems involving Doubling, halving, grouping and sharing

Coconuts Calculations – Basic addition and subtraction facts to 20.

Animated Song: Hoopy Zoo – Adding one more to five

Fingers and Toes – Counting to ten

Wobbly Tower – Counting backwards from ten

Tools – Tree Tool (Partitions), Dice Tool, Number Line, Counting Tool, Number cards, Number Square and Function Machine.

Record book

Pages 6–16

Key vocabulary

- **More**
- **Less**
- **Add**
- **Plus**
- **Take away**
- **Subtract**
- **Minus**
- **Altogether**
- **Subitise** – recognise an amount without counting
- **One more/less**
- **Double**
- **Near double**
- **Odd**
- **Even**

Misconceptions

Potential misconception	Addressing the misconception
As language is still developing, the children may not understand the instruction in the language.	• Model language clearly and address any misuse of words. Ensure you expose the children to a range of language.
The children do not count the objects with one to one correspondence.	• The children must be secure with all the counting principles.
The children miscount.	• The children need to know the counting sequence well. Ensure they touch all the objects when counting to guarantee one to one correspondence. Encourage the children to move objects if they need to in order to ensure none are missed or double counted.
When using a number line, the children begin counting on the number they are adding or subtracting with.	• The children should jump on or back from the number they start with. Concrete experiences transferred on to a number line will help with this.
Mistakes when performing addition and subtraction involving zero.	• The children see zero therefore think the answer is zero. For example, 7-0=0 or 8+0=0. The children are just seeing the zero and not thinking about the calculation. Give the children lots of practical experiences of facts involving zero. Use number lines and ask them to discuss what happens.

Teaching ideas

Unit 3.1 Conceptual subitising

Stations

- **Flash Patterns** – Quickly show the children a card with a regular pattern. Then mix in irregular patterns. Start with small amounts and increase to ten then beyond ten. Question the children, **How many? How did you know? What did you see? Which were easier?** (Small amounts and regular patterns should be easier at first.)
- **Pick a Card** – Place tens frame dot patterns (see Photocopiable resource 3.2) face down on the table. Ask the children one at a time to turn over a card and describe to the group. For example, there are seven dots in the pattern five and two. I can also see three spaces. As more cards are turned over the children can discuss and compare with each other. For example, 'My amount was bigger than yours by one' or 'We had the same amount but my pattern was three and four dots.'
- **Finger Patterns** – Make finger patterns for the numbers one to ten in different ways and ask the children to describe them. Encourage the use of mathematical language. For example, 'I put up two fingers on this hand and four on the other because two and four makes six.' Compare different ways the total was made and highlight doubles.
- **Rekenrek Slide** – Ask the children to show an amount up top. Try to do this in one big push. Then the same amount but using the two bars. Discuss the different combinations. For example, show me six up top. Show me six using top and bottom.

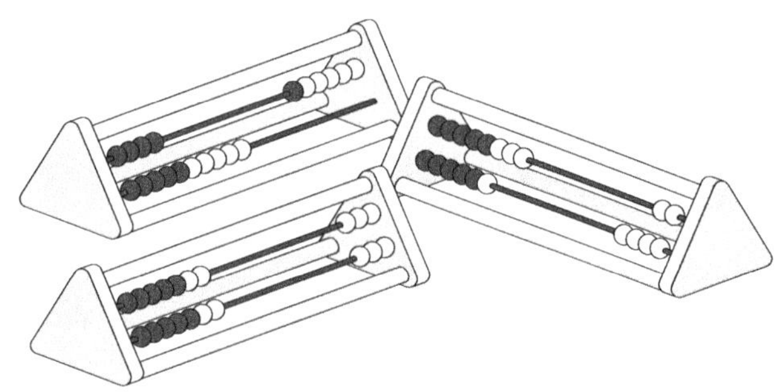

- **Magic Number Dice** – Pick a number as the magic number. Roll the dice taking turns around the group. The children quickly shout out their number and pass the dice on. While this is happening a sand timer can be turned over. Can the children find the magic number before the sand runs out? The dice tool can be used as an alternative for this activity.
- **Subitise Pairs Match** – Have irregular dot patterns cards (see Photocopiable resource 3.4) in rows face down. Make sure there are pairs of the same amount in different patterns. The children take it in turns to turn over two cards and describe what they see. If the amounts match they keep the pair. If not, they return them face down in place. Play moves on. Once all the pairs have been found the winner is the person with the most pairs.
- **Subitise Ordering** – Have irregular patterns (see Photocopiable resource 3.6) in larger amounts (7–20/30). Give each child three random cards and ask the children to place the cards in order from the smallest to the biggest. This should be done by subitising and not counting. Ask the child to explain what they did and why. Then ask the other children in the group if they agree. Once the children are confident with this, more cards can be placed in order.

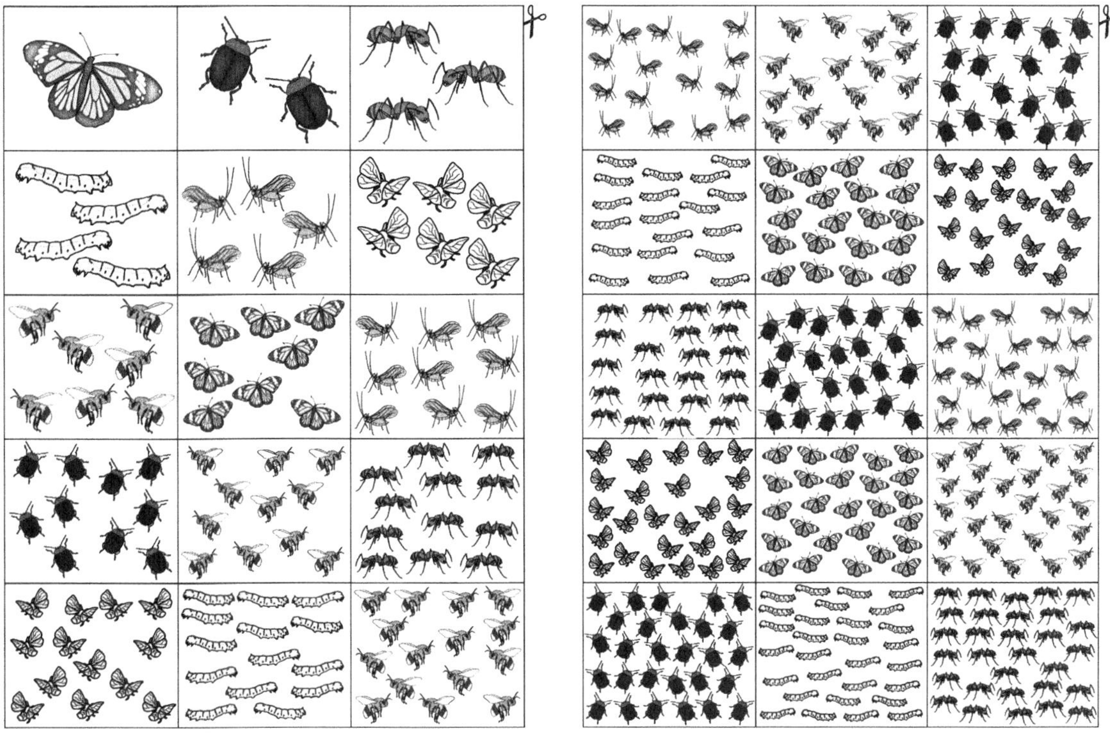

Photocopiable resource 3.6 – Irregular Patterns to 30.

- **Five and Ten Reference Points** – Using finger patterns and dot patterns, ask the children to describe the patterns using 5 and 10 as a reference point. For example, 'I knew to put one finger down to make nine because nine is one less than ten' – 10 as a reference point; 'I saw five on the top and three on the bottom. Five and three makes eight' – 5 as a reference point. 'Two empty boxes. Ten take away two makes eight' – 10 as a reference point.

Unit 3.2 Odd and even number patterns

Stations

- **Digit Cards** – ask the children to lay out the digit cards zero to ten. Then ask the children to separate into two rows by skip counting in twos. The even numbers 0, 2, 4, 6, 8, 10 should be in one row and the odd numbers 1, 3, 5, 7, 9 should be in another row. Get the children to discuss what they see and chant the numbers. This is fun to do in different voices, for example, a loud voice, a quiet voice, a low voice or a robot voice. This can later be extended to the numbers to 20. Then patterns can be discussed such as 2 and 12 and 4 and 14.

0 2 4 6 8 10 12 14 16 18 20

1 3 5 7 9 11 13 15 17 19

This can be done using the Number cards in the digital pack, turning over odd and even number patterns.

- **Twenty Number Track** – Using a number track to 20 (see Photocopiable resource 3.7) ask the children to skip count in twos colouring in each number they land on (ensure they use a light colour or highlighter so they can still see the numbers). Discuss the pattern and numbers. Practise chanting the even number sequence both forwards and backwards following the pattern with their finger as they say the number. This activity can also be repeated with the odd numbers.

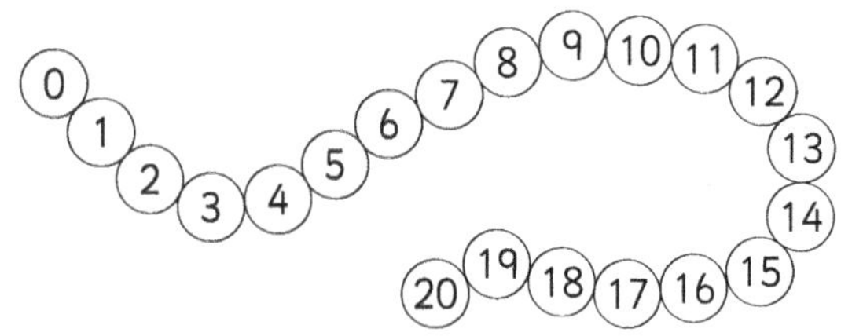

Photocopiable resource 3.7 – Twenty Track.

- **Count and Clap** – Ask the children to count from zero to thirty. When saying odd numbers clap hands together and even numbers clap hands on knees. This can also be done counting backwards from twenty. Point out all the odd numbers hands are together and even on knees. To mix it up only say the odd or even numbers aloud.
- **Odd and Even Grid** – Give the children a grid from 1 to 30 (see Photocopiable resource 3.8) in rows of ten. Ask them to colour the odd numbers yellow and the even numbers green. Discuss the patterns they see. The children should notice when going down a column the ones number is always the same and always even or odd. This can be done using the Number Square in the digital pack. Using colour tool in different colours to highlight the odd and even sequences.

1	2	3	4	5	6	7	8	9	10
11	12	13	14	15	16	17	18	19	20
21	22	23	24	25	26	27	28	29	30

Photocopiable resource 3.8 – 1–30 Grid.

- **Groups of Two** – Ask the children to get into groups of two and count how many there are in twos. If there is an odd one, discuss that that is because we have an odd number.
- **Human Line Up** – Give the children all a number from 0 to 10, then extend to 20. The children should be asked to line up in order. Then ask only the odd or even children to step forward. The children should notice it is every second person. Then ask the children to shout out their number either in the forwards or backwards counting sequence.
- **Make the Number** – Using multi-link ask the children to make the numbers. The numbers should be joined in pairs where possible. This will clearly show if the number is odd or even. The children can lay the numbers to ten, then to twenty in a line in order that they can see the odd, even, odd, even, odd, even sequence.

- **Counting Stick** – Using an unmarked counting stick in ten sections. Tell the children to imagine the number 0 on one end and 20 on the other. Then the children have to follow your finger (move your finger high and slow so that it is clear where it is going and gives the children thinking time) as they imagine the missing numbers on the number line and count the even sequence. This can be done starting at 0 or 20 for the forwards or backwards counting sequence. Once confident, try hiccup counting where the finger jumps both forwards and backwards. Repeat for the odd number sequence. Then use the counting stick to identify a number in the sequence. Point to a marker and ask the children what the number is. They may be able to imagine it or may need to drop back and count part or all of the sequence. Discuss how they knew. For example, if pointing to 8, the child may say that half way is 10 and it is the number before.

- **1–30 Board** – Ask the children to lay amounts for each number. This can be done with cubes, Diennes or counters. Encourage the children to lay the counters out in pairs so that the odd and even sequence can be easily discussed. Point out that 1, 11 and 21 are odd and ask the children to try and explain why. **Would 31 be odd? Why?** Repeat for other odd and even numbers to try and determine a rule for odd and even numbers.

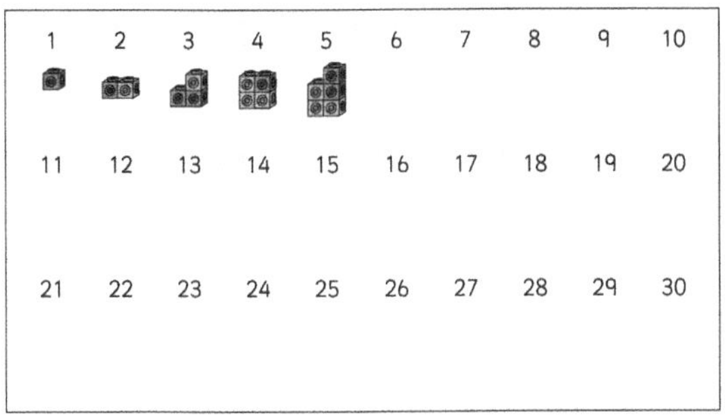

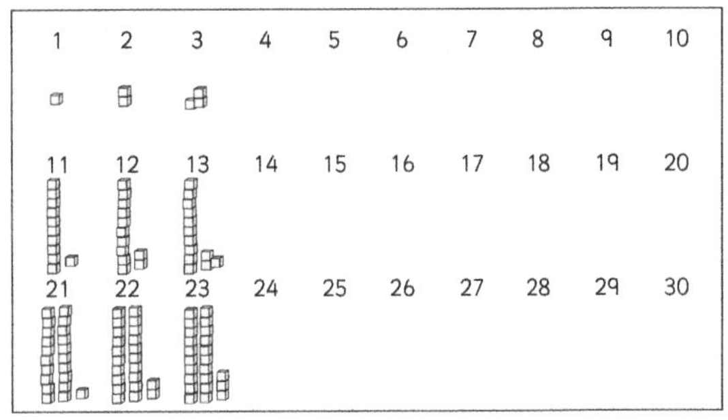

- **100 Board** – If you have a large moveable 100 board, block out the odd or even numbers so that only an even or odd pattern is showing. Practise counting aloud the forwards and backwards sequences. Turn a number the wrong way and ask a child what the missing number is. **How do you know?** Discuss how the pattern helped them to find the missing number. This can also be done using the Number Square in the digital pack.

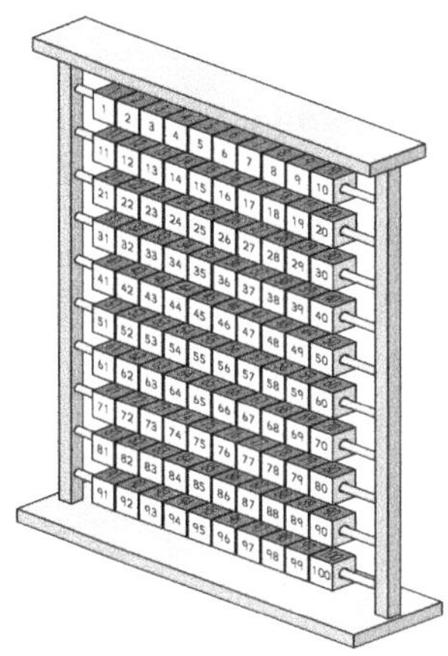

- **Stairs** – Using interlocking cubes make stairs to 10, then 20. Look at the odd and even pattern. Pull apart to show just the odd or even number sequences.

Unit 3.3 Doubles number facts within 10

Stations

- **Rainbow Towers 2** – Choose two different colours of multi-link and make two towers of the same height. Discuss how these are the same and check by counting and measuring. Join the counters together to find the total.
- **Double Finger Flash 2** – Ask the children to show you double 4 on their fingers or show me 8 using two hands with an equal amount of fingers on each hand.
- **Doubles to 10 Tower Challenge 2** – Using different colours of multi-link can the children find all the doubles to ten. Skip count in twos up and down the towers. Ask the children to tell you the doubles fact. For example, three and three makes six or double three is six.
- **Rekenrek Doubles 2** – Ask the children to show you double three by pushing the beads over using top and bottom. **What is the total?** Also ask: **Can you show me six using the top and bottom making sure the beads are equal?**
- **Doubles Clapping Rhyme** – The children sing their doubles facts while clapping and snapping their fingers. One plus one is two snap snap. Two plus two is four snap snap...
- **Partners** – Ask the children to join hands showing the doubles facts. Ask each group to shout out the doubles fact they are representing.

1 + 1 = 2 2 + 2 = 4 3 + 3 = 6

- **Interlocking Chains** – Use interlocking chains to show the doubles facts and say/record the facts.
- **Are Doubles Odd or Even?** – Give the children access to cubes/counters and ask them: **Are doubles facts odd or even?** Get the children to make their doubles facts and discuss the answers. They should notice they are all even. If they get an odd number, it was near double. Discuss what this means and explore some together.
- **Double the Fun** – Play the interactive game in the digital pack where the children can solve double problems.

Unit 3.4 Fact families to and within 5

Stations for addition

- **Addition Facts To and Within 5** – Give the children zero, then one, two, three, four and five counters. Work with the children to manipulate the counters to find all the addition facts to and within 5. Record the facts together on a whiteboard. Use 'and' and 'makes' or the correct symbols.

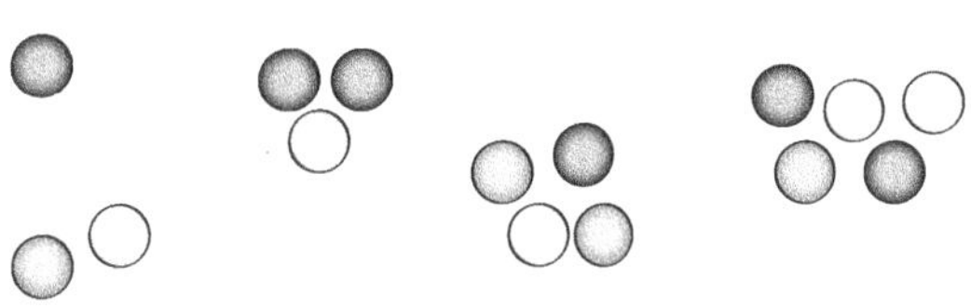

- **Paper Plate Investigation** – Ask the children to work in pairs. Give them two paper plates and a bundle of counters up to 5. Ask the children to find all the ways they can split the counters between the plates. Ensure the children can explain orally what they are doing. Also supply an individual whiteboard and pen so that the children can record the facts if they wish to. Discuss the facts they found. Did they get them all? Highlight the fact that there is always one more fact than the number they are working with.
- **Balance Scales** – Use multi-link in balance scales to show the addition facts. For example, put three joined cubes and two joined cubes in one side and five cubes in the other. Ask the children to verbalise the fact 3 add 2 makes 5.

Stations for subtraction

- **Subtraction Facts within 5** – Give the children zero, then one, two, three, four and five counters. Work with the children to manipulate the counters to find all the subtraction facts within 5. Record the facts together on a whiteboard. Write the words at first before introducing the symbols.
- **Five Fingers** – Cut out a paper hand and glue only the palm to a piece of card, ensuring the fingers are unstuck. Ask the children to fold over the fingers and record the subtraction facts for 5. Then put the fingers back up showing the corresponding addition.

Stations for addition and subtraction

- **Linking Facts within 5** – When the children have investigated both addition and subtraction facts for numbers up to 5, explore each number. Use concrete materials for the addition and then show how manipulating the materials backwards makes the subtraction sum. Highlight the link between addition and subtraction.
- **Tower of Cubes** – Give the children towers of cubes from 2 up to 5. Ask the children to break the towers into two. Explain the sum to the group as a subtraction. Then put the two parts back together and explain it as an addition. Record all the facts for the children, both additions and subtractions showing how they are linked together.
- **Picture Stories** – The children each need an individual whiteboard and pen. Say a number story (this could be linked to a topic such as dinosaurs) and the children draw the story and find the answer. Examples of this are: 'There were five *Apatosaurus*. A *T. rex* came along and ate three of them.' **How many are left?** 'I went shopping and bought four apples and one strawberry.' **How many pieces of fruit are in my basket?**
- **Write the Sum** – Draw a picture number story on the board, for example, a fruit bowl with three apples and two oranges. The children can verbalise the problem and write the sum using words or symbols. The children should explain how they knew to add or subtract noticing addition sums end with the big amount (whole) and subtraction sums start with the big amount (whole). The counting tool can be used to illustrate a word problem.
- **Part-Part-Whole** – Give the children a part-part-whole bar diagram and ask them to fill it in with the appropriate numbers. Using bar diagrams that fit with multi-link will help the children who are still relying on concrete materials. The children can also be given diagrams with some numbers filled in and they have to put in the missing numbers. It may be one of the parts or the whole.

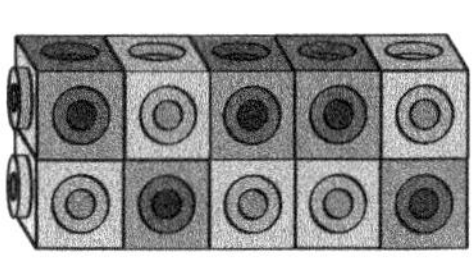

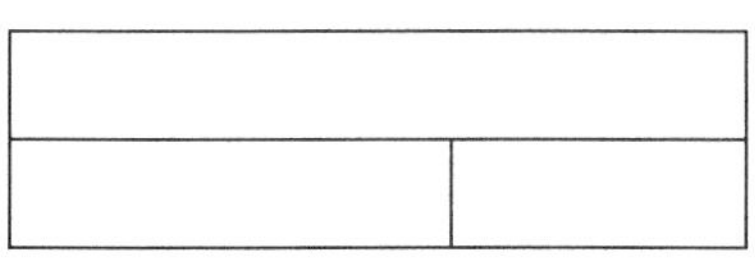

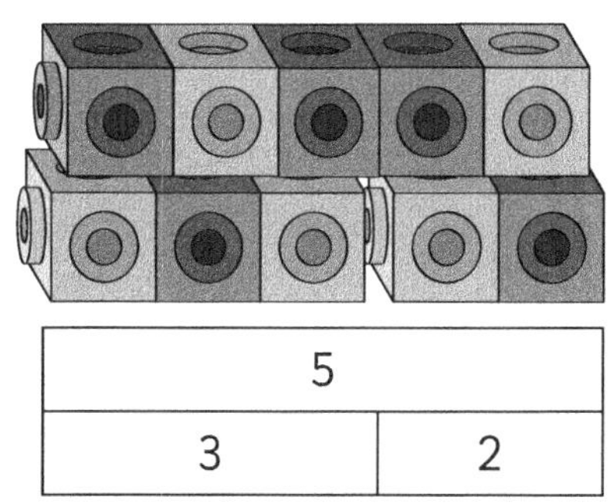

Unit 3.5 Mathematical symbols involved in addition and subtraction

Whole class

- **Symbols** – The symbols (+, –, =) can be introduced as the activities for number facts are repeated. Always highlight the language used and link to the symbols. Vary the language so that the children are familiar with a variety of words for the symbols and not just add or take away. Equals or *same as* should be used for the = symbol. Once the symbols have been introduced, allow the children to use them when they feel ready and comfortable when recording work.

Stations

- **Introduce Symbols** – Perform a range of concrete experiences of addition and subtraction using words to record and then show how symbols can replace the words. Introduce the symbols and encourage the children to record sums using the symbols. The children will copy your example of recording number stories with the corresponding symbols before becoming independent at this. For example, six red cubes and three blue cubes can be recorded as 6 and 3 equals 9. This can then be changed to 6 + 3 = 9.
- **Add or Subtract** – Read out a range of number problems using different language. The children hold up a plus or minus card for the operation they think that problem involves. Discuss why they think that.

- **Think Board** – Using the template (see Photocopiable resource 3.9) encourage the children to record a sum using symbols, pictures, words and materials. One part could be filled in and the children complete the other parts, or they could be given a blank think board and come up with their own sum. This works well if the children work in differentiated pairs so that a more able child can support a less able child.

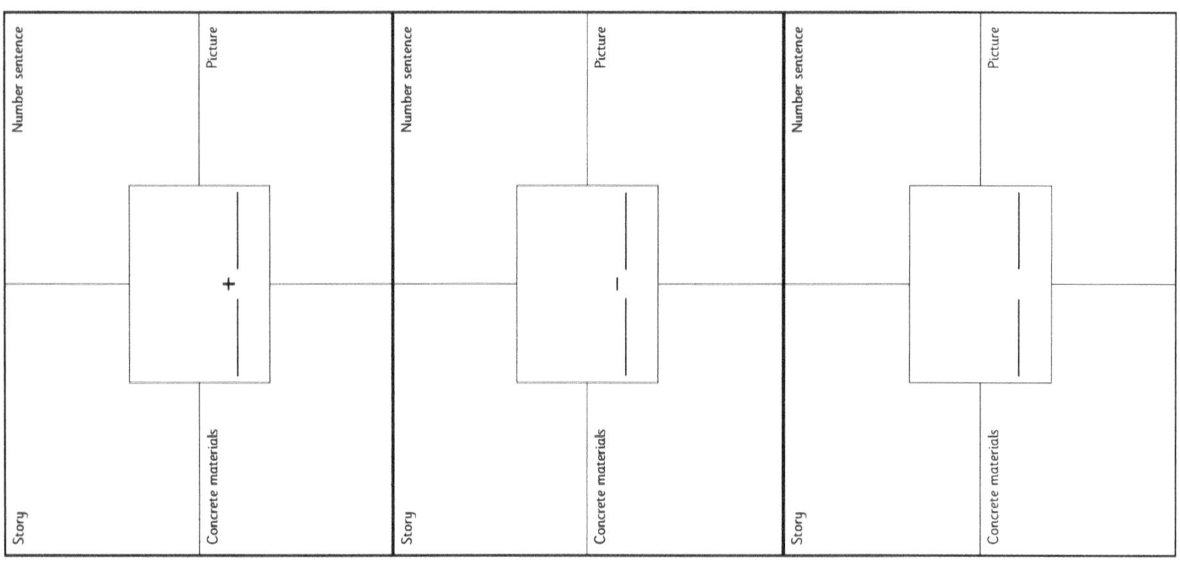

Photocopiable resource 3.9 – Think Board.

Unit 3.6 Addition and subtracting using count all concrete materials and imaging

Stations for addition

- **Grab Mat** – Using a grab mat the children grab two handfuls of cubes then count them all to find the total.
- **Empty Tens Frame** – The children can be given addition calculations. They place counters on the tens frame then count and record the answer.
- **Rekenrek Addition** – The children use the rekenreks to show additions by moving the beads along and counting the total.
- **Ladders** – Roll the dice and the children go and stand on that number. Then roll the dice again. The children add on that number by jumping along the ladder. The dice tool in the digital pack can be used.

Stations for addition and subtraction

- **Draw a Picture** – Ask the children to add and subtract by listening to a word problem and drawing the appropriate picture. They could write the answer or record as a complete sum. Examples of this are: 'Finlay has ten beads, six are red.' **How many are white?** 'Isla had six sweeties. She ate three.' **How many are left?** 'There were some cakes on a plate. Amman ate two. There is one left.' **How many cakes were on the plate?** 'In the line there are three boys and six girls.' **How many children are in the line?**
- **Counters** – The children use counters to complete addition and subtraction sums. The children can record the answer or the complete sum.
- **Digit Card and Counters** – The children are given an addition or subtraction sum. The start number is a digit card and the counters are the second number. The children then add or subtract from the number using imaging.

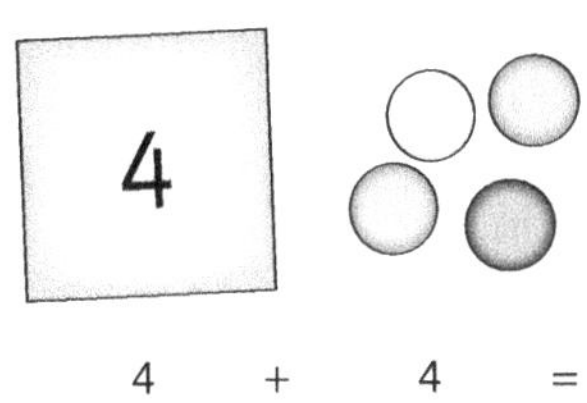

- **Number Line** – The children use a number line to add or subtract. With a larger number line, show the children how to jump forwards or backwards from the start number using a toy frog.

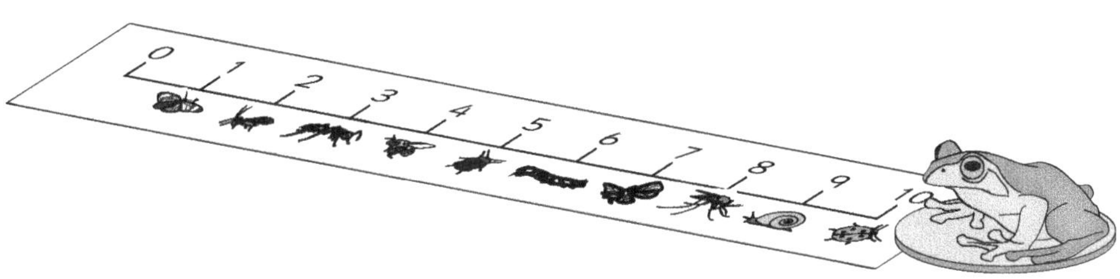

- **Head and Fingers** – For addition and subtraction the children put the number in their head (the children can do this figuratively by touching their head and saying the number). Then the children can use their fingers to help them count on or count back to the second number to find the answer.
- **Interactive Games** – Fishy Facts and Ants on the Bus interactive games in the digital pack that can be played to help children recall addition and subtraction facts.

Stations for subtraction

- **Counters and Cups** – Lay out a number of counters. Get the children to count them. Ask the children to shut their eyes while you cover some of the counters with a cup. When the children open their eyes, ask: **How many counters are under the cup? Why do you think that?** Lift the cup and check. Verbalise and/or record the corresponding subtraction sum.

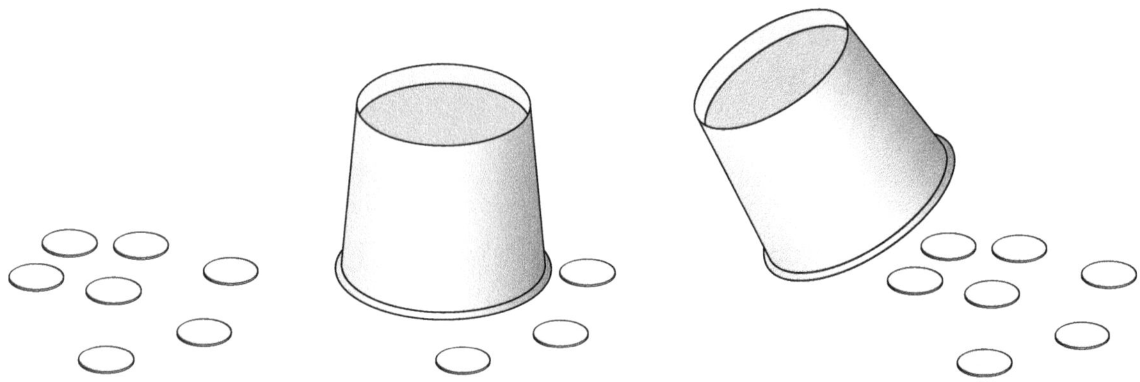

- **Magic Box** – Have 10 teddies in a box (this number can vary). Tell the children there are 10 teddies in the box. You are taking four out (different numbers can be chosen). Remove four and place them on the table beside the box. Ask the children: **How many teddies are in the box?** Ask the children to explain their thinking. Then remove the lid and check if the children were correct.

Unit 3.7 Continuing a sequence (next 2, 3, 4, numbers forwards to 30 and backwards from 20)

Whole class

- **Counting Stick** – Pick a number to put on the start or end of the counting stick. The children should imagine the number there. Using the markers on the counting stick the children can count on or back from the given number.
- **Pick Two Numbers** – Put some numbers in a feely bag. Ask a child to come and pick two numbers. One is the start number, one the finish number. The children can then be told or decide if they want to count forwards or backwards.
- **Number Track** – Using a number track. Ask the children to put their finger on a number. Tell them to count forwards 2, 3 or 4 or backwards 2, 3 or 4.
- **Pick a Number** – Ask a child to pick a number out of a bag. Then pick an arrow card (forwards or backwards). Then roll a dice with only 2, 3 or 4. The children then follow the instructions they chose counting on or back 2, 3 or 4.
- **In the Middle** – Write a number on the board. Choose a child to come and write the numbers that come after and/or before.
- **Number Chain** – Give the children all a number card. Put one card in the middle of the floor. The children with the numbers on either side have to come and join on. Then the next numbers join on and so on.

Unit 3.8 Counting on and back (from one unscreened numeral to locate a numeral on a screened numeral track)

Whole class

- **100 Square** – Only show the range of numbers the children are working on by turning the blank side of the numbers beyond the range you want the children to focus on. The children pick a number and count on or back. This can be made more challenging by screening the numbers and only showing the starting number.
- **Pendulum Count** – Place a sponge ball on the end of a string to make a pendulum. Swing the pendulum to keep pace when counting aloud. The children can choose a start number to count on or back from on a blank number track. The numbers they choose to start and stop at can be written on the number track. The numbers can also be chosen randomly from a bag.
- **Hidden Numbers** – A number track or grid can be shown with some numbers covered. The children have to say the hidden numbers. The numbers can be a sequence or random. The track or grid can be forwards or backwards. The Number Card tool or Number Square tool can be used from the digital pack.
- **Which Way?** – If you have a moveable 100 board, turn all the numbers away to show a blank 100 board. Turn around one number and ask the children to say it aloud. Then tell the children they have to count to another number which you choose. Do not show them yet. The children have to tell you if they are going to use a forwards counting sequence or a backwards counting sequence and explain why they think that. The number can then be revealed, and the children can see if they were correct. Then the children count on to the number. **How many jumps did they take to get there?**

Unit 3.9 Counting jumps on a number line

Whole class

- **Froggy Counting** – Jump forwards and backwards along a number line or a hundred square using the frog toy/laminated picture to lead the way and keep the children together. Start and end at different numbers each time. Discuss the number patterns seen. Frog helps to focus the children on jumping forwards or backwards on to the next number. Ask the children what happens when we count forwards or backwards. If counting in twos use a rabbit toy/laminated picture as it has bigger legs, therefore jumps further. Therefore, use a kangaroo toy/laminated picture for skip counting in five and a spider toy/laminated picture for skip counting in tens down the web (column).
- **Human Number Track** – The children line up holding the digit cards in order. Going forwards and backwards down the number line the children each shout out their number in order. The children can form a sort of Mexican wave as they count holding their number high as they shout it out.
- **Pass the Bean Bag** – The children stand with a number in front of them. Choose a number to take the bean-bag. The children then choose a number to stop at. They then pass the bean bag forwards or backwards down the line shouting out their number.

- **Circle Count** – The children sit in a circle their number in front of them. The child with one starts by saying one and then the children continue the count around the circle until you shout stop. This can be done for a forwards or a backwards counting sequence. To challenge the children, every time you clap your hands the count changes from a forwards counting sequence to a backwards counting sequence.
- **Chalk Number Line** – The children draw a giant chalk number line in the playground. The children then take it in turns to travel along the number line. The other children decide which number to start on and how many jumps their friend should take. This can be done both forwards and backwards. All the children shout aloud the numbers they jump on.

Stations

- **Card and Dice** – Give each child an individual number line. The children choose a card which gives them the number to start on and then roll the dice to see how many jumps forwards they take. Which number do they land on? This game can also be played for a backwards counting sequence.
- **Counters** – The children place two counters on different numbers on the number line. Ask them which numbers they picked. How many jumps will it take to get from one counter to the other?

Unit 3.10 Addition using concrete materials to count on

Stations

- **Count on Counters** – The children pick two bundles of counters. They say how many are in each bundle. They then place their hand on the biggest pile saying how many and then count on the second pile touching each one as they do to find the total. The children can also do this starting with the smaller pile and compare the result. Ask the children if it was easier to start with the small pile or big pile and why.
- **Count on Pipe Cleaners** – Give the children a bundle of pipe cleaners with a band around them and the total written on it and some individual pipe cleaners. Ask the children to find how many pipe cleaners there are altogether. The children will hopefully start with the bundle total and count on the individuals to get the total.
- **Pencils** – Tell the children, 'Here are five pencils. I add two more.' **How many pencils do I have now?** Allow the children to add pencils to tell you the total. Repeat for different amounts.
- **Draw the Sum** – 'There are five cakes in a bag and three more on the plate.' **How many cakes are there altogether?** Provide the children with a whiteboard and pen to draw the number story and find the answer. Repeat for similar addition stories.
- **Counting Tool** – The pictures can be dragged on to the screen by the children and counted. The numbers can then be added to find the total. This can be found in the digital pack.

Unit 3.11 Addition using counting on using screened concrete materials

Stations

- **Screened Piles** – Tell the children there are five counters under the paper. They can see four more beside it. Ask the children: **How many counters are there altogether?** Encourage the children to count on from five although some children may not trust the count and need to count from one. Repeat for other amounts.
- **Boxes** – Put some teddies in a box and have some on the side. Tell the children 'There are three teddies in the box.' **How many teddies are there altogether?**
- **Hidden Counters** – In pairs, one child hides some counters in their hand. Their partner lays some more counters out on the table. The partner says 'I have six in my hand.' Then they count on to find the total.
- **Under the Cup** – One child hides some cubes under a cup and puts the digit card for the amount on the top. He then places some more at the side. The partner tells how many there are altogether and the first child confirms if they are correct. Repeat for other amounts for addition within 10.

Unit 3.12 Addition using counting on by imaging

Stations

- **Addition Act Out** – Get the children to act out some simple number problems. For example, 'There were three boys standing in line and two more came and joined them.' **How many are there now?** Repeat for other amounts.
- **Addition Stories** – Tell the children a story. On their whiteboard they can write the answer or the whole sum. For example, 'Nuria put four apples in her bag and two oranges'. **How many pieces of fruit does Nuria have in her bag?**
- **How Many in the Boxes?** – Show the children two boxes and tell them how many items are in each in box. Ask: **How many are there altogether?**
- **Part-Part-Whole Diagram** – The children are given a sheet with the two part numbers in the diagram and they must put in the whole amount.
- **Addition Bingo** – Ask the children to write six numbers on their board. Read out addition sums either as a word problem or a fact. If the answer is one of the numbers on the children's board, then they score it off. The winner shouts Bingo once all their numbers are scored off.
- **Function Machine** – In the digital pack, the function machine tool can be set to perform one step addition calculations.
- **Peg It Up** – Have a number line on a washing line. The children pick a peg with a sum on it. The children work out the answer mentally and put their peg on the correct number.

Unit 3.13 Subtracting using concrete materials to count back

Stations

- **Count Back Counters** – The children pick a bundle of counters. The children say how many are in the bundle. They then take some away and count how many are left. The children should explain what they are doing and record as a sum.
- **Broken Towers** – Give the children a tower of blocks up to 10. They take some away. **How many are left?** The children say and record the sum.
- **Funky Fingers** – Tell the children to hold up 10 fingers. One child rolls a dice and shouts out the number. The children put down that number of fingers. The children then say the corresponding take away sum. Repeat for different amounts.
- **Draw the Sum** – 'There are five cakes on the plate. Amman ate four.' **How many cakes are left?** Provide the children with a whiteboard and pen to draw the number story and find the answer. Repeat for similar subtraction stories.
- **Counter Count Down** – The children work in pairs or small groups and each has 10 counters. In turn the children roll a dice and place that number of counters in a pile in the middle. They say the subtraction they did. Play continues around the group. The winner is the first person to have no counters left.

Unit 3.14 Subtracting using counting back using screened concrete materials

Stations

- **Track Back** – The children draw a chalk number line in the playground. One child stands on a given number. Another child throws the dice and shouts out the number. The child then steps back down the number track to find the answer.
- **Jumping Out of the Box** – Tell the children, 'There were ten teddies in the box.' Point to some teddies at the side. Ask the children: **How many teddies are left in the box?**
- **What's in the Cup?** – Lay a number of counters on the table and ask the child to count them. The child shuts their eyes and you hide some under the cup. The child then opens their eyes to see what is left. They then say how many are under the cup. Lift the cup to check and discuss how the child knew.

Unit 3.15 Subtracting using counting back by imaging

Stations

- **Subtraction Act Out** – Get the children to act out some simple number problems. For example, 'There were eight girls standing in line and two went away.' **How many girls are there now?** Repeat for other amounts.
- **Subtraction Stories** – Tell the children a story. On their whiteboard they can write the answer or the whole sum. For example, 'Nuna put four apples in her bag. She ate one.' **How many pieces of fruit does she have in her bag now?**
- **Part-Part-Whole Diagram** – The children are given a sheet with the two part numbers in the diagram and they must put in the whole amount.
- **Subtraction Bingo** – Ask the children to write six numbers on their board. Read out subtraction sums either as a word problem or a fact. If the answer is one of the numbers on the children's board, then they score it off. The winner shouts 'Bingo' once all their numbers are scored off.
- **Function Machine** – In the digital pack, the function machine tool can be set to perform one-step subtraction calculations.
- **Washing Line Sums** – Use a washing line to make a number line. The children pick a peg with a subtraction sum on it. The children work out the answer mentally and peg their peg on the correct number showing the answer.

Unit 3.16 Fact families for 5 and 10

Stations

- **Families** – Using dominoes and five and tens frames, record the facts for 5 and 10. Remember to highlight the commutative aspect clearly seen in the resources. For example, if the domino had 3 spots on one side and 2 on the other, the fact family would be $3 + 2 = 5$, $2 + 3 = 5$, $5 - 2 = 3$ and $5 - 3 = 2$. For a tens frame with three dots, the fact family would be $3 + 7 = 10$, $7 + 3 = 10$, $10 - 3 = 7$ and $10 - 7 = 3$.
- **Facts for 5** – Give the children five counters for finding fact families of 5. Ask the children to manipulate the counters to find all the facts and record them. Repeat for all the other numbers to 10.
- **Bead Line** – Give the children a pipe cleaner with beads for the fact family total. The children push the beads on the pipe cleaner to show all the bonds. Show how addition and subtraction facts are linked. For example, the picture shows $3 + 5 = 8$, $5 + 3 = 8$, $8 - 5 = 3$ and $8 - 3 = 5$.
- **Facts are Facts** – Through hands-on experimentation with a collection of objects, make sense of the notion that there are basic facts. Gather these facts through discussion with the children and prove that they will always be the fact no matter which objects are used and which arrangement they are in. For example, give one child 9 cubes and another 9 counters. Ask them to split the cubes and counters into 5 and 4. They both have nine whether it is cubes or counters. **Did you lay your counters out differently?** Perhaps one had the four in a line and the other had them in a square (2 on top of 2). 'It still makes 9.' And so on.

Unit 3.17 Partitions to 10

Stations

- **Tens Frames** – Discuss the number of spots and number of empty spaces. Record partitions to ten (see Photocopiable resource 3.2).
- **Rekenreks** – Using the top line only. Ask the children to push some beads across. How many are left? Record the partitions to 10.
- **Finger Fiddle** – Ask the children to record partitions to 10 using their fingers. Hold up one amount on her fingers and the children hold up the partner amount. For example, if you hold up 3, the children hold up 7. Ask one of the children to verbalise the bond, proving the fact using the fingers. As always, use mistakes as learning opportunities. Choose a child showing you the right number of fingers to prove the bond to 10 and one/some with the wrong number of fingers to join with yours and show how it does not make 10.
- **Matching Pairs** – Give the children number cards to 10. Ask the children to find their partner to make 10. For example, 6 finds and matches with 4.
- **Feely Bag** – In a feely bag, place the multi-link towers of one colour from 1 to 10. On the table have the same towers in a different colour. The children then pull a tower out of the bag and match it to one on the table so that together they make 10. For example, if the children pull out 3, they match it to the tower of 7 on the table.
- **Pairs to Ten** – Stand in the middle of the circle of children. Throw a ball to one of the children in the circle and shouts out a number. The child catches the ball and throws it back saying the number that pairs with the teacher's number to make 10.
- **Clap to 10** – The children sit in pairs. The first child claps some claps (between 0 to 10). The second child finishes the claps to make 10. The children then verbalise the pairs to 10 they clapped.
- **Digit Cards** – In pairs each child needs the digit cards 0–10. One child places down a digit card and the other child finds the partner to make 10. They swap so the other child puts down the first number.

- **Two Plates** – The children are given two plates and ten counters. How many ways can they split the counters to make 10?
- **Tower Splits** – The children are given a tower of ten multi-link. How many ways can they split the tower to make 10. For example, 3 add 3 add 4. Discuss the many varied options and record using symbols.
- **Squeezy Buttons** – Fill a sandwich bag with some clear liquid soap. Pop in ten colourful buttons. Seal the bag with strong tape and tape the top and bottom of the bag to the table. On the front of the bag in permanent marker draw a big circle at the top with two lines coming off to two smaller circles. The children start with all the ten buttons in the big circle. They then squeeze them into partitions that make 10. They verbalise or record the partitions. Did they find them all?

- **Bar Diagrams** – Use bar diagrams with two bars to 10. The bottom bar has to be split to show the partition to 10. The line should be relevant to the size of the part and the numbers filled in to show the partitions. The children can then discuss, and record, the related addition and subtraction facts.

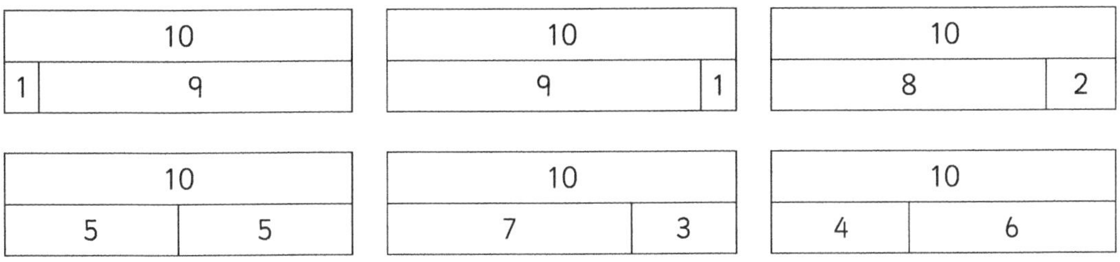

Unit 3.18 Adding and subtracting mentally using known facts

Stations

- **Number Problems** – Read aloud a number story which the children answer mentally. This can be done with no hands up, giving all the children thinking time before asking one child to share the answer or by giving all the children an individual whiteboard to record the answer. The answer should be discussed, and the children asked to share their strategy. Compare all the different strategies used. This is very important as the children will be encouraged to try different mental strategies. For example, 'Nuria has seven red stickers and two blue stickers.' **How many stickers does Nuria have?** 'Amman had seven balloons, two burst.' **How many unburst balloons has he now?**

- **Whiteboard Challenge** – Give the children a set block of time to record all the facts for numbers from 0 to 10: **The answer is … What is the question?** For example, 'The answer is 7, what is the question?' At the end of the time the children gather back together to share the facts they found and discuss the strategies they used. Discuss any connections made. The children need to prove they were correct, therefore learning from any mistakes made. This can be done individually or in pairs.

- **Strategy Boards** – Give the children a number. The children then write a number fact sum that achieves that number. They then show how that sum can be calculated using a number line, picture and partitioning to solve it. This shows their thinking and any links that are made in the numbers.

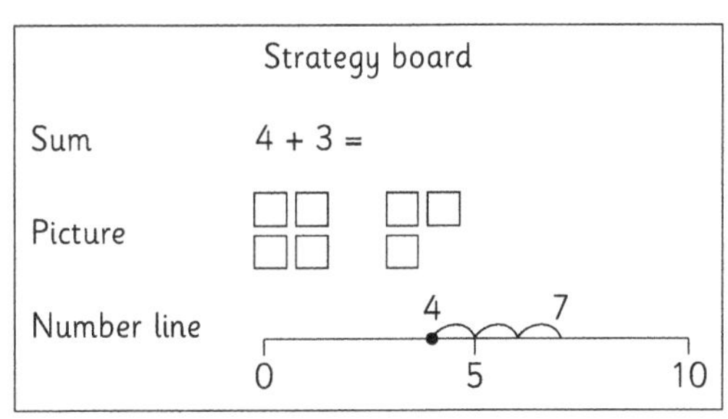

- **Add/Subtract Zero or One** – Use concrete materials and number lines to investigate facts involving zero and one. Ask the children to explain their thinking. For example, 'Have six counters on a plate add zero.' **How many counters are there now?** Discuss the fact that because we did not add any more counters, the amount did not change, so we still have six. Repeat starting with zero then add six and discuss that as we started with zero when we added six there will only be six. Do the same for subtraction. When adding or taking away one, use the number line to show how we jump on or back to the neighbouring number.

- **Doubles** – Children should recall all the doubles facts mentally to 10. Hold up a number card with the double answer on it. The children shout out the number that doubles to make it. Then turn over the card for the children to check by having the answer on the back of the card.

- **Double and Near Doubles** – Once the children are confident with their doubles facts and have noticed all the answers are even numbers. Discuss what a near double sum would look like. Use Rekenreks or concrete materials to create them from the double. Highlight where the numbers are on a number line, calling them next door neighbours. Once the near double sums have been created ask the children to find the answer. Compare the answers to the related doubles facts. Discuss what happened and why. Once all the near double facts have been found, look at the answers and see if the children notice all the answers are odd. Discuss why this is.
- **Partitioning** – Introduce how known facts to 5 can be used to add bigger numbers. This should be done with concrete materials. For example, 5 + 3 could be seen as 2 + 3 + 3. The use of Tree tool in the digital pack can support this learning.
- **Odd Plus Odd Makes Even** – Give the children concrete materials and ask them to investigate and prove this theory.
- **Even Plus Even Makes Even** – Give the children concrete materials and ask them to investigate and prove this theory.
- **Odd Plus Even Makes Odd** – Give the children concrete materials and ask them to investigate and prove this theory.

Unit 3.19 Solving simple missing number problems

Stations

- **Missing Number Story Problems** – Give the children a start or change unknown number problem. The children should be given an individual whiteboard to solve the problem by drawing or imagining the situation. For example, 'Nuria has ten beads, seven are red.' **How many white beads does Nuria have?** (Change unknown.) 'There were some cakes on a plate. Amman ate two of them and now there are four left.' **How many cakes were there to begin with?** (Start unknown.) Discuss the strategies the children used.

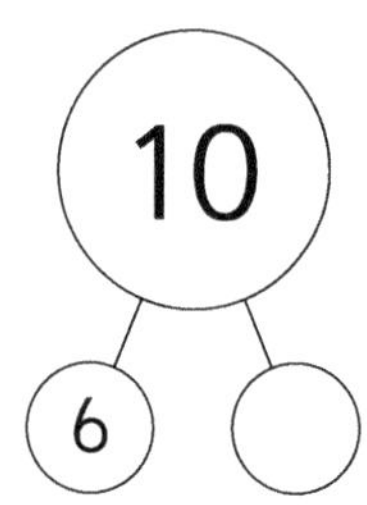

- **Missing Part-Part-Whole** – Give the children a part-part-whole diagram with one of the parts missing. **What is the missing number? How do you know?**

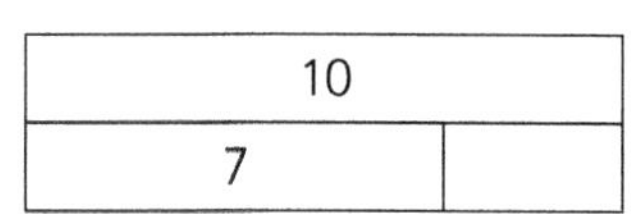

- **Missing Bar Diagram** – Give the children a bar diagram with the whole amount and one of the parts numbers. The children have to find the other parts number. **What is the missing number? How do you know?**
- **Secret Number** – In pairs, one child picks a secret number. The child adds the number the partner chooses and gives the answer. The partner then has to say what the missing number is and explain how they knew. For example, Child A has the secret number in his head. Child B tells him to add 5. Child A says the answer is 9. Child B works out the missing number must be 4 because 5 add 4 makes 9 and 9 take away 5 leaves 4.
- **What's Missing?** – In pairs, one child shuts their eyes. The other child counts out some counters and hides a few of them under a cup. The first child opens their eyes and is told there are nine counters. **How many are under the cup?** The child needs to count how many are seen and work out how many are under the cup to answer the question.

What to observe

- Children can confidently subitise amounts, saying the number without counting. Ask the children to explain their thinking. Note if the children are recognising partitions or using 5/10 reference points.
- Children can remember the odd and even sequences and are beginning to identify odd and even numbers.
- Children can mentally recall their doubles facts within 10.
- Children can recall fact families to 5 and within 5.
- Check the children can understand the language of addition and subtraction by giving the children word problems to record as sums using the appropriate symbols. Examples could be: 'I have three lollipops in a bag, I put in two more lollipops. How many lollipops are in the bag now?' 'There are six cookies on a plate, I ate two of them. How many cookies are left on the plate?'
- Check that children can perform calculations, noting if they add and subtract by relying on concrete materials to count all, by imaging or by counting on or back.
- Children can count on in the range of 30 and back in the range of 20 up to four numbers from a given number.
- Children can count on and back crossing decades with both unscreened and screened numbers.
- Children can use a number line accurately to count on and back from an unscreened number to a screened number.
- Children can add by counting on using concrete materials.
- Children can add by counting on using screened concrete materials.
- Children can add by using counting on by imaging.
- Children can subtract by counting back using concrete materials.
- Children can subtract by counting back using screened concrete materials.
- Children can subtract by counting back using imaging.
- Children have a good recall of fact families to and within 5 then 10.
- Children are confident with partitions to 10.
- Children can add and subtract mentally using known facts. Ask them to explain the strategies used.
- Children can solve simple missing number addition and subtraction problems.

4 Fractions

An Introduction to Fractions A

Section overview

Units of learning	Review units	Prior learning	Benchmarks/Es&Os
Unit 4.1 **Equal/fair shares**		• From their lives the children will have an ever expanding level of vocabulary and a range of personal experiences to draw upon. Such as sharing, cutting up food, making groups and so on.	**MNU 0-07a** • Splits a whole into smaller parts and explains that equal parts are the same size. • Uses appropriate vocabulary to describe halves. • Shares a group of items equally into smaller groups. I can share out a group of items by making smaller groups and can split a whole object into smaller parts.
Unit 4.2 **Breaking a whole into fair parts**	**An Introduction to Fractions A** Unit 4.1		
Unit 4.3 **Fair shares may have leftovers**	**An Introduction to Fractions A** Units 4.1 and 4.2		
Unit 4.4 **Language of fractions**	**An Introduction to Fractions A** Units 4.1, 4.2 and 4.3		
Unit 4.5 **Halves**	**An Introduction to Fractions A** Units 4.1, 4.2, 4.3 and 4.4		

Conceptual context

This concept is based on the learning that one whole object or amount can be split into equal parts. This can be done physically by grouping or sharing a whole amount. The children need to understand that they will get a fraction of the whole. In a fraction of an amount the children should understand that sometimes this cannot be done fairly and this will lead to leftovers.

Children will be familiar with sharing, grouping and fractions through situations and events they experience in their daily lives. It is important to draw on as many of these real-life experiences as possible to make the learning relevant while addressing any misconceptions that may have arisen. For example, if the apple they were given and told was a half had not been cut equally, then it should not be called a half. The children need to be exposed to different types of sharing, grouping and fractions by maximising naturally arising situations such as getting into groups for an activity or sharing at snack time. Highlighting fairness is important and language should be used consistently. Modelling the use of sharing, grouping and fractions should be maximised showing the real-life context and significance of the concept.

Resources

- Counters
- Small toys
- Paper plates
- Cups
- Paper shapes and strips
- Fraction picture cards
- *Sharing a Shell* by Julia Donaldson
- Wooden fraction puzzles

Digital

- Fraction wall

Record book

Pages 17–18

Key vocabulary

- **Sharing** – splitting an amount or object into groups or parts
- **Grouping** – taking a whole amount and distributing into sets
- **Equal** – of the same value
- **Fair** – all parts or groups are the same/equal
- **Fraction** – parts of a whole
- **Whole** – the complete object or amount
- **Half** – one part of something divided into two equal parts
- **Leftovers** – an amount left over when sharing equally

Misconceptions

Potential misconception	Addressing the misconception
Units 4.1, 4.2, 4.3, 4.4 and 4.5 The children may think that something cut in two is a half.	• When showing a half, it is important to show – by comparing – that both parts must be equal for them to be halves.
Units 4.1, 4.2, 4.3, 4.4 and 4.5 The children may believe that sharing is just giving out the group of items.	• When sharing, highlight the fact that the items must be shared out equally.
Units 4.1, 4.2, 4.3, 4.4 and 4.5 The children may believe that grouping is just putting the items into a number of groups.	• When grouping, highlight the fact that all the groups must have an equal amount.

Teaching ideas

Unit 4.1 Equal/fair shares

Whole class

- **Fair Shares** – Count an amount (this is more meaningful if it is something the children want, e.g. food or toys). Tell the children you are going to share it between two of the children. Give one of the children one and the other the rest. Ask them: **Is it fair? Why not? What would make it fair?** Now share the amount again showing how to give one to each at a time and keep repeating until all the amount has been shared. Check that it is fair by counting and/or comparing. Highlight the words equal and fair shares. Repeat this activity often with different objects.
- **Equal Groups** – Ask the children to get into groups of four. Ask the children to check to see if all the groups have the same amount. To do this, compare by standing next to each other. This activity can be repeated for groups of different amounts.
- **Sharing** – Sharing socially introduces the concepts of sharing with others. Stories such as *Sharing a Shell* by Julia Donaldson is a nice introduction. Then the children should discuss times when they share. When they share pencils, toys, and food at snack time, the children should have their attention drawn to the fact they are sharing and how to do this fairly.

Stations

- **Counters and Cups** – Give a cup to each child in the group. Have an amount of counters in the middle. Ask one child to share out the counters until there are none left. Make sure this is done one at a time around the group so that no children are missed or given extra. Once all the counters have been given out, ask: **How many counters do you have in your cup?** Go round the circle noticing that each child has the same amount.
- **Building Bricks** – Have a pile of building bricks in the middle. Tell the children you are going to share them out so that everyone gets a fair amount. Get the children to help do this and then ask them to check that it has been done fairly. This can be done by comparing or counting. Once they are sure everyone has the same amount, ask the children to build something with their bricks. Look at all the different models made with the same amount of bricks.
- **Pom-Pom Tweezers** – Set up an activity with a bundle of pom-poms, giant tweezers and bowls. Ask the children to share the pom-poms fairly/equally between the bowls. Check this was done by comparing or counting. This can be done again with a different number of pom-poms and bowls. The use of tweezers is also great for developing the fine motor skills involved in early writing.

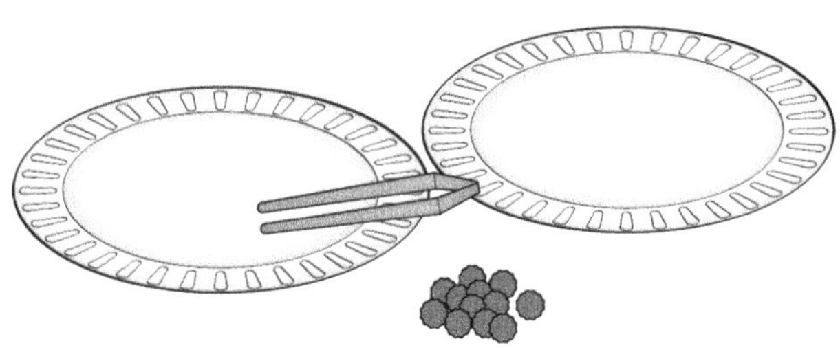

- **Garages** – Tell the children there are 12 cars to be parked equally in 3 garages. The children can do this using garages made from cardboard boxes and toy cars. Ask them: **How many cars are in each garage?**

- **Train Carriages** – Give two children an engine each. Tell them they are to be train drivers. They must have trains of the same length. The children must share out the given carriages fairly and check by comparing.

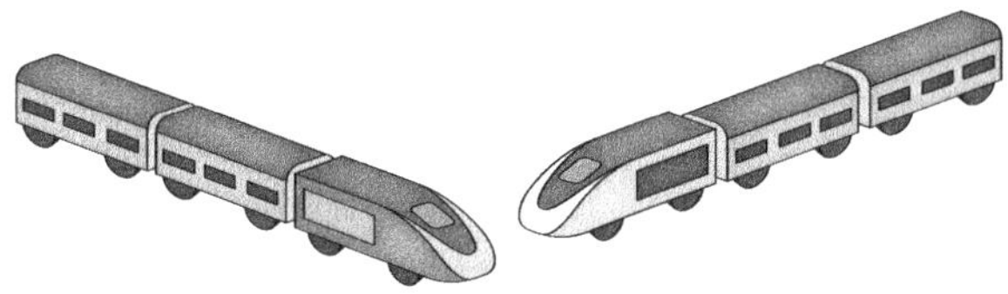

- **Necklaces** – Give three children a string each and say, 'There are 18 beads to put on your necklaces but they must be shared fairly.' Discuss how they could do this. Let the children make their necklace. **Did you share the beads fairly? How do you know? How do you feel if one person has more than the other? How can we fix it?** Repeat with other amounts.

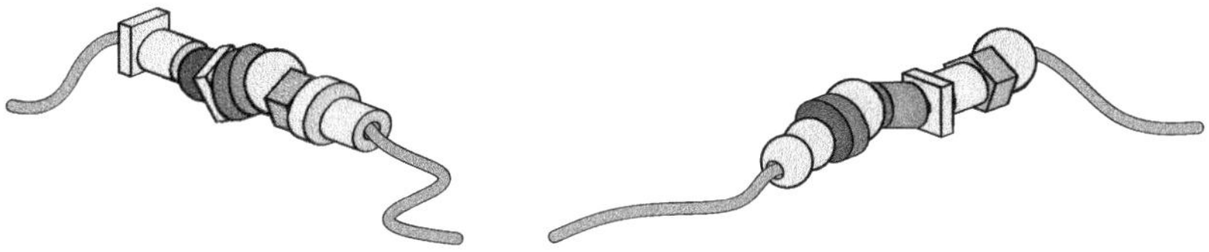

Unit 4.2 Breaking a whole into fair parts

Whole class/Stations

- **Fruit and Vegetables** – Look at a whole fruit or vegetable then cut it in half. Talk about finding the middle. Ask the children: **Which way is best to cut this vegetable so that it is in half?** Point out that a half must be two equal parts. Look at a melon (oval shape), which can be cut in half two ways: along the length or round the middle. The halves look very different, but both are still halves. Look at a carrot, which if cut across the width gives two pieces of different sizes but if cut across the length gives halves as the two pieces are of equal size. When cutting, make some cuts that are not in the middle so that you can discuss how some parts are bigger and smaller. Point out that unequal cutting gives just two parts; it is not cut in half.

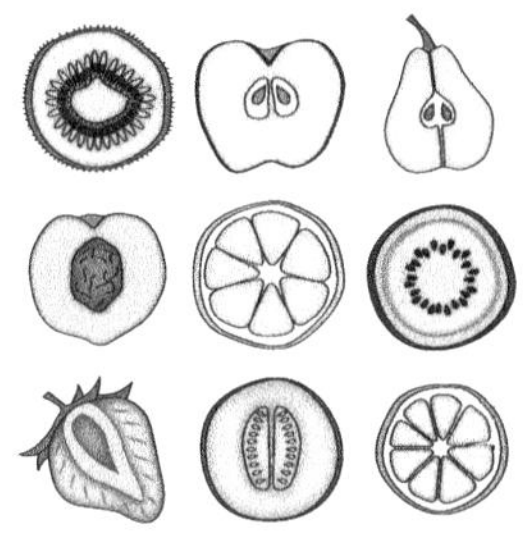

- **Towers** – Give the children an even number of bricks and ask them to make towers. Then ask the children to show you the middle of their tower and break into two equal parts. Check by comparing.

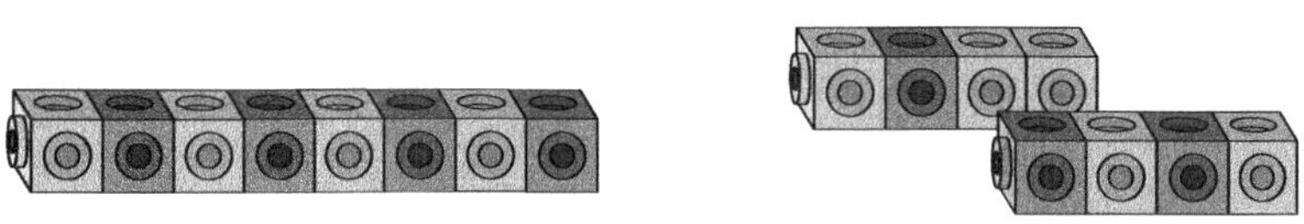

- **Paper Strips** – Give the children a strip of paper in their favourite colour. Ask them to fold their strip into fair parts. Get ideas from the children how to do this. Then tear it on the fold. **How many parts did you get from your strip? Were all the parts the same size? How do you know?**

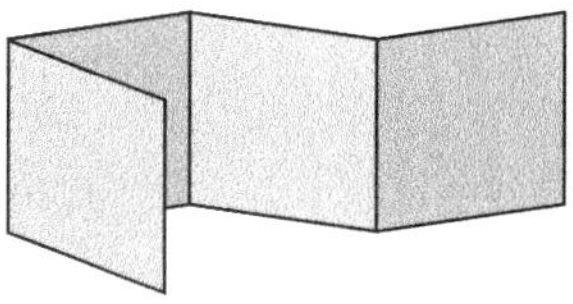

Unit 4.3 Fair shares may have leftovers

Whole class/Stations

- **Sharing** – Give the children seven toys and ask them to share them between two children. **Does it share equally? How many does each person get? How many are left over?** When the children are sharing, let them realise there must be leftovers as giving to one child and not another would not be fair. The children pick up on this very quickly. Repeat for other amounts which do not share equally.

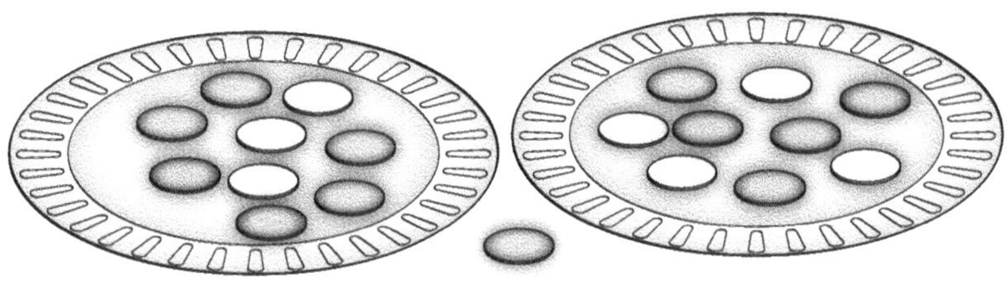

- **Groups** – Give the children a bundle of 10 pencils and ask them to put them into groups of three. Let the children work out that to be three in a group there must be some left over. **How many are in each group? How many are left over?** Repeat for other amounts which do not group equally.
- **Handfuls** – Take a handful of objects. **How many are there?** Share the objects between the group. **Does it share equally? How many does each person get?** If there are extras, **How many are left over?** Discuss how sometimes you can share fairly and sometimes there are leftovers.

Unit 4.4 Language of fractions

Whole class

- **Language** – The language of fractions and the concepts will arise naturally in the nursery setting. These should be highlighted by focusing on the language used. Make a point of showing half an apple when preparing a snack with the children or showing how to share fairly when giving out the toys or putting children into groups for a game.
- **Wall Display and Investigation Station** – As fraction work is covered, a wall display showing the concepts can be added to with an investigation station – a table or tuff tray with lots of examples of halves to explore such as, plastic fruits or pizza the children can cut in half, paper strips and 2D shapes to fold in half, towers of cubes to break in half, fraction wooden puzzles, photographs of fractions around us, word cards and so on. The children can find fractions around the setting to photograph and print out for the display.

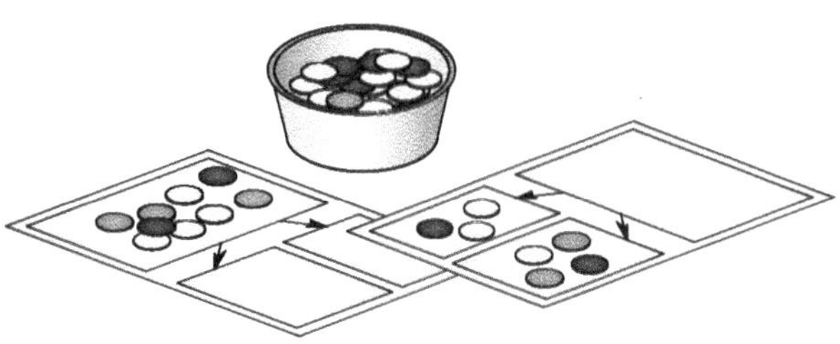

Unit 4.5 Halves

Whole class/Stations

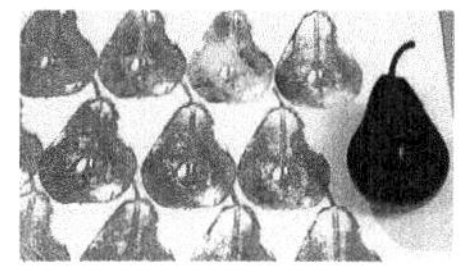

- **Printing** – Cut fruits and vegetables in half. Discuss how the cut had to be made in the right place to give two pieces of the same size. Use the fruit and vegetable halves to do some printing.

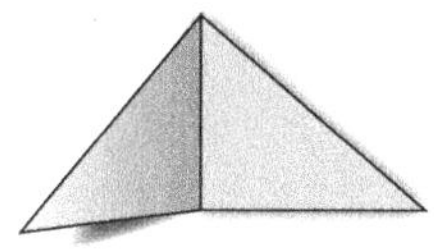

- **Paper Shapes** – Have a selection of 2D shapes of various sizes made from paper. Name and discuss the 2D paper shapes. Ask the children to show you where a half would be. Ask them to explain their thinking. Remind the children each half should be the same size. Ask the children: **How we can check it is a half?** Encourage them to check, measuring half by folding the shape over on top of itself. The children can then cut the shape in half gluing the two halves on to a piece of paper of a different colour.
- **Paper Strips** – Ask the children to fold a paper strip in half. Ask them how they did it. Emphasise that both pieces should be the same size for it to be a half. Show them how measuring can be done by folding it on top of itself.
- **Halves Mat** – Count an even amount of counters into the top box of the halves mat (see Photocopiable resource 4.1). Share the counters between the two bottom boxes. **How many are in each half?**
- **Lego** – Give the children an eight piece of Lego and show how two four pieces of Lego fit underneath exactly. Discuss how this is half. Print with the eight piece in one colour and then once dry print over the top with the four piece. This shows how the two halves are the same as the whole.
- **Ladybirds** – Using paper plates and sticky dots. Get the children to draw a ladybird with two wings. Give the children a strip of 2, 4, 6, 8 or 10 sticky dots. Tell the children to share the dots fairly between the ladybird's wings. **How many dots on each wing?**
- **Lego Towers** – Give the children an even numbers of bricks. Ask them to snap the tower in half. Discuss how they knew where half was.
- **Role Play Area** – Set up a pizza shop or a fruit shop with lots of plastic food the children can cut in half.

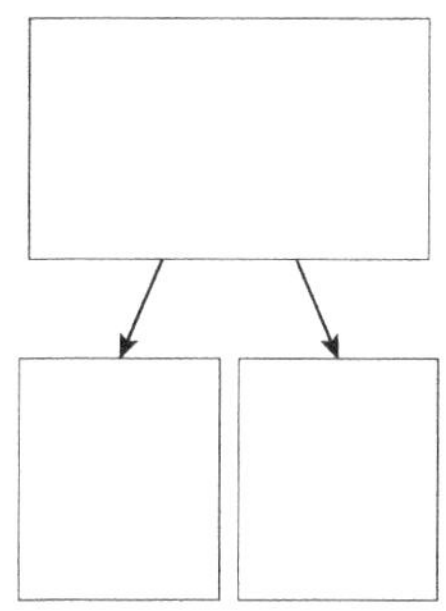

Photocopiable resource 4.1 – Halves Mat.

What to observe

- Children who are confident using the appropriate vocabulary. Use the key vocabulary list as a guide. Question the children to uncover the children's thinking process. An example of this may be a child that says, 'I shared the grapes fairly with my friend.' Ask, **How do you know you shared them fairly?** If the child can answer that they both got the same amount, then they are showing confidence and understanding.
- Children who are beginning to talk about fractions when they encounter them in their life.
- Children who can use sharing and grouping in their play.
- Children who know a half is two equal parts.

Fractions and Fair Shares B

Section overview

Units of learning	Review units	Prior learning	Benchmarks/Es&Os
Unit 4.1 **Beginning to use the language of fractions**	**An Introduction to Fractions A** Unit 4.4	• I can use the language of fractions.	**MNU 0-07a** • Splits a whole into smaller parts and explains that equal parts are the same size. • Uses appropriate vocabulary to describe halves. • Shares out a group of items equally into smaller groups. I can share out a group of items by making smaller groups and can split a whole object into smaller parts.
Unit 4.2 **Breaking a whole into equal parts**	**An Introduction to Fractions A** Unit 4.2	• I understand a whole can be broken into parts.	
Unit 4.3 **Making a whole from halves**	**An Introduction to Fractions A** Unit 4.5	• I know halves are two equal parts.	
Unit 4.4 **Understanding the terms 'halfway' and 'half of'**	**An Introduction to Fractions A** Unit 4.5	• I can use the language of fractions. • I know halves are two equal parts.	
Unit 4.5 **Identifying half of a whole**	**An Introduction to Fractions A** Unit 4.5	• I can use the language of fractions. • I know halves are two equal parts. • I can make halves by sharing or grouping. • I can compare amounts by measuring and counting.	
Unit 4.6 **Knowing if two parts are not equal, they cannot be called halves**	**An Introduction to Fractions A** Unit 4.5	• I know halves are two equal parts.	
Unit 4.7 **Recognising the notation for one half and one quarter**	**An Introduction to Fractions A** Unit 4.5 **Fractions and Fair Shares B** Unit 4.4	• I know halves are two equal parts. • I know quarters are four equal parts.	
Unit 4.8 **Sharing into quarters**	**An Introduction to Fractions A** Unit 4.1	• I can use the language of fractions. • I know sharing must be done fairly. • I can make halves by sharing or grouping. • I can compare amounts by measuring and counting. • I know quarters are four equal parts.	

Units of learning	Review units	Prior learning	Benchmarks/Es&Os
Unit 4.9 **Finding the quarter of a whole**	**An Introduction to Fractions A** Unit 4.5	• I know quarters are four equal parts.	
Unit 4.10 **Understanding that one quarter can be made by halving a half**	**An Introduction to Fractions A** Unit 4.2 and 4.5	• I know quarters are four equal parts.	
Unit 4.11 **Beginning to make a whole from quarters**	**Fractions and Fair Shares B** Unit 4.3	• I know quarters are four equal parts.	
Unit 4.12 **Finding half of a quantity**	**An Introduction to Fractions A** Unit 4.1	• I can use the language of fractions. • I know sharing must be done fairly. • I can make halves by sharing or grouping. • I can compare amounts by measuring and counting.	
Unit 4.13 **Finding a quarter of a quantity**	**An Introduction to Fractions A** Unit 4.1	• I can use the language of fractions. • I know sharing must be done fairly. • I can make quarters by sharing or grouping. • I can compare amounts by measuring and counting. • I know quarters are four equal parts.	
Unit 4.14 **Problems where the group of items does not share equally and there are 'leftovers'**	**An Introduction to Fractions A** Unit 4.3	• I can share items fairly.	

Conceptual context

This concept is based on learning that one whole object or amount can be split into equal parts. This can be done physically by grouping or sharing a whole amount. The children need to understand that they will get a fraction of the whole. In a fraction of an amount, the children should understand that sometimes this cannot be done fairly, and this will lead to leftovers which are called remainders.

The children should have started to develop the language of sharing, grouping and fractions. They should be beginning to understand that this should be done fairly and that when it can't be done fairly, there will be leftovers. This will have been achieved through engaging with the activities in section A throughout preschool and drawing on real-life experiences.

The children should now consolidate and extend learning in this concept through developing the language and attempting more formal sharing, grouping and fraction activities. Quarters should also be introduced. During this unit, children's real-life experiences should continue to be drawn upon and naturally arising events should be highlighted and maximised in the children's learning.

Resources

- Counters
- Small toys
- Paper plates
- Cups
- Paper shapes and strips
- Fraction picture cards
- Fraction word cards
- Beebot
- Mini whiteboard
- Teddies

Digital

- Fraction wall

Record book

Pages 17–18

Key vocabulary

- **Sharing** – splitting an amount or object into groups or parts
- **Grouping** – taking a whole amount and distributing into sets
- **Equal** – of the same value
- **Fair** – all parts or groups are the same/equal
- **Fraction** – parts of a whole
- **Whole** – the complete object or amount
- **Half** – one part of something divided into two equal parts
- **Quarter** – one part of something divided into four equal parts
- **Remainder** – an amount left over when sharing equally

Misconceptions

Potential misconception	Addressing the misconception
Units 4.1, 4.2, 4.3, 4.4 and 4.5 The children may think that something cut in two is a half and cut in four is a quarter.	• When showing a half or a quarter it is important to show – by comparing – that the parts are be equal.
Unit 4.8, 4.9 and 4.14 The children may believe that sharing is just giving out the group of items.	• When sharing, highlight the fact that the items must be shared out equally.
Unit 4.11 The children may believe that grouping is just putting the items into a number of groups.	• When grouping, highlight the fact that all the groups must have an equal amount.

Teaching ideas

Unit 4.1 Beginning to use the language of fractions

Whole class

- **Language** – The language of fractions and the concepts will arise naturally. These should be highlighted; focusing on the language used and building on previous learning experiences. A role play area such as a shop or café is ideal for revising the language of fractions through play.
- **Wall Display and Investigation Station** – As fraction work is covered, a wall display showing the concepts can be added to with an investigation station – a table or tuff tray with lots of examples of fractions to explore such as, plastic fruits or pizza the children can cut in half, paper strips and 2D shapes to fold in half, towers of cubes to break in half, fraction wooden puzzles, photographs of fractions around us, word cards, a halving mat and so on.

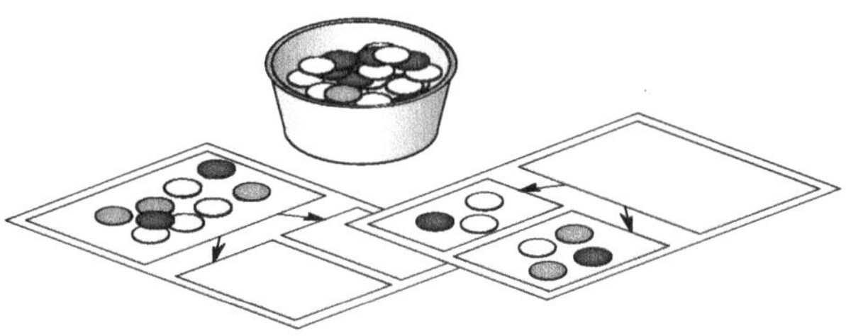

Unit 4.2 Breaking a whole into equal parts

Whole class/Stations

- **Fruit and Vegetables** – Look at the whole fruit or vegetable, then cut it in half. Talk about finding the middle. Ask the children: **Which is the best way to cut this so that it is in half?** Point out that a half must be two equal parts. Look at a melon (oval shape), which can be cut in half two ways; along the length or round the middle. The halves look very different, but both are still halves. Look at a carrot, which if cut across the width gives two pieces of a different size but if cut across the length gives halves as the two pieces are of equal size. When cutting, make some cuts not in the middle so that you can discuss how some parts are bigger and smaller. Point out that unequal cutting gives just two parts; it is not cut in half. Ask the children: **Are there other ways we can cut the fruit/vegetables keeping the parts equal?**
- **Towers** – Give the children an even number of bricks and ask them to make towers. Then ask the children to show you the middle of their tower and break into halves. Check by comparing. **Can you break the tower again but keep the parts equal?**

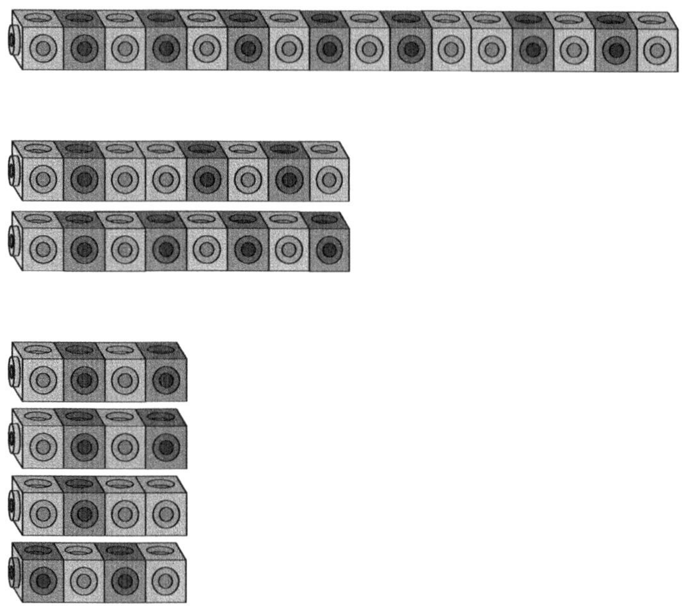

- **Paper Strips** – Give the children a strip of paper in their favourite colour. Ask them to fold their strip into fair parts. Get ideas from the children how to do this. Then tear it on the fold. **How many parts did they get from their strip? Are the parts equal? How do they know?** The children could also name the parts as halves and quarters.
- **Equal/Not Equal** – Give the children a variety of shapes with a line either showing halves or two different parts or cut into three equal parts or three different parts or four equal parts or four different parts. Sort them and discuss their reasoning.

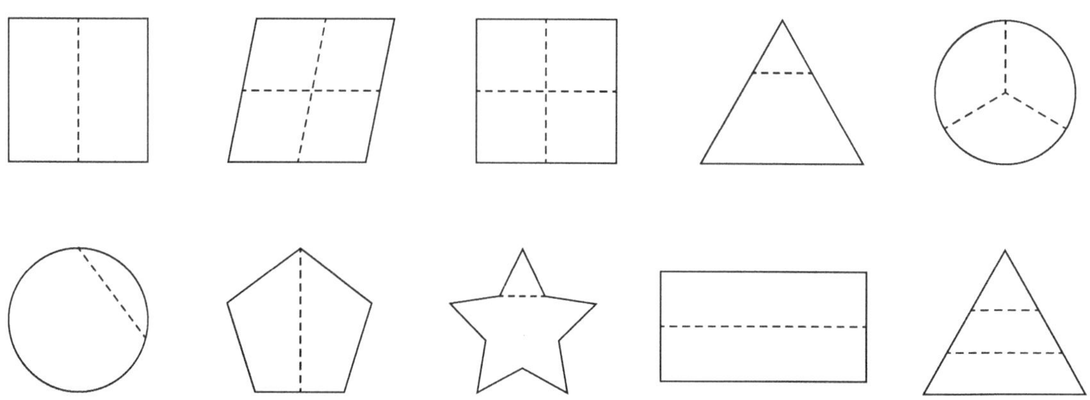

Unit 4.3 Making a whole from halves

Whole class/Stations

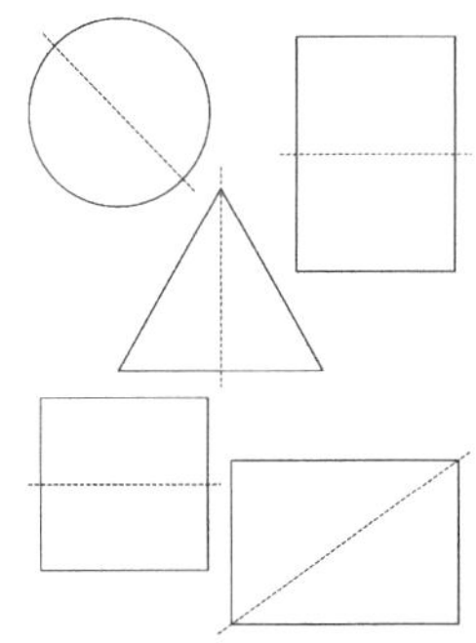

Photocopiable resource 4.2 – Colour Half.

- **Colour Half** – Give the children a shape with half marked (see Photocopiable resource 4.2). Colour one half red and one half green. **How much of the shape is coloured red?**
- **Fruit and Vegetables** – Cut the fruit and vegetables in half and put back together.
- **Paper Shapes** – Cut out two 2D shapes, from a range of 2D shapes, exactly the same in different colours. Cut one in half. Glue one half on the front left of the whole shape and the other half on the back right. Discuss how the two halves make a whole.
- **Two Groups** – Count how many children there are in the whole group. Make sure it is an even number in the group. Split the group into two groups. Discuss how the amounts in the two groups are the same as the whole amount of children.
- **Find My Half** – Cut in half lots of 2D shapes – circles, ovals, triangles, rectangles, squares, pentagons and hexagons – all made from stiff cardboard. Make each 2D shape a different size, shape or colour. Keep half of each shape laid out on the table and put the other half in a feely bag. The children take turns to pull a half out of the bag and find its matching partner.

Unit 4.4 Understanding the terms 'halfway' and 'half of'

Whole class/Stations

- **Halfway** – Lay a measured piece of rope on the ground. Ask the children to make a chalk mark and write their name where they think halfway is. Fold the rope to check and discuss their accuracy.

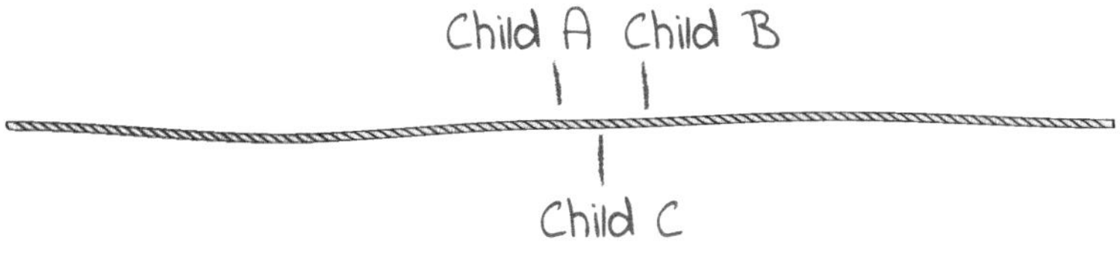

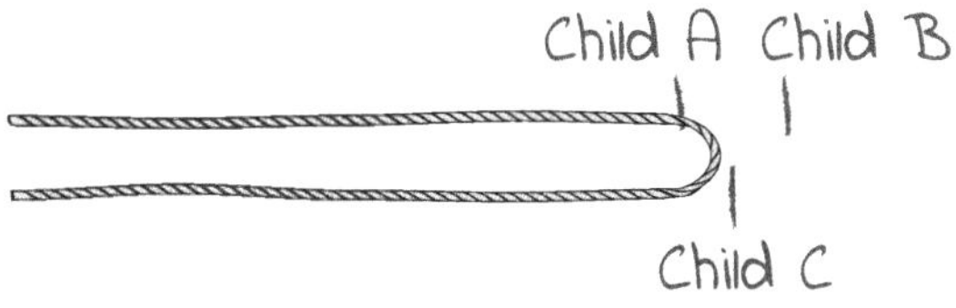

- **Beebot** – Using a Beebot grid, lay two objects on the grid a distance apart. Ask the children to program Beebot to stop halfway between the objects.
- **Half Of** – Give the children a number of objects. Ask them to split them in half. Check by comparing the objects side by side.

- **Paper Strips** – Fold strips of paper in two to show halves or in four to show quarters. Ask the children how many equal parts they have and explain that they are called halves and quarters.
- **Flat Shapes** – Fold 2D shapes to show halves or quarters. Ask the children to check that the parts are equal and remind them only then can they be called halves or quarters.
- **Towers** – Build towers from multi-link and split into halves and quarters. Ask the children to check that the parts are equal and remind them only then can they be called halves or quarters.

Unit 4.5 Identifying half of a whole

Whole class/Stations

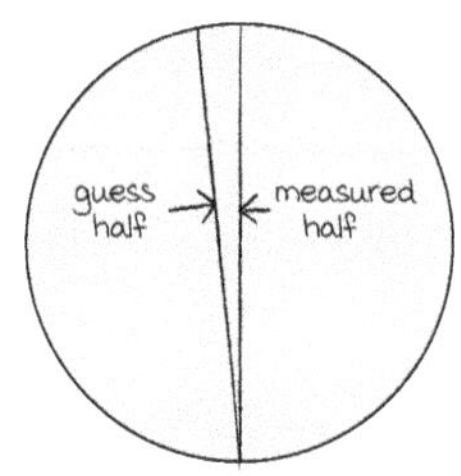

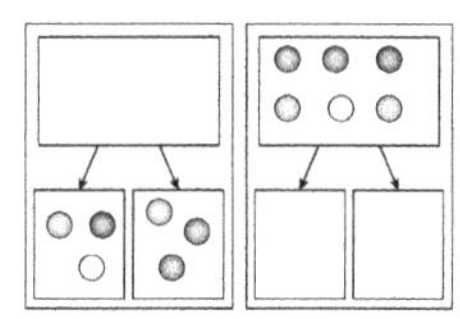

- **Shape Match** – Give the children a range of 2D shapes with their corresponding halves all jumbled up in the middle of the table. Match the halves to the whole shape by laying the halves on top of the shape.
- **Draw the Half Line** – Give the children a cardboard 2D shape. Ask them to use chalk to draw on a line to make halves. Fold the shape to check their accuracy.
- **Halves Mat** – Count an even number of counters into the top box. Share the counters between the two bottom boxes. How many are in each half?
- **Draw to Find Half** – Ask the children to draw an amount of pictures, for example stars, on their mini whiteboard. Then ask them to draw a ring around half. Count to check.
- **Share** – Take an even number of objects and ask the children to share between two. **How many do you each get?**

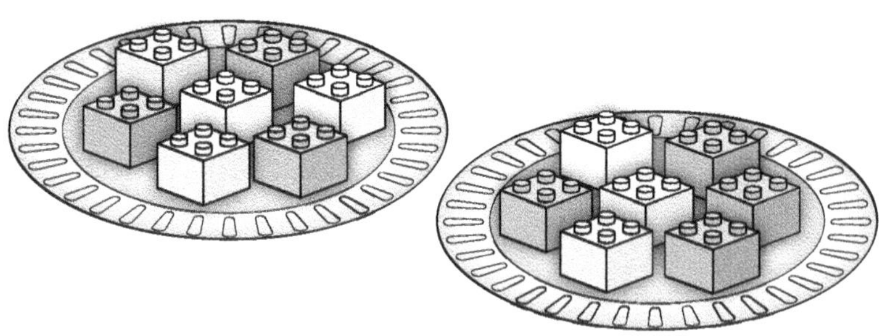

- **Draw the Other Half** – Give the children half a shape. Keep it simple with well-known 2D shapes such as triangles, squares, circles and rectangles. Challenge with more complicated shapes such as pentagons, hexagons and octagons. Ask them to draw the other half.

Unit 4.6 Knowing if two parts are not equal, they cannot be called halves

Whole class/Stations

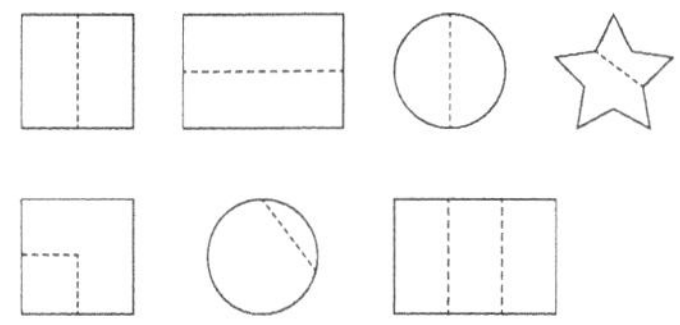

- **Halves/Not Halves** – Give the children a variety of shapes with a line either showing halves or two different parts or shapes cut into three/four equal parts. Sort them and discuss their reasoning.
- **Woolly Wonder** – Give the children a length of wool. Ask them to cut the wool where they think halfway would be. Measure the halves to check they are indeed a half. Ask the children to justify their answer.
- **Two Piles** – Give the children a pile of counters. Ask them to split the pile into two equal groups without counting. Then match up the groups to check if they are indeed equal. Qualify if it is half or not and explain reasons why.

Unit 4.7 Recognising the notation for one half and one quarter

Whole class/Stations

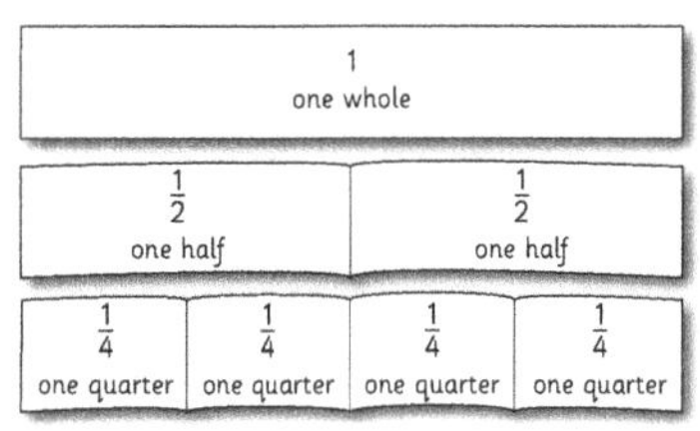

- **Paper Display** – Start with three strips of equal length. Lay one above the other. With the children, write one whole and the number one on the top strip. Fold the second strip in half and write *one half* on each side of the fold and lay below the first strip. On the third strip fold it into four equal parts and write one quarter and ¼ on each quarter then lay below. Discuss the notation and where the numbers come from in simple terms.
- **Lego Notation** – Using an eight piece, 2 four pieces and 4 two pieces of Lego. Build them up one on top of the other and ask the children to write 'one whole', 'one half' and 'one quarter' on the appropriate pieces.

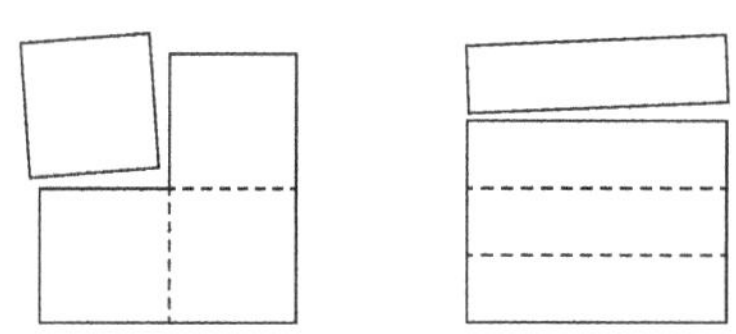

- **Colour Halves and Quarters** – Give the children a range of shapes with halves marked on with a line and one half displayed on each half. Ask them to colour one half of each shape. Repeat the activity for quarters; the children colour one quarter of each shape.

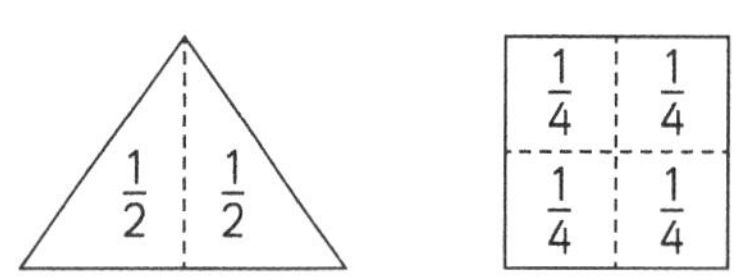

Unit 4.8 Sharing into quarters

Whole class/Stations

- **Share** – Pick out four children. Place a number of objects (multiple of 4) in the middle. Ask the children to share the objects out fairly by each taking one around the circle until there are none left. **How many does each get?** Explain that each child has quarter of the whole amount.
- **Paper Plates** – On the table have four paper plates and tubs of objects (multiples of 4). Ask the children to share the objects between the four plates. How many are on each plate? Repeat with another tub of objects. Relate back to each plate being a quarter of the whole amount as it was shared equally among four.
- **Cups** – Give the children a pile of counters (multiple of 4). Ask them to share them between the four cups. How many in each cup? Relate back to each cup being a quarter of the whole amount as it was shared equally among four.

Unit 4.9 Finding the quarter of a whole

Whole class/Stations

- **Fold the Shape** – Ask the children to fold their shape (circle, triangle, square and rectangle) to show quarters. Ask: **Which shapes could we do this with and why? Were they quarters? Why?**
- **Find the Missing Quarter** – Cut out lots of 2D shapes. Make each shape a different size, shape or colour. Cut one quarter out of each shape. Ask the children to find and match the missing quarter.
- **Colour the Quarter** – Give the children shapes marked in quarters. Ask them to colour one quarter of each.

Unit 4.10 Understanding that one quarter can be made by halving a half

Whole class/Stations

- **Towers** – Ask the children to build a tower with 16 cubes. Ask them to break it in half. Discuss and count how many in each half as well as comparing the length to check it is a half. Then ask the children to break each half in half. What do they find? Count and compare the four sections. Discuss that half of a half is a quarter.
- **Paper Strips** – Ask the children to fold a paper strip in half. Check they do this accurately. Then ask them to fold it in half again. Ask them to unfold their paper strip and explain what they see. They can mark their strip with the notations 'one half' and 'quarter and quarter' on the other side. Highlight the fact that one half is the same as two quarters. Therefore, a quarter is half of a half.
- **Chocolate** – Look at a whole chocolate bar and count the segments (Make sure there are 8). Ask a child to break the bar in half. Agree that this has been done correctly and discuss how many segments in each half. Then take one half and break it in half again. Place the broken half next to the other half. Ask the children: **What fraction do we have now? How many segments in a quarter?** Compare the two quarters to the half.
- **Empty Number Line** – Mark a number line with 0 and 1. Add a line to mark the half and quarter points. Ask the children to mark a half on the top of the number line and the quarters underneath. Point out how one half is the same as two quarters.

Unit 4.11 Beginning to make a whole from quarters

Whole class/Stations

- **Circles** – Give the children various sizes of circles cut into quarters and mixed up. Ask the children to find all the correct quarters to make the circles.
- **Colour Quarters** – Give the children a shape with quarters marked. Colour one quarter red, one quarter blue, one quarter yellow and one quarter green. **How much of the shape is coloured?**
- **Cake** – Cut the cake into quarters and put it back together. Discuss.
- **Paper Shapes** – Cut out 2D shapes, such as squares and rectangles, in different colours. Cut them in quarters. Look at how the same shape can be cut into quarters in different ways. For example, a square can be four small squares or four rectangles for quarters. Discuss how the four quarters go back together to make a whole. It is interesting for the children to investigate and find out which shapes this works for and which it does not. For example, a circle can only be quartered one way.
- **Four Groups** – Count how many children are in the group. Make sure it is a multiple of four in the group. Split the group into four equal groups. Discuss how the amounts in the four groups are the same as the whole amount of children.

Unit 4.12 Finding half of a quantity

Whole class/Stations

- **Sharing** – Share an even amount of objects between two. **How many does each person get?** Point out that each child gets half of the whole amount.
- **Grouping** – Put an even number of objects into two groups. **How many in each group?** Point out each group has half of the whole amount.
- **Solve the Problem** – Tell the children you have 12 balls (have the balls there for the children to manipulate). Ask the children to halve the balls between two people. **How many is half?** Repeat with other amounts. Discuss how a half can be a different number as it is relevant to the starting amount.
- **What Do You See?** – Lay out two groups of an amount of objects. Ask the children to describe what they see in fraction terms. **How much is in each half? How much is the whole amount?** Repeat for different sizes of groups.

Unit 4.13 Finding a quarter of a quantity

Whole class/Stations

- **Sharing** – Share a multiple of four objects between four. **How many does each get?** Point out each child gets quarter of the whole amount.

- **Grouping** – Put a number (multiple of 4) of objects into four groups. **How many in each group?** Point out each group has quarter of the whole amount.
- **What Do You See?** – Lay out four groups of an amount of objects. Ask the children to describe what they see in fraction terms. **How much is in each quarter? How much is the whole amount?** Repeat for different sizes of groups.
- **Solve the Problem** – Tell the children you have 12 teddies (have the teddies there for the children to manipulate). Ask the children for a quarter of the teddies. **How many is a quarter? How did you know?** Repeat with other amounts.

Unit 4.14 Problems where the group of items does not share equally and there are 'leftovers'

Whole class/Stations

- **Grouping** – Put several objects into groups. **How many in each group? How many left over?**
- **What Do You See?** – Lay out two/four groups of an amount with some leftovers. Ask the children to describe what they see. Repeat for different sizes of groups.
- **Cake Shop** – Ask the children if we can share seven cakes between two people fairly. Ask the children to justify their answer. Have 'cakes' and two plates on hand for the children to manipulate and prove their answer. Repeat for other amounts.

What to observe

- Children who are confident using the appropriate vocabulary. Use the key vocabulary list as a guide. Question the children to uncover the children's thinking process. An example of this may be asking a child to share all the teddies fairly between them and their three friends, 'I shared the fourteen teddies equally with my friends.' The teacher asks: 'Did you all get a fair share?' If the child can answer that they all got the same amount but there were some left over and explain why, then they are showing confidence and understanding.
- Children who are beginning to talk about fractions and use them in their daily life.
- Children who know halves are two equal parts.
- Children who know quarters are four equal parts.
- Children who measure/count to check accuracy.
- Children who recognise notations for a half and a quarter.
- Children who can use sharing and grouping in halves and quarters.
- Children who understand that you cannot always share an amount equally.

5 Money

An Introduction to Money A

Section overview

Units of learning	Review units	Prior learning	Benchmarks/Es&Os
Unit 5.1 **Money in play situations**		• From their lives the children will have an ever expanding level of vocabulary and a range of personal experiences to draw upon.	**MNU 0-09a** • Identifies all coins to £2. • Applies addition and subtraction skills and uses 1p, 2p, 5p and 10p coins to pay the exact value for items to 10p. I am developing my awareness of how money is used and can recognise and use a range of coins.
Unit 5.2 **Naming coins 1p–10p**	**An Introduction to Money A** Unit 5.1	• The children will have experiences of coins and shopping from their lives and have started to investigate this in their play.	
Unit 5.3 **Ordering coins by value 1p – 10p**	**An Introduction to Money A** Unit 5.2	• I can recognise coins from 1p to 10p.	

Conceptual context

This concept is based on understanding the value of coins and how they can be used in life to pay for goods. The children need to understand that there are different coins each worth a different amount. The children will have to understand the equivalency of amounts made with different coins.

Children will be familiar with money through their daily lives and it is important to point out all the ways of paying for goods in today's society including cash, cards, phones and digital banking. However, it is equally important to stress that money must be earned and even if cash is not being used, the money must be available.

Resources

- Real money
- Plastic money
- Plastic debit cards
- Toy till
- Role play resources
- Purses/wallets
- Laminated giant coin pictures
- Feely bag
- Sorting trays
- Washing line and pegs
- Play dough
- Sand
- Ink and roller
- Wax crayons

Digital

- Money

Record book

Page 19

Key vocabulary

- **Coins** – metal discs used to represent the lower value of money
- **Copper** – 1p and 2p coins traditionally made of copper; also slang for small value coins
- **Silver** – 5p, 10p, 20p and 50p coins, silver colour
- **Notes** – higher value money made of paper/plastic and in a different form to coins
- **Value** – important in highlighting the value of each coin and getting the children to understand that one coin does not always mean one in value
- **Worth** – key word in highlighting the value or worth of one coin
- **Pence** – symbol 'p' used to show the context of money
- **Pound** – symbol '£' used to show the context of money

Misconceptions

Potential misconception	Addressing the misconception
Units 5.2 and 5.3 The children do not understand the value of the coins. They see each coin as one object not noting their different value.	• When introducing the coins talk about their different shapes, sizes, colours and values. Highlight equivalency to the one pence coin.
Unit 5.3 The children may believe that size is linked to value.	• Stress that size is not linked to value. Highlight the bigger numbers and relate to one pence.

Teaching ideas

Unit 5.1 Money in play situations

Whole class

- **Role Play** – Provide role play areas where the children can use coins in their play to purchase different items/services. Shops, cafés, hairdressers, post offices and so on are all great for allowing children to become immersed in the language and process of shopping. Make sure the children have purses/wallets, real coins, priced items, shopping bags and a till so that they can experience all aspects of the shopping process. It is even better if you can visit a real shop and allow the children to buy something first, so they have a real-life experience to draw on in their play. Adults can scaffold the play modelling the language and process.
- **Money Display** – Set up an interest table with coins and cards (such as sample loyalty cards, debit and credit cards etc.) for the children to explore. It is a great idea to have coins from other countries too to discuss and compare.

Unit 5.2 Naming coins 1p–10p

Stations

- **Coin Exploring 1** – Use a real purse, open it and discuss the contents, both cards and coins. Explain how each can be used to buy objects/services. Discuss each coin in turn, asking the children to explain the coin noticing the size, shape, colour and any numbers or pictures they find. Focus in on the 1p, 2p, 5p and 10p coins. Compare different coins of the same value, for example, two ten pence coins. **What is the same and different?** They will both have the number ten and the queen's head but a different picture on the back. Notice that all the ten pence coins are silver and the same size.
- **Coin Match 1** – Lay out laminated cards of each coin from 1p to 10p. Get the children to say what each coin is. From a large pile of different coins, the children take it in turns to pick a coin and match it to the correct picture.

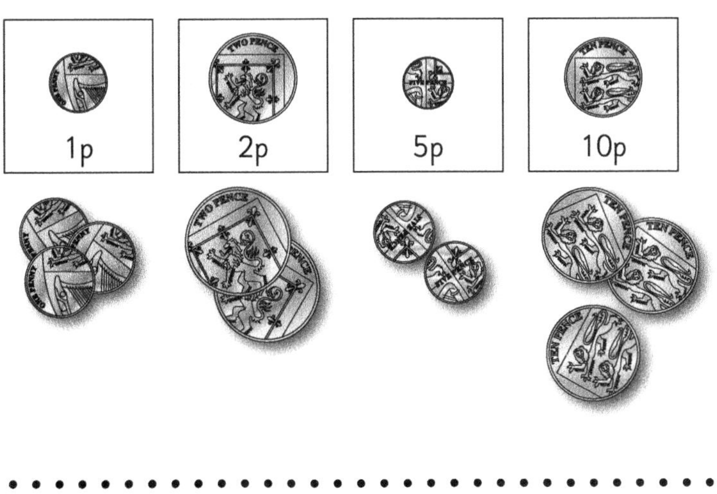

- **Feely Bag 1** – Place a number of 1p, 2p, 5p and 10p coins in a feely bag. Each child in turn places their hand in the feely bag and pulls out a coin. **Which coin did you get?**
- **Coin Rub 1** – The children can use real coins to make wax rubbings. The children should say what coin they have and place it under a piece of paper. Rub over the coin with the side of a wax crayon. Turn the coin over and rub over the other side. Discuss both sides of the coin.

- **Coin Printing 1** – Using ink and a roller, roll the ink over a real 1p, 2p, 5p or 10p coin and print it on paper. Ensure the children know which coin they are printing.

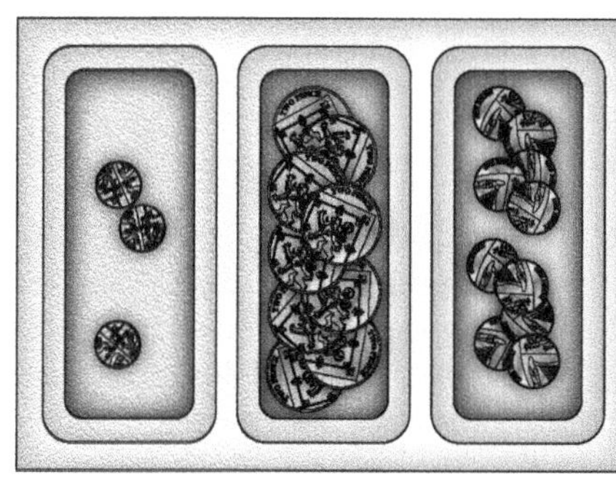

- **Coin Sorting 1** – Use a sorting tray to sort a large pile of coins. Ensure the children are naming each coin and discussing their value. Use only 1p to 10p coins.
- **Play Dough Press 1** – Press the coins into play dough and look at the print it makes. Use only 1p to 10p coins. **Which coin did you use?**
- **Coin Number Match 1** – Have 1, 2, 5 and 10 large digit cards. Ask the children to place the coin on the correct number.

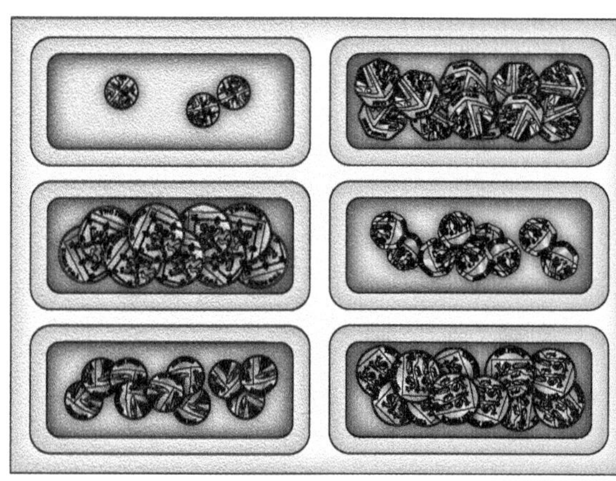

- **Washing Line Coins 1** – Hang laminated pictures of 1p, 2p, 5p and 10p coins on the washing line. Ask the children to close their eyes. Turn one coin card round. When the children open their eyes, ask: **Which coin is hiding?**

Unit 5.3 Ordering coins by value 1p–10p

Stations

- **Coin Value 1** – Look at the coins 1p, 2p, 5p and 10p. The children should discuss each one and show their value in one pence coins. Note a pot of 1p coins is required.
- **Treasure Hunt 1** – Hide 1p, 2p, 5p and 10p coins in the sand. Ask the children to dig out the coins and lay them in order of their value.
- **Washing Line Value 1** – Have a pile of laminated 1p, 2p, 5p and 10p coin cards. Ask the first child to pick a card, name the coin and place it on the washing line. The next child picks a card and places it on the washing line in order of value stating if its value is more or is less than the one before. When all the coins are on the washing line from the least to the most value, discuss. Do the same process but order the coins from the most value to the least value.
- **Match Totals 1** – Have laminated cards with 1p, 2p, 5p and 10p totals in pennies. Ask the children to match the actual coins to the correct value shown on the laminated cards.
- **Coin Mix Up 1** – Lay out the 1p, 2p, 5p and 10p coins in order of their value from least to most or vice versa. Discuss. Ask the children to close their eyes and the teacher mixes them up. When the children open their eyes, ask: **What is the mix up? How do you know?** The children can then correct the order.
- **Purse Put Away 1** – Give the children some 1p, 2p, 5p and/or 10p coins. Ask the children to give you back their coin with the most value to put in the purse. This can be varied by asking for the coin with the least value. As the children have different coins, one child's most/ least value may be different to another's. Discuss.

What to observe

- Children who are confident using the appropriate vocabulary and processes involved in money.
- Children who can identify and name 1p, 2p, 5p and 10p coins.
- Children who can discuss the value of coins and order them accordingly.

Money and Shopping B

Section overview

<table>
<tr><th>Units of learning</th><th>Review units</th><th>Prior learning</th><th>Benchmarks/Es&Os</th></tr>
<tr><td>Unit 5.1
Naming all coins</td><td>An Introduction to Money A
Unit 5.2</td><td>• I can recognise coins from 1p to 10p.</td><td rowspan="4">MNU 0-09a
• Identifies all coins to £2.
• Applies addition and subtraction skills and uses 1p, 2p, 5p and 10p coins to pay the exact value for items to 10p.

I am developing my awareness of how money is used and can recognise and use a range of coins.</td></tr>
<tr><td>Unit 5.2
Knowing the value of all coins</td><td>An Introduction to Money A
Unit 5.3</td><td>• I can order 1p–10p coins by value.</td></tr>
<tr><td>Unit 5.3
Using 1p–10p coins to buy items (applying correct language for money)</td><td>Money and Shopping B
Units 5.1 and 5.2</td><td>• I can name all coins.
• I know the value of all coins.</td></tr>
<tr><td>Unit 5.4
Using addition and subtraction for money purchases</td><td>Money and Shopping B
Units 5.1, 5.2 and 5.3</td><td>• I can name all coins.
• I know the value of all coins.
• I can use coins to buy items applying the correct language.</td></tr>
</table>

Conceptual context

This concept is based on understanding the value of coins and how they can be used in life to pay for goods. The children need to understand that there are different coins each worth a different amount. The children will have to understand the equivalency of amounts made with different coins. The children will have to use coins to make amounts, being encouraged to use as few coins as possible. The children will have to count amounts. The children will also have to use addition and subtraction in the context of money when making purchases.

Children will be familiar with money through their daily lives and it is important to point out all the ways of paying for goods in today's society, including cash, cards, phones and digital banking. However, it is equally important to stress that money must be earned and even if cash is not being used the money must be available. The children should have started to develop the language of and the processes involved in money. This will have been achieved through engaging with the activities in section A throughout preschool and drawing on real-life experiences.

The children should now consolidate and extend learning in this concept through developing the language and extending their knowledge and experience of coins with higher values. During this unit children's real-life experiences should continue to be drawn upon and naturally arising events should be highlighted and maximised in the children's learning.

Resources

- Real money
- Plastic money
- Plastic debit cards
- Toy till
- Role play resources
- Purses/wallets
- Laminated giant coin pictures
- Feely bag
- Sorting trays
- Washing line and pegs
- Cake cases with amounts written inside
- Ink and roller
- Play dough
- Wax crayons
- Money fan

Digital

- Money

Record book

Page 19

Key vocabulary

- **Coins** – metal discs used to represent the lower value of money
- **Copper** – 1p and 2p coins traditionally made of copper; also slang for small value coins
- **Silver** – 5p, 10p, 20p and 50p coins, silver colour
- **Notes** – higher value money made of paper/plastic and in a different form to coins
- **Value** – important in highlighting the value of each coin and getting the children to understand that one coin does not always mean one in value
- **Worth** – key word in highlighting the value or worth of one coin
- **Pence** – symbol 'p' used to show the context of money
- **Pound** – symbol '£' used to show the context of money

Misconceptions

Potential misconception	Addressing the misconception
Units 5.2 and 5.3 The children do not understand the value of the coins. They see each coin as one object not noting their different value.	• When introducing the coins talk about their different shapes, sizes, colours and values. Highlight equivalency to the one pence coin.
Units 5.1 and 5.2 The children may believe that size is linked to value.	• Stress that size is not linked to value. Highlight the bigger numbers and relate to one pence.
Units 5.2, 5.3 and 5.4 The children may focus on the amount and use values which don't exist. An example of this may be if asked for 6p they make it with 3p and 3p.	• The children are focusing on their number bonds and forgetting the context of coins. Spend time refreshing the child's understanding of the coins available and their value and equivalency.

Teaching ideas

Unit 5.1 Naming all coins

Stations

- **Coin Exploring 2** – Use a real purse, open it and discuss the contents, both cards and coins. Explain how each can be used to buy objects/services. Discuss each coin in turn, asking the children to describe the coin, noticing the size, shape, colour and any numbers or pictures they find. Focus on the 1p, 2p, 5p and 10p coins. Compare different coins of the same value, for example, two ten pence coins. **What is the same and different?** They will both have the number ten and the queen's head but a different picture on the back. Notice that all the ten pence coins are silver and the same size. Look at different coins. Show the children how the coins can make different values, for example, a 2p has the same value as two 1p coins.
- **Coin Match 2** – Lay out laminated cards of each coin. Get the children to say what each coin is. From a large pile of coins, the children take it in turns to pick a coin and match it to the correct picture.
- **Feely Bag 2** – Place a number of coins in a feely bag. Each child in turn places their hand in the feely bag and pulls out a coin. They have to say which coin they got.
- **Coin Rub 2** – The children can use real coins to make wax rubbings. The children should say what coin they have and write it, remembering the 'p' for pence. Then place the coin under a piece of paper. Rub over the coin with the side of a wax crayon. Turn the coin over and rub over the other side. Discuss both sides of the coin.
- **Coin Printing 2** – Using ink and a roller, roll the ink over a real coin and print it on paper. Ensure the children know which coin they are printing.
- **Coin Sorting 2** – Use a sorting tray to sort a large pile of coins. Ensure the children are naming each coin and discussing their value.
- **Play Dough Press 2** – Press the coins into play dough and look at the print it makes. **Which coin did they use?**
- **Coin Number Match 2** – Have a 1, 2, 5, 10, 20 and 50 large digit card. Ask the children to place coins on the correct numbers.

- **Washing Line Coins 2** – Have laminated pictures of 1p, 2p, 5p, 10p, 20p, 50p, £1 and £2 coins hanging on the washing line. Ask the children to close their eyes. Turn one coin card round. When the children open their eyes, ask: **Which coin is hiding?**
- **Flash Cards** – Hold up laminated flash cards of all coins. The children name the coin from the picture.
- **I'm Thinking of a Coin …** – Describe the coin orally, the size, shape and colour. Give clues about its value; for example, it is worth more than a 5p but less than a 20p. The children hold up the coin. **What coin is it? How did they know?**
- **Fastest Finger** – Give each child one of each coin to lay on the table in front of them. The teacher says a coin and the children point to it as quickly as they can.

Unit 5.2 Knowing the value of all coins

Stations

- **Coin Value 2** – Look at all the coins. Discuss each one and show their value related to other coins.
- **Treasure Hunt 2** – Hide all the coins in the sand. The children dig out the coins and lay them in order of their value.
- **Washing Line Value 2** – Have a pile of laminated coin pictures. The children pick a card, name the coin and place it on the washing line. The next child picks a card and places it on the washing line in order of value, stating if its value is more or is less than the one before. Once all coins are laid from least to most value, discuss. Do the same process but order the coins from the most value to the least value.
- **Match Equivalent Amounts** – Lay out amounts using various combinations of coins. The children match the actual coin. Discuss all the different ways you could make the amount.
- **Coin Mix Up 2** – Lay out all the coins in order of their value from least to most or vice versa. Discuss. Ask the children to close their eyes and the teacher mixes them up. when the children open their eyes, ask: **What is the mix up? How did you know?** The children then correct the order.

- **Purse Put Away 2** – Give the children some coins. Ask the children to give you back the coin with the most value to put in the purse. This can be swapped for the coin with the least value. As the children all have different coins, one child's least value may be different to another's. Discuss.

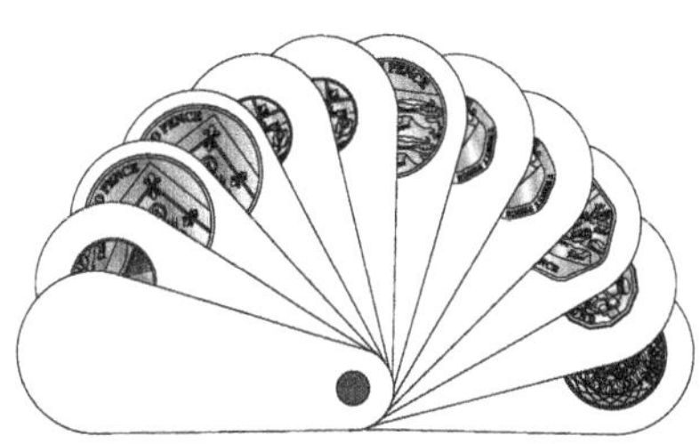

- **Show Me –** Give the children a set of all the coins. Ask them to show you the coin with the most value, the least value, a value more than 2p but less than 10p, and so on. Then ask the children to put all the coins in order.
- **Money Fans** – Ask the children to find the coin with the most value, least value, the coins with a greater value than 50p or smaller value than 20p and so on.

Unit 5.3 Using 1-10p coins to buy items

Stations

- **Pennies 1** – Use one pence coins to make amounts to 10p.
- **Pennies 2** – Lay out amounts using only pennies. Ask the children: **What is the value of these coins?**
- **Introduce 2p coin** – Introduce the 2p coin then use a combination of 1p and 2p coins to make amounts.

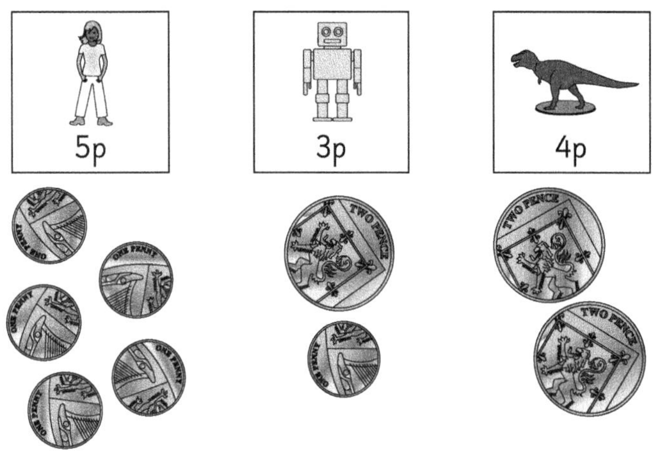

- **Using the 2p Coin** – Lay out amounts using 1p and 2p coins. Ask the children: **What is the value of these coins?**
- **Introduce 5p Coin** – Introduce the 5p coin. Find all the different ways to make 5p using combinations of 1p and 2p coins.
- **Make Amounts Using the 5p Coin** – Give the children a picture card with amounts up to 10p. Ask them to make the amount using combinations of coins. Look at the different ways the children made the amount. Compare.
- **Count Amounts Including 5p** – Lay out amounts using 1p, 2p and 5p coins. Ask the children: **What is the value of these coins?**
- **Introduce 10p Coin** – Introduce the 10p coin. Find all the different ways to make 10p.
- **Cake Cases** – Give the children a cake case with an amount written in the bottom. The children put that amount into the cake case.
- **Shops 1** – Set up a class shop and price the items up to 10p. Each child chooses an item to 'buy' and gives the designated shopkeeper the correct money.

Unit 5.4 Using addition and subtraction for money purchases

Stations

- **Shops 2** – Set up a class shop and price the items up to 10p. The children choose two or three items to buy and find the total cost. The children pay and work out what change, if any, they should get back from the shopkeeper.
- **Toys** – Show the children a set of picture cards of toys with their prices. Each child chooses two cards and finds the total amount. Ask the children: **How did you find the total?** Look for them to tell you they added. Then get the children to lay out the coins to pay for the item.

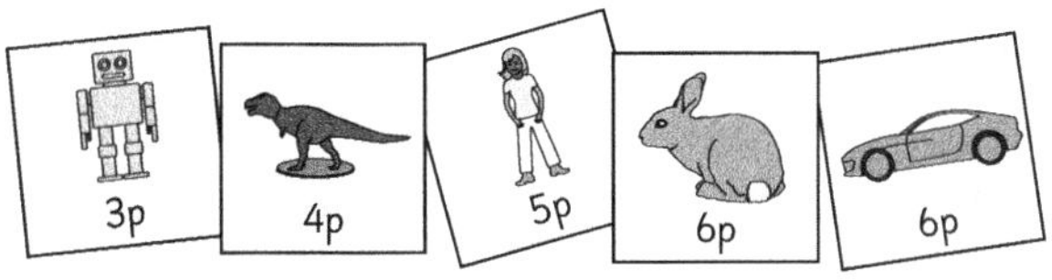

- **Can I Buy?** – Hold up a coin. Then hold up a priced item. Ask the children: **Can you buy this item? How do you know?** At first the children may say no if it is not the exact amount not realising that if they have more, they could be given change. Discuss.
- **What's My Change?** – Tell the children you have 10p to spend. You buy a banana for 7p. **What is my change?** The children select the correct coins to lay out for change. Discuss how they worked out the answer. Repeat for other amounts.
- **Pocket Money Challenge** – Lay out some toys with various prices less than 10p. Tell the children they have 10p pocket money. Ask them what they would buy and why. Ask the children to explain their thinking. Look at who got the most toys for their money. **Who spent the same but got different items? Who spent exactly 10p?**
- **Purse Challenge** – Lay out two laminated purses. Give the children a range of 1p, 2p and 5p coins. Ask the children to put exactly 10p in each purse but they must have five coins in each purse.

What to observe

- Children who are confident using the appropriate vocabulary related to money and can name all the coins.
- Children who can give the value of all coins.
- Children who can use 1p to 10p coins to buy items.
- Children who can use addition and subtraction for money purchases.

6 Time

An Introduction to Time A

Section overview

Units of learning	Review units	Prior learning	Benchmarks/Es&Os
Unit 6.1 **Daily routines**		• From their lives the children will have an ever expanding level of vocabulary and a range of personal experiences to draw upon. The children may be familiar with counting sleeps down to a special event or know some of the days of the week relating to when they do things in their lives. The children may be familiar with some time devices such as the clock in their house or a family member's watch, although more likely today, the digital clock on their phone. All these experiences will be different depending on the encounters each child has had. Staff should be aware of this and ensure they consolidate and build on the child's previous learning.	**MNU 0-10a** • Links daily routines and personal events to time sequences. • Names the days of the week in sequence, knows the months of the year and talks about the four seasons in relevant contexts. • Recognises, talks about and where appropriate, engages with everyday devices used to measure or display time, including clocks, calendars, sand timers and visual timetables. • Reads analogue and digital o'clock times (12 hour only) and represents this on a digital display or clock face. • Uses appropriate language when discussing time, including before, after, o'clock, hour hand, minute hand. I am aware of how routines and events in my world link with times and seasons, and have explored ways to record and display these using clocks, calendars and other methods.
Unit 6.2 **Using language to describe time**			
Unit 6.3 **Introducing days of the week**			
Unit 6.4 **Seasons**			
Unit 6.5 **Introducing months**			
Unit 6.6 **Describing and using time devices**	**An Introduction to Time A** Unit 6.2	• I can use the language of time.	

Conceptual context

Children will be familiar with time through situations and events they experience in their daily lives. It is important to draw on as many of these real-life experiences as possible to make the learning relevant. The children need to be exposed to different types of time and have their attention drawn to naturally occurring change in the world around them. Language is vitally important in developing the concept of time and should be exploited at every opportunity to extend the child's vocabulary and understanding.

Modelling the use of time and timers in routines should be maximised showing the real-life context and significance of the concept. Attention should also be drawn to the cyclical pattern of time so children become familiar with repetitive patterns and routines. All timing devices should be highlighted to children and they should be given the chance to explore these with a variety of first hand experiences.

Resources

- Sand timers
- Clocks – analogue, digital
- Watches
- Calendars
- Diaries
- Seasonal pictures
- Photographs of daily events
- Day/night pictures
- Washing line
- Hula hoop
- Globe
- Torch
- Pom-poms
- Containers
- Giant tweezers
- Tuff tray
- Interlocking cubes
- Play dough
- Role play area (an area where the children can take on the role of a character and act out or perform)

Digital

- Clock

Record book

Pages 20–21

Key vocabulary

- **Morning**
- **Afternoon**
- **Day**
- **Night**
- **Today**
- **Tomorrow**
- **Yesterday**
- **Day**
- **Week**
- **Month**
- **Year**
- **Season**
- **Minute**
- **Hour**
- **Before**
- **After**
- **Earlier**
- **Later**
- **Digital** – clock using numerals only
- **Analogue** – clock with moving hands

Misconceptions

Potential misconception	Addressing the misconception
Units 6.1, 6.2, 6.3, 6.4 and 6.5 As language is still developing, the children may over-extend the use of words. An example of this may be saying tomorrow, meaning another day.	• Model language clearly and address any misuse of words.

Teaching ideas

Unit 6.1 Daily routines

Whole class

- **Daily Calendar** – as part of the daily routine, discuss the day, date and weather linking to the season. This also introduces the children to ordinal numbers. A great way to bring the days of the week, months and seasons in to daily use. Ask the children what day comes next (tomorrow) or what day was yesterday to extend their vocabulary. Highlight key events of the day.

Stations

- **Visual Timetable** – Take photographs of the main events in a day and display them for the children to discuss and order. When the children are familiar with the daily routine, take the photographs off the display. Ask the children to put them back on in the correct order, giving the opportunity to use lots of language such as before/after/next. When the children become more confident, this can be built up to a weekly timetable.

- **My Favourite Day** – Using a template with morning, afternoon and evening (see Photocopiable resource 6.1). Discuss with the children what their favourite day might look like. Have a number of prompt cards for them. These may include a picture of a water park, a funfair, a circus, a play park, a cinema and so on. Have blank pieces of paper to hand so the children can also draw their ideas. Stick the cards on to the template and discuss with the children their favourite day highlighting key vocabulary.

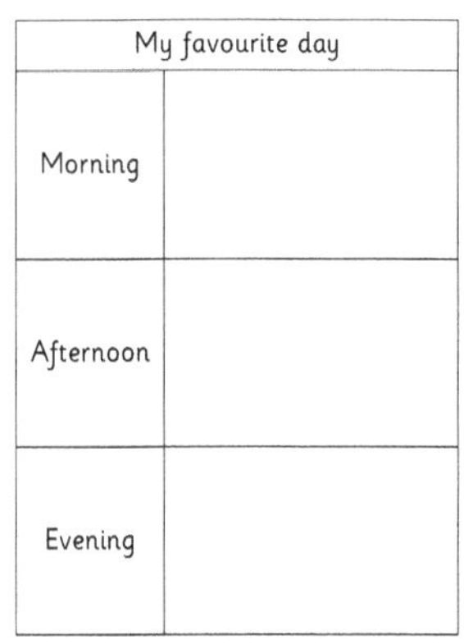

Photocopiable resource 6.1 – My Favourite Day.

Unit 6.2 Using language to describe time

Whole class

- **Start-Stop Game** – In the gym hall or outside area, ask the children to travel in different ways, stopping and starting on command. The children can skip, hop, jump, gallop, walk backwards, crawl and so on. This game also develops their co-ordination and listening skills. You can also bring in moving quickly or slowly. All developing the language of time.
- **Time-related Games**

  What's the Time Mr Wolf? – One child stands at one end of the space with their back to the other children. The rest of the class are at the opposite end of the space and shout 'What's the time Mr Wolf?' The wolf replies with an o'clock time and the children take that number of steps towards the wolf. For example, if the wolf says three o'clock, the children take three steps forward. When the wolf replies dinner time, the children run back to where they started. The wolf tries to catch a child. That child then becomes the wolf. A great game for the language of time and counting.

  The Clock on the Tower – The children stand in a circle and are numbered 1 to 4/5/6 (depending on their ability and class size) around the circle. One child stands in the middle of the circle and says 'The clock on the tower strikes the hour ...' and they clap the number. All the children with that number run round the outside of the circle, back through their original space and into the middle. The first person to get to the person in the middle stays there for the next turn. For example, the clock on the tower strikes the hour ... four claps. All the number fours run round the outside of the circle, through their space and into the middle. The winner then repeats the sentence with a different number. This game introduces the language of time as well as counting.

Stations

- **Day and Night** – Through discussions with the children, create a display of pictures, art and objects associated with day and night. Highlight specific language. Create a dark space the children can explore with a torch, perhaps finding nocturnal animals (cuddly/plastic toys or laminated pictures on card) hidden in the space. Following the children's interest, this may be a suitable time to get the globe and torch out to show how the Earth spinning around while travelling around the Sun gives us day and night. Another linked activity is a DIY human sundial. The children can go outside at different times of the day to look at their shadows and how they change throughout the day. Try to spot patterns over a few days and encourage the children to think about why the changes are happening. This a great way to understand about the Earth's rotation and how we can use the Sun's position in the sky to help tell the time.

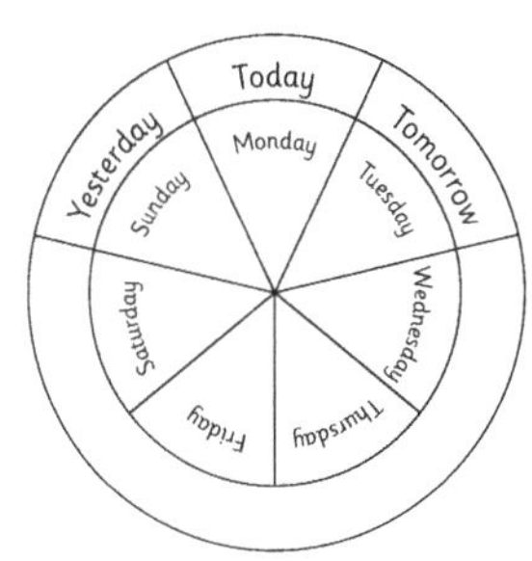

- **Yesterday, Today, Tomorrow Spinner** – Create a spinner wall display which can be used to prompt discussion. Use pictures to show what the children are doing each day. Use the display for discussion to show how the days change and move in a cyclical pattern.
- **Washing Line Sequencing** – Use a washing line to sequence real events such as cleaning teeth, getting dressed or making a sandwich. Familiar stories could also be used such as 'The Three Billy Goats Gruff', 'Little Red Riding Hood', *Room on the Broom* by Julia Donaldson, *The Very Hungry Caterpillar* by Eric Carle or *I Am Bat* by Morag Hood. Use photographs or pictures to stimulate discussion. Ask the children to decide which order to peg them up in.

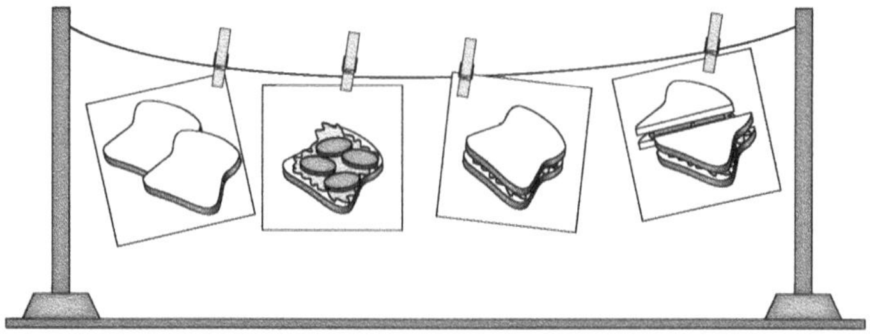

- **Hickory Dickory Dock** – Use the traditional rhyme to highlight the vocabulary used about clocks. Continue the rhyme for two, three, etc., asking the children to think of suitable rhyming words. An example may be, 'The clock struck two the mouse turned blue'. The children can draw pictures to match their rhyme and display them in order.
- **What Next?** – Focusing on the words before and after, place several pictures face down in a circle. Ask a child to turn over a picture and explain what is happening. The pictures could be related to their nursery routine or familiar routines such as brushing teeth (getting the toothbrush, putting toothpaste on the brush, brushing teeth, rinsing toothbrush). As each picture is turned over, ask, **What do you do after?** The game can then be repeated changing the question to, **What do you do before?**

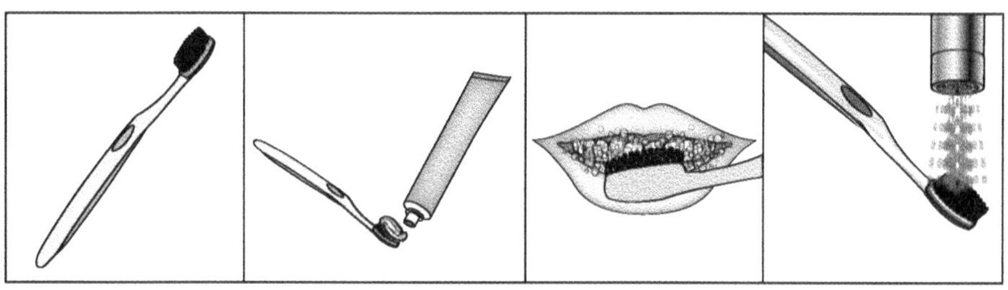

- **Prop Box Routines** – Have a prop box full of props related to the topic of discussion. An example may be a pair of pyjamas, a toothbrush and a book. Ask the child to discuss the items and think about what time of the day they might be used. They should explain their reasoning. The child can then use the props to 'act out' the routine explaining what they do first, next and last. Other examples include breakfast cereal and clothes or a shopping basket, purse with money, food and shopping bags.

Unit 6.3 Introducing days of the week

Whole class

- Days of the week song – sung to the tune of The Addams Family:

  'Days of the week, clap, clap.

  Days of the week, clap clap.

  Days of the week, days of the week, days of the week clap, clap.

There's Sunday and there's Monday, There's Tuesday and there's Wednesday.

There's Thursday and there's Friday and then there's Saturday.

Days of the week, clap, clap. Days of the week, clap clap.

Days of the week, days of the week, days of the week clap, clap'

Stations

- **Washing Line** – Peg up cards with the days of the week names on them. Use photographs to show what the children do on each day. The washing line can also be used as a countdown to special events happening later in that week. When the children become more familiar with the days of the week, Mr/Ms Trickster can 'steal' one of the days of the week cards and hide it around the nursery. **Who can spot which day is missing? Who can put it back in the right place? Which day does it come before/after or between?** You could also jumble the order and ask the children to put the cards back in the correct order.
- **Human Days of the Week Line** – Ask seven children to stand in line. Give them a card each with the days of the week on it. Ask each child in turn to shout out their day of the week. Repeat several times so that the children become very familiar with the order of the days of the week. If other children in the group are watching, they can join in. The children can say the days in different voices, for example loud, quiet, robot. When the children are very familiar with the order, start in different places to highlight the cyclical aspect of the days. Ask a child holding one of the day cards to turn around. Ask the other children: **Which day is missing?** Ask a child to swap with a certain day. When the children are confident with this, it can be extended to the day before/after. The children are not expected to read the days of the week but to become familiar with the order. Some children may start to look at the words and pick out the first letter used.
- **Stories** – There are many stories which highlight the days of the week. *The Very Hungry Caterpillar* by Eric Carle and *Jasper's Beanstalk* by Nick Butterworth and Mick Inkpen are good examples. When they are familiar with the stories, the children can sequence pictures from the books into order related to the days of the week and talk about what happened each day.

Unit 6.4 Seasons

Stations

- **Dressing Up Box** – Have a box of children's clothes from all seasons available. Ask the children to choose clothes for a season. **Why did you choose those clothes?**
- **Seasons Walk** – Each season take the children outside for a season senses walk. Focus on what they can see, hear, smell, touch and taste. Take photographs to create a wall display and collect examples of objects for an interest table. Keep the photographs in order that the children can compare different seasons.
- **Season Event Calendar** – Create a calendar which shows main events each season. For example, Hallowe'en and Bonfire Night in Autumn. This can be added to throughout the year and is more meaningful if it contains pictures of the children, such as, Santa visiting in winter.
- **Season Matching** – Use big cards with the season words written on them. Cut out seasonal pictures or use seasonal photographs to look at. As a group, discuss the images and which season they belong to. Ask the children to explain their thinking. Add the pictures to the correct season on the display.

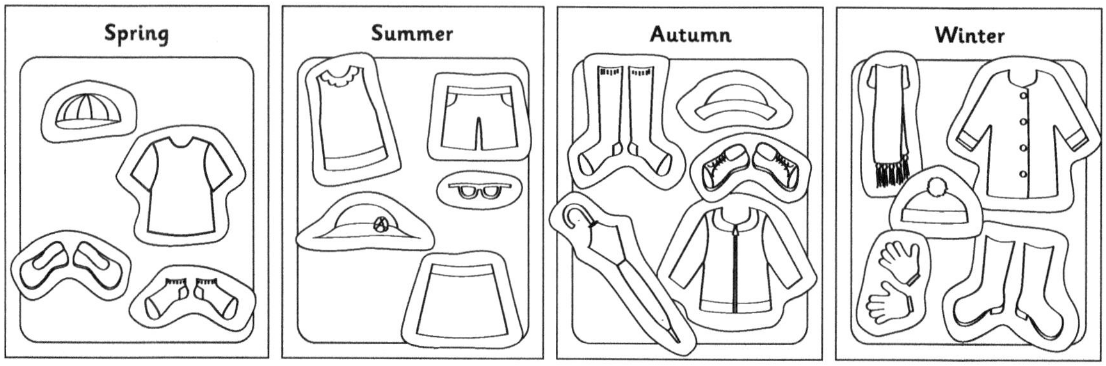

- **Season Mix Up** – Following on from above once the pictures have been sorted for each season, ask the children to close their eyes. Move one of the pictures to the wrong season. When the children open their eyes ask them to find the mixed up picture. **How did you know? Why is it in the wrong place?**
- **Seasonal Trees** – Use finger painting to make trees for each season. Display a tree from each season around a hoop to show the cyclical aspect of the seasons. Discuss which season comes before/after and between. The children can jump around the hoop saying each season they jump on. The children can stand on their favourite season and share why it is their favourite. As an extension with more able children, ask the children to stand on a season then question the children about which season is before/after. The children can physically move to the appropriate season before they answer.

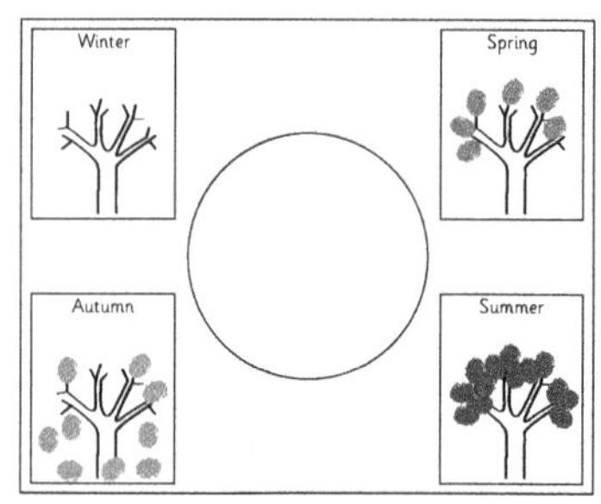

Unit 6.5 Introducing months

Whole class

- **Birthday Display** – Create a display of all the months. Attach a photograph of each child to their birthday month. Discuss how many children have a birthday each month. **Whose birthday is first?** The months can also be linked to seasons by pointing out who has a birthday in summer and so on.
- **Month March** – Line the children up in order of their birth month starting with January. March around the space with the teacher at the front of the line calling out the month and the children repeating. When the children are more secure with the sequence of the months, everyone can say the months together in order as they march. As an extra challenge, the children can raise their hand when it is the month their birthday is in as they march.
- **Month Mexican Wave** – Line the children up in order of their birth month starting with January. Each month group calls out their birth month and stands up, sits down again as they shout it out. Go down the line from January to December to create a Mexican wave effect. When the children are proficient at this you can start with a different month highlighting the cyclical aspect of the months. Ask the children: **Which month is before/after?**

Unit 6.6 Describing and using time devices

Stations

- **How Long?** – Measure how long it takes the children to do an activity using claps. This can be something simple like tidying up or changing their shoes. Vary the tempo of the claps to see if the children notice that sometimes it takes lots of claps to complete the task and sometimes just a few. Develop the discussion to move towards agreeing a 'standard' unit of measure so that the measurement is fair. This could be a metronome or the second hand on the clock.
- **Tuff Tray Experience** – Set up a tuff tray with a wide variety of sand timers, clocks, watches and digital timers. Allow the children to explore and play with them. Be on hand to answer any questions the children have and to explain the devices' features, their uses and how they work. Encourage the children to talk about where they have seen things like this before by relating to their own experiences.
- **Different Clocks** – Look at an analogue clock and a digital clock. Discuss that they both tell the time but in different ways. Discuss the features pertinent to each. Have pictures of different types of clocks from life, these could be famous clocks such as Big Ben, clocks from the nursery, cooker clocks from home, mobile phone display and so on. Ask the children to place the pictures beside the correct clock – analogue or digital. Don't be frightened about using 'big' words with the children, they love it!
- **Clock Shop** – In the role play area, set up a clock shop full of different timing devices for the children to play with.
- **Timed Activity** – In the tuff tray, set up a sand timer and a tub with pom-poms, an empty tub and some giant tweezers. Ask the children to transfer all the pom-poms to the empty tub before the sand runs out. This activity also develops fine motor skills such as the pinch grip for future writing.
- **Tower Race** – Set up a tuff tray with a big pile of interlocking building blocks and a sand timer. Ask the children to race against their friends to see who can build the tallest tower in the time it takes for the sand to run through.
- **Timed Tower Comparison** – This can be extended by using a longer timer and shorter timer. Point out that in more time they were able to make bigger towers and in less time the towers were shorter.
- **Timer Claps** – Count how many claps it takes the children to do an activity. You can get two children to do the same activity and compare the results. For example, **How many claps for Isla to walk over and touch the light switch? How many claps for Amman to do the same? Who was fastest? How do we know?**
- **Beat the Timer** – Using a sand timer, give the children an activity to do and see if they can beat the timer. For example, this may be setting the table for snack time or tidying the toys away. The children could also try to build a tower taller than the teacher's or fill the bucket full of sand.

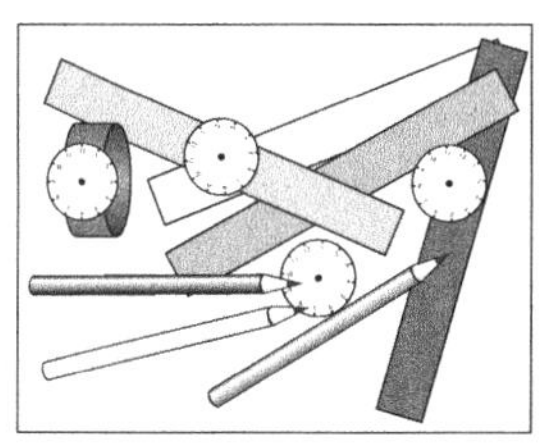

- **Paper Watches** – After looking at a variety of real watches and noting their features, the children can make their own paper watches and decorate them.
- **Giant Clock** – Use a tuff tray as the clock face. Make play dough numbers and place them around the tray in the appropriate places. If the children need support making the numbers, then put laminated cards of the numbers onto the tray ready for the children to cover over with the play dough. The children can then roll more play dough into sausage shapes to make the long and short hands.
- **Laminated Clock Faces** – Provide the children with laminated clock faces where they can write over the numbers with whiteboard pens to become familiar with the features and numbers on an analogue clock.

- **Natural Clock** – In the outdoor area collect natural objects and use them to make an analogue clock.
- **Human Clock** – The children can make a giant human clock. They can form their bodies into the shapes of the numbers and use their bodies as the hands. Use a camera to take pictures to print out and display.

What to observe

- Children who are confident using the appropriate vocabulary. Question the children to uncover the children's thinking process.
- Children who are beginning to talk about time and the part it plays in their life.
- Children who are using the days of the week in sequence.
- Children who can talk about the months of the year.
- Children who can discuss the four seasons and features related to each.
- Children who can start identifying timing devices around them.

Telling the Time B

Section overview

Units of learning	Review units	Prior learning	Benchmarks/Es&Os
Unit 6.1 **Sequencing days of the week**	**An Introduction to Time A** Unit 6.3	• I can say the days of the week.	**MNU 0-10a** • Links daily routines and personal events to time sequences. • Names the days of the week in sequence, knows the months of the year and talks about the four seasons in relevant contexts. • Recognises, talks about and where appropriate, engages with everyday devices used to measure or display time, including clocks, calendars, sand timers and visual timetables. • Reads analogue and digital o'clock times (12 hour only) and represents this on a digital display or clock face. • Uses appropriate language when discussing time, including before, after, o'clock, hour hand, minute hand. I am aware of how routines and events in my world link with times and seasons, and have explored ways to record and display these using clocks, calendars and other methods.
Unit 6.2 **Sequencing months of the year**	**An Introduction to Time A** Unit 6.5	• I can say the months of the year.	
Unit 6.3 **Time to the hour**	**An Introduction to Time A** Units 6.2 and 6.6	• I can use the language of time. • I can recognise timing devices.	
Unit 6.4 **Time to the half hour**	**An Introduction to Time A** Unit 6.6 **Telling the Time B** Unit 6.4	• I can tell the time to the hour.	
Unit 6.5 **Draw the analogue time**	**An Introduction to Time A** Unit 6.6	• I can use the language of time. • I know the features of an analogue clock. • I can tell the time to the hour. • I can tell the time to the half hour.	
Unit 6.6 **Draw the digital time**	**An Introduction to Time A** Units 6.2 and 6.6	• I can use the language of time. • I know the features of a digital clock. • I can tell the time to the hour. • I can tell the time to the half hour.	

Conceptual context

The children should have started to develop the language of time and to have become familiar with days, months and seasons. The children should have some understanding of the cyclical aspect of time and be familiar with some timing devices. This will have been achieved through engaging with the activities in section A throughout preschool and drawing on real-life experiences.

As the children develop skills in literacy, they will begin to connect words with the days, months and seasons. The children's understanding of units of time will develop and refine through further experience. The children will extend their knowledge of time and timers into analogue and digital clocks and be able to tell the time to the hour then half hour. During this unit children's real-life experiences should continue to be drawn upon and naturally arising events should be highlighted and maximised in the children's learning.

Resources

- Sand timers
- Clocks – analogue, digital, laminated
- Washing line
- Hoop
- Motorised clocks – large, individual
- Lollipop stick clocks
- Tuff tray
- Play dough

Digital

- Clock

Record book

Pages 20–21

Key vocabulary

- **Today**
- **Tomorrow**
- **Yesterday**
- **Day**
- **Month**
- **Year**
- **Season**
- **Second**
- **Minute**
- **Hour**
- **O'clock**
- **Half past**
- **Before**
- **After**
- **Earlier**
- **Later**
- **Analogue clock** – clock with moving hands
- **Digital clock** – clock with numerals only
- **Big hand** – minute hand
- **Little hand** – hour hand

Misconceptions

Potential misconception	Addressing the misconception
Units 6.3, 6.4 and 6.6 The children use the familiar base ten to base their calculations.	• Time is not calculated on base ten. It is important that the children understand that time does not fit in to the metric scheme of measures and that its units are built up on several different bases. The children have to become familiar with the units based on sixty and twelve.

Teaching ideas

Unit 6.1 Sequencing days of the week

Whole class

- **Daily Calendar** – Days of the week and months of the year songs can be sung daily before completing the calendar. The daily calendar should have: Today is … The month is … The weather is … The date is __/__/____. Underneath a laminated grid showing the days of the week across the top allows the children to write in the date on the correct day showing how the days progress through the month and when to move on to a new month. This provides scope for lots of discussion on the calendar and how to interpret one. A real calendar (Month to view) placed beside the homemade one for reference is also beneficial. At the end of the month you can ask the children a range of questions such as, **Which day did May begin? How many days are in May? What day was the 5th of May? How many Tuesdays were in May? What day was the last day in May?** This activity can start small at the beginning of the year and be developed as the children progress throughout the year.
- **Visual Timetable** – The children should use a visual timetable to show the day's events. There should also be a weekly timetable highlighting weekly events and special activities. This promotes the use of time specific language as well as recognising the days of the week words.

	9.00	10.30	12.30	3.00
Monday			Gym	
Tuesday	Assembly			
Wednesday			Gym	
Thursday		Music		
Friday	Nursery visit			

Stations

- **Washing Line** – The children should now become familiar with recognising the days of the week words. Knowing the sequence of the days of the week well will help. When the children know their initial sounds, they can read the words by looking at the initial sound. M for Monday, T for Tuesday, W for Wednesday, Th for Thursday and F for Friday. The weekend words both start with S therefore the children will have to look at the first two sounds, Sa for Saturday and Su for Sunday. These should be written in a different colour ink to draw the child's attention to the appropriate letter. When the children's phonics are more developed reading the whole word will be possible. The children can then peg the days of the week cards up in the correct order. As an extension, start the order from a different day of the week. Turn a card around and ask the children to say which one it is.
- **Sequence Cards** – Write the days of the week on strips of cards and cut up. Ask the children to put the cards in the right order. At the beginning the children can do this activity in pairs to support each other. One partner can close their eyes while the other partner hides a day behind their back. **Which one is missing? How did you know?**

Unit 6.2 Sequencing months of the year

Whole class

- **Daily Calendar** – As mentioned earlier, using the daily calendar and extending to months when the children are ready is a great way to include time in to the daily routine.
- **Birthday Display** – The children can add photographs of themselves or write their name on the months of the year to create a birthday display. As with the days of the week, start to look at the words.
- **Human Month Line Up** – Give 12 children a month card each. Shout out January and the child with that card lines up first. Then February and so on until all the months are in order. Keep chanting the months in order while the child with that card holds it up. Other children can swap with the child holding a month. For example, 'Finlay go and hold September.' When the children can do this, you can repeat with the month before/after/between. Ask the children: **What is the third month? The month after …? The month before …? The month between …? The month Santa comes?**

Stations

- **Month/Season Sort** – Lay out the months of the year around a hoop and talk about the features. Pick out the months of the year cards and discuss which season they match to. Lay these in an outer circle around the hoop. The children can then talk about the cyclical aspect. The children can illustrate each month with seasonal pictures.

Unit 6.3 Time to the hour

Whole Class

- **One Minute Activities** – Use a one minute sand timer to see how many jumps, hops, claps, etc., the children can do within one minute. Point out that on an analogue clock the 60 little marks around the clock face are each one minute. There are 60 minutes in one hour. Explain this on an analogue clock pointing out the movements of the hands. Look at the second hand and explain that there are 60 seconds in a minute. Point out the same features on a digital clock.
- **One Minute!** – Using an analogue clock with a second hand ask the children to close their eyes then raise their hand when one minute is up. It is good to do this activity in pairs with one partner keeping an eye on the second hand. The children can discuss how early or late they were before swapping over.
- **Visual Timetable** – Make a new daily visual timetable to include o'clock times. The children can place all the activities they do at the right time. When doing activities in the class, refer to the time and get the children to look at the class clock. For example, it is 11 o'clock, time to go to assembly. Or tell the children we are going to assembly at 11 o'clock and ask them to tell you when it is time to go.

Stations

- **Stop Watch** – Explain the features of a stop watch and the concept of sixty seconds in a minute, sixty minutes in an hour. Choose a child and ask the child to say stop when they think an amount of seconds have passed. Then show them the stop watch and see if their timing was accurate or not. The children can also do this activity in pairs once they are proficient with the use of timers.
- **Sand Timer** – The children can build their own sand timer using two plastic bottles, sticky tape and sand. The children can then use their timers to time themselves doing various activities.

- **Paper Plate Clock** – After looking at a range of analogue clocks and discussing their features, the children can make their own clock using a paper plate, paper fastener and pre-cut hands.
- **What's the Time?** – Use a large/digital analogue clock. Tell the children when the big hand points straight up it is o'clock. Link the big hand to previous work on minutes. 'The little hand is the hour hand. The number (hour) it points to tells us the o'clock time.' Keep language consistent. Use a motorised clock and show how the hands change when moving on to the next hour. Keep setting the clock to different o'clock times. Give the children thinking time, then ask a child to tell you the time. Discuss their answer. **How did you know?** Or show where they went wrong. Link the times to what they might be doing. For example, if it is nine o'clock, the school bell would be ringing for the start of the day. This is the ideal time to point out there are two nine o'clock times in day. Introduce a.m. and p.m.

- **Set the Time** – In pairs one child says an o'clock time. The other sets a mini clock to the given time. Check and swap.
- **Match the Time 1** – Give the group a number of clock faces set to o'clock times and cards with the corresponding o'clock times written in words (4 o'clock) (see Photocopiable resource 6.2). The children take it in turns around the group to match the cards.

Photocopiable resource 6.2 – Match the Time 1.

- **Washing Line O'Clock Order** – Give the children a range of o'clock analogue clock faces (from one o'clock to twelve o'clock or just a selection). One child picks up a clock and says the time then pegs it on to the washing line. The next child picks up a time. Says the time then places it on the washing line either before or after the time(s) already up. The children continue to do this until all the times are pegged up in order.
- **One Hour Earlier/Later** – Have a range of analogue clocks showing all the o'clock times. Have the labels 1 hour earlier, now and 1 hour later prepared and laid out on the carpet. Ask a child to find a time and place it below the now label. Then ask another child to find the time one hour earlier and place it below the 1 hour earlier label. Then another child to find one hour later. Discuss their answers looking at an analogue clock or give them the clock to support their thinking. This can be done with just before or after then mixing it up to challenge the children.

- **Digital Time to the Hour** – Introduce the digital display for o'clock times. Show digital o'clock times and ask the children to say the time. The children could then also show the matching analogue time on mini clocks.
- **Matching O'Clock Times** – Give the children a range of o'clock times in analogue and digital forms. The children take it in turns to match the corresponding clocks saying the time to their group. This activity can also be done as a memory game. The cards should be placed face down in rows and turned over to see if they match. If the children get a matching pair, they keep them; if not, they turn the cards back and play moves on. The game continues until all the times have been matched.

Unit 6.4 Time to the half hour

Whole class

- **Half a Clock** – Make an analogue clock and cut it in half (two equal parts). Discuss half of 60 is 30 and relate this to the minutes on the clock. The children could also colour each half of the clock in two different colours. Marking 60 and 30 minutes on the clock.
- **What's the Time?** – Using a large/digital analogue clock. Tell the children when the big hand points straight down it is half past. Link the big hand to previous work on minutes highlighting thirty minutes in half an hour. The little hand is the hour hand. The number (hour) it has just gone past tells us the half past time. Keep language consistent. Use a motorised clock and show how the hands change when moving on to the next hour and half hour. Keep setting the clock to different half past times. Give the children thinking time then ask a child to tell you the time. Discuss their answer. **How did you know?** Or show where they went wrong. Link the times to what they might be doing. For example, if it is half past seven, time for bed. Point out there are two half past seven times in a day. At half past seven in the morning, the alarm may be going off to get up. When the children are secure, mix up o'clock and half past times.
- **Visual Timetable Part 2** – Make a new daily visual timetable to include o'clock and half past times. The children can place all the activities they do at the right time. When doing activities in the class, refer to the time and get the children to look at the class clock. For example, it is 11 o'clock, time to go to assembly. Or tell the children we are going to gym at half past nine and ask them to tell you when it is time to go.
- **Digital Time to the Half Hour** – Introduce the digital display for half past times. Show digital o'clock times and ask the children to say the time. The children could then also show the matching analogue time on mini clocks.

Stations

- **Set the Time** – In pairs one child says a half past time. The other sets a mini clock to the given time. Check and swap.
- **Match the Time 2** – Give the group a number of clock faces set to half past times and cards with the corresponding times written in words (half past 4) (see Photocopiable resource 6.3). The children take it in turns around the group to match the cards.
- **Washing Line Time Order** – Give the children a selection of o'clock and half past analogue clock faces. One child picks up a clock and says the time then pegs it on to the washing line. The next child picks up a time. Says the time then places it on the washing line either before or after the time(s) already up. The children continue to do this until all the times are pegged up in order. This activity can be extended with the children then placing pictures of what they do at certain times throughout the day beside the corresponding time.

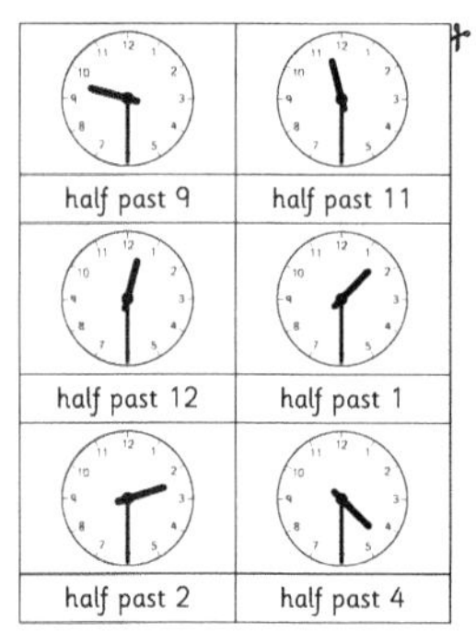

Photocopiable resource 6.3 – Match the Time 2.

- **Matching Half Past Times** – Give the children a range of o'clock times in analogue and digital forms. The children take it in turns to match the corresponding clocks saying the time to their group. This activity can also be done as a memory game. The cards should be placed face down in rows and turned over to see if they match. If the children get a matching pair, they keep them; if not, they turn the cards back and play moves on. The game continues until all the times have been matched.

Unit 6.5 Draw the analogue time

Whole class

- **Draw the Hands** – On an interactive whiteboard have a number of clocks with the hands missing. Ask the child to draw on the hands for the time. Ensure you can clearly see the big and little hand.
- **Show Me** – Say a time. The children set their individual mini clock to that time and hold it up for the teacher to check on the 'show me' command.

Stations

- **Draw the Time** – Give the children individual laminated clocks with missing hands. Ask them to draw the hands on for the given time.
- **Play Dough Times** – Give the children laminated clock faces and a piece of play dough. Ask them to show the time making sure they have the big hand and little hand in the correct place.
- **Time Dice** – Roll a dice with digital o'clock times. The children have to show the time the dice lands on using their laminated clock. This can be a clock where the children set the hands or a laminated one where the children draw on the hands. Repeat this activity for half past times before mixing both.

Unit 6.6 Draw the digital time

Stations

- **Write the Matching Time** – Give the children individual laminated clocks set to a time. In the digital blank space below ask them to write the time.
- **Play Dough Digital** – In a tuff tray have a range of o'clock and half past times shown in words and analogue form. The children match the two and in play dough make the corresponding digital time.
- **Lollipop Clocks** – Say an o'clock time. The children draw the hands on the analogue side of their lollipop clock and write the digital time on the other. They then hold up their lollipop clocks for checking, flipping over to show both sides.
- **Digital Order** – Give the children a range of analogue time cards with the corresponding digital time missing. The children write in the missing digital time then order the time cards on the washing line.
- **Time Story** – Talk about your day saying the times and events. The children write the digital time and draw the event beside it. An example of this would be, at 8 o'clock I got in my car. At half past eight, I arrived at school. The children would write 8.00 on their whiteboard with a picture of a car and 8.30 with a picture of the school. At 12.30 the children ate their lunch and at 4.00 they went to the shop.

What to observe

- Children are familiar with the language of time and using these terms correctly.
- Children can sequence the days of the week.
- Children can sequence the months of the year.
- Children can name the seasons and discuss their features.
- Children are confident with the features of an analogue clock. That they understand the hours and minutes and their corresponding hands.
- Children can show the same time in different forms – words, analogue and digital.
- Children can confidently show o'clock times.
- Children can confidently show half past times.

7 Measurement

An Introduction to Measurement A

Section overview

Units of learning	Review units	Prior learning	Benchmarks/Es&Os
Unit 7.1 Experiencing situations where measurement is used		• From their lives the children will have an ever expanding level of vocabulary and a range of personal experiences to draw upon.	**MNU 0-11a** • Share relevant experiences in which measurement of length, heights, mass and capacities are used, for example, in baking. • Describe common objects using appropriate measurement language, including tall, heavy and empty. • Compare and describe lengths, heights, mass and capacities using everyday language, including longer, shorter, taller, heavier, lighter, more and less. • Estimate, then measure, the length, height, mass and capacity of familiar objects using a range of appropriate non-standard units. I have experimented with everyday items as units of measure to investigate and compare sizes and amounts in my environment, sharing my findings with others.
Unit 7.2 Comparative language	**An Introduction to Measurement A** Unit 7.1	• The children will have experiences of measure from their lives and have started to investigate this in their play using descriptive language.	
Unit 7.3 Opposites	**An Introduction to Measurement A** Unit 7.2	• I can use comparative language.	
Unit 7.4 Using non-standard units	**An Introduction to Measurement A** Unit 7.1	• I can understand when measure is used.	

Conceptual context

This concept is based on understanding the need for measure in relation to length, height, weight, mass and capacity. The children should develop the language involved and understand that measure can give them a better understanding of the properties of objects so they can make comparisons and decisions.

Initially, children have a relational understanding of measurement and develop language to describe differences in size (big/small, etc.). Preschool children use their perception to decide on size, who is taller or 'Can I fit under that table?' They then begin to use direct comparison. Direct comparison is a valid but limiting form of measurement. Only items that can be moved or lifted can be compared. Children often need help to understand that the ends must be lined up for a fair comparison. Often called matching, children can use this method to order items in size by direct comparison.

'Conservation' of measurement is a key aspect of a child's understanding. Young children often believe that if an object is moved its size changes. They need plenty of experience to observe that length, area, volume, mass and capacity can be conserved.

The next step that children take is to use a third object to compare two others. This can be achieved by using string or body parts, for example. This helps children to understand that measurement not only *compares*, but that this can be achieved by using some 'unit'. This is still a comparison skill, but children are building a repertoire of units or objects that they can use in the future to compare *and* estimate. By developing the skills of comparison, both direct and indirect, children are developing a conceptual sense that measurement helps to determine size.

Once we begin to use indirect measurement, we soon realise that there needs to be some element of uniformity in measurement to help us communicate with others. Non-standard units of measurement help children to compare objects and share their answers. The following key principles are introduced:

- Measurements are expressed by counting the total number of units used;
- The unit must not change during the measurement activity;
- The units of measure used are not absolute but are chosen for their appropriateness.

Children will become familiar with these measurement principles through activities such as baking and through play such as playing with toys in the water tray. Language should be introduced and developed during these activities. The children should be exposed to lots of comparisons, opposites and ordering as described in the following activities.

Resources

- Opposites sorting cards
- Texts involving measure such as 'Goldilocks and the Three Bears', 'Jack and the Beanstalk' and *The Smartest Giant in Town* by Julia Donaldson.
- Role play area and resources
- Baking resources
- Different sized clothes and shoes
- Balance scales
- Tubs – various sizes
- Funnels
- Spoons
- Water tray
- Dried rice
- Lentils
- Paper strips
- Ribbon/string
- Cubes
- Objects - various weights
- Natural objects
- Wool/string
- Tuff tray

Digital

- Treasure Measure
- Capacity, Weighing

Record book

Pages 22–23

Key vocabulary

- **Length** – the size of an object or the distance between two points
- **Weight** – how much matter its in an item including the effect of gravity, therefore the weight of an object will change depending on its location
- **Mass** – how much matter is in an object, its mass will never change
- **Capacity** – the amount an object can hold
- **Small**
- **Medium**
- **Large**
- **Bigger than**
- **Smaller than**
- **Holds more**
- **Hold less**
- **Longer than**
- **Shorter than**
- **Full**
- **Empty**
- **Half full**
- **Big**
- **Little**
- **Long**
- **Longer**
- **Short**
- **Shorter**
- **Tall**
- **Small**
- **Thick**
- **Thin**
- **Heavy**
- **Light**
- **High**
- **Low**

Misconceptions

Potential misconception	Addressing the misconception
Unit 7.2 The children do not understand the language of measure.	• When introducing measure highlight and explain the language used and take every opportunity to consolidate the language in daily experiences.
Unit 7.4 The children may notice one characteristic and not realise the specific characteristic linked to the type measure, for example, a tall and thin container may lead them to say it holds more than a short, wide container as they are focussing on height not capacity.	• Give the children lots of experience of different objects and keep their focus on the specific area of measure and their related characteristics.
Unit 7.4 The children do not use non-standard measures accurately, for example, for length and capacity.	• Teach the children the correct way to measure, for example, for length ensure the children measure the whole object from end to end. When moving the measure ensure they mark it correctly or for capacity ensure they fill the measure object each time. Discuss the fact it may not be an exact amount and therefore must use the word 'about'.

Teaching ideas

Unit 7.1 Experiencing situations where measurement is used

Whole class

- **Role Play** – Provide role play areas where the children can use different sizes and measures. For example, in the shop add scales to weigh the fruit and vegetables or in the science lab include lots of containers to fill with different substances. Set up a role play area linked to stories such as *The Smartest Giant in Town* by Julia Donaldson or 'Jack and the Beanstalk' with giant plates for the giant and small plates for the children and so on. Adults can scaffold the play modelling the language.
- **Stories** – Read stories involving size such as 'Goldilocks and the Three Bears'. A role play area can be set up to show the different sizes while the children act out and retell the story. A prop bag can be made to go along with the story with small, medium and large spoons and bowls, chairs and beds as well as the toy characters and a box of porridge.
- **Dress Up Box** – Have a dressing up box with all styles and sizes of clothes and shoes for the children to explore. Encourage the children to use appropriate language in their play.
- **Cooking** – The children should be involved in baking (cheese scones with 2½ cups of self-raising flour, 2½ cups of grated cheese and 2 cups of milk) or cooking (soup with cups of stock, vegetables and lentils, a pinch of salt, pepper and herbs) to experience measure and the need for accuracy.
- **Growing** – Notice how tall the children are and how they have grown and changed. Compare their heights against each other. Plant some seeds and notice how the plants grow taller. Compare their plant to another. Bean seeds are great as they grow quickly. Also, a nice link to the 'Jack and the Beanstalk' story.

Unit 7.2 Comparative language

Stations

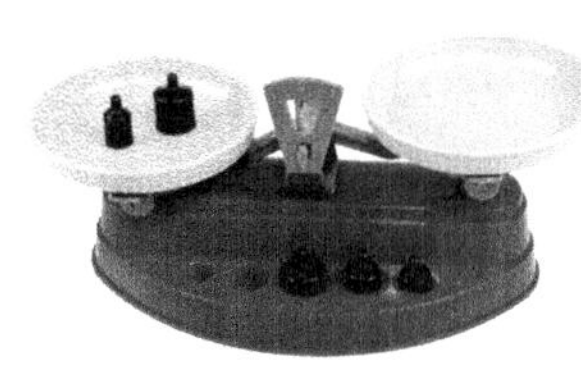

- **Balance Scales 1** – Set up a station with balance scales and a variety of objects of differing weights. Use the scales to find out which is heaviest/lightest. Ask the children to guess first by holding and comparing the objects. Test in the scales to see if they were correct.
- **Capacity** – Set up a station with various containers, spoons, scoops, funnels and so on. Use dried coloured rice or lentils and let the children explore the capacity of the containers. Ask: **Which container holds most/least? How do you know?** Ask the children to make the container full, empty or half full. **Which containers hold the same amount? How do you know?**

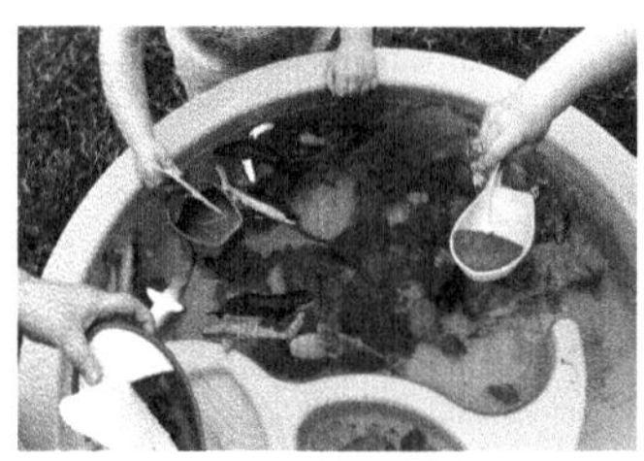

- **Water Play** – In a water tray have lots of different sized containers, tubes and funnels. Let the children explore the containers. Ask: **Which container holds most/least? How do you know?** Ask the children to make the container full, empty or half full. **Which containers hold the same amount? How do you know?**

- **Messy Wool** – Have a bundle of different coloured wool cut into different lengths. Ask a child to pull out two different colours and compare. **Which is longest/shortest?** Continue round the group. **Can you find something around the room longer or shorter than your piece of wool?**
- **Height** – Pick two children and compare their height. If the children are sensitive about this, use dolls instead.
- **Hand/Footprints** – Get the children to print their hand or foot. Cut them out carefully. Pick two and compare. **Which one is the biggest/smallest?** The hand/footprints can be ordered from smallest to biggest.

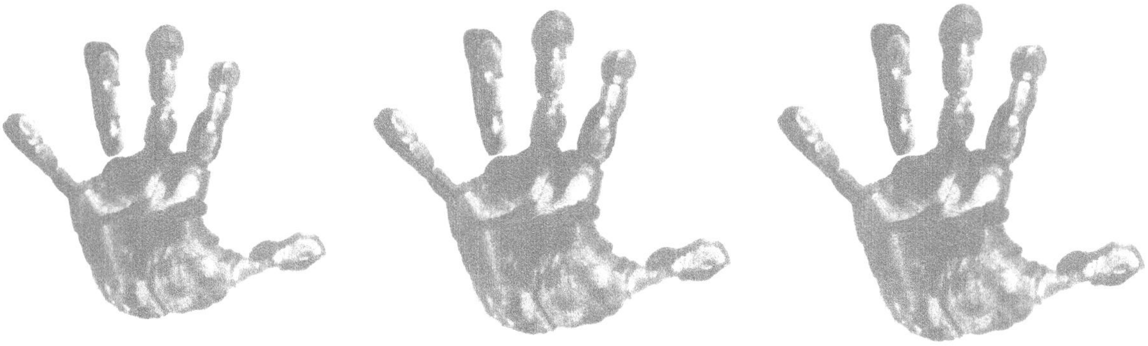

- **Washing Line** – The children hang up the clothes in order of size from the biggest to the smallest. These can be dolls clothes from various sized dolls. The children can then select the correct size of clothes to dress each doll.
- **Line Up** – Put toy dinosaurs in order from the longest to the shortest or smallest to tallest and so on. Other groups of objects can be used such as cups, spoons, zoo animals and so on.

Unit 7.3 Opposites

Stations

- **Picture Cards** – Have lots of opposite picture cards in the middle of the circle. The children pick a card and find the opposite one. Discuss the opposites. Examples are big/little, long/short, tall/small, thick/thin, heavy/light, empty/full, high/low.
- **Pick an Object** – Have lots of different objects in the middle of the group. The objects should be linked to opposites like a feather, a door stop, a thick pencil and a thin pencil, a long string and a short ribbon, a big teddy and a little teddy, an empty fancy box and a see-through jar filled with glitter and so on. Ask the children to pick up an object and describe it. Discuss what the opposite might be. Pick it up and compare.
- **Sorting Hoops** – Have two hoops and write labels for each which are opposite. Find items around the room and outside area which would go in each hoop. Discuss the objects found. For example, if the labels read heavy and light, in the heavy hoop the children might find and put in a stone, a big book, a chair, the fruit bowl and so on. In the light hoop they might find and put in a pencil, a piece of paper, a feather, a leaf and so on.

- **Point To** – Go outside and ask the children to point to a tall flower and a small flower, a long stick and a short stick, a heavy stone and a light stone and so on.
- **Moving Opposites** – In the gym or outdoor space get the children moving in opposites. Run fast then slow, walk forwards then backwards, skip in a straight line then a wiggly line, jump high then low or do a long jump then a short jump and stop then go and so on.

Unit 7.4 Using non-standard units

Stations

- **Balance Scales 2** – The children pick an object and place it in one side of the scales. Then count how many cubes weigh the same. The same item can then be weighed again using another unit of measure, such as buttons or marbles. Discuss.
- **Ribbons** – Use paper clips to measure the length of different ribbons. The ribbons can be measured again with cubes. Discuss.
- **Chalk Children** – Ask the children to lie down on the ground and draw round them with chalk. Then use feet to measure the height. Different children can do this and discuss how the same chalk child can measure a different number of feet.
- **Handspan Printing** – Ask the children to lie down on a piece of a paper and draw round them. The children then put paint on their hands and print their hands along the length of the outline. The children then count to see how many handspans tall they are.
- **Containers** – Let the children find out how many cups of water, sand or lentils it takes to fill a container. The children can then guess how many spoons it will take to fill the same container. **Were they correct?** Repeat for lots of different containers and compare them to each other. **Which held the most/least?**
- **Length** – Set up a table with various objects; these may be random like pencils, toy cars, a book and so on or themed such as different sized dinosaurs. Have tubs of measuring objects for the children to choose from such as paper clips, cubes or buttons. The children can then measure the lengths of the objects and discuss.
- **Outdoor Measures** – The children can find objects to measure and choose suitable natural objects to measure them. For example, they might choose a stick to measure the length of the shed or leaves to measure the height of a flower. **Why did they choose that object to measure with?**
- **Biggest Step** – The children can use chalk to mark their step or how far they jump. They can then measure their step or jump with a stick from the garden. Use the chalk to mark the end of the stick each time and move it forward. Compare their number with others. **Who stepped/jumped the furthest? How do they know?**
- **How Big?** – Find out how big a dinosaur or giraffe or elephant really is. Then use a stick (metre stick but just call it a stick) to measure and draw out the creature in the playground using chalk. This can also be done in the park using items such as sticks, stones and acorns that the children find lying around.

What to observe

- Children who can discuss and use measurement in daily activities and play.
- Children who are confident using the appropriate vocabulary when discussing size and measurement and can make comparisons.
- Children who can explain opposites relating to size and measurement.
- Children who can use non-standard measures to measure length, weight, mass and volume.

Early Measurement B

Section overview

Units of learning	Review units	Prior learning	Benchmarks/Es&Os
Unit 7.1 **Using descriptive language for measurement**	**An Introduction to Measurement A** Units 7.2 and 7.3	• I can use the language of measure. • I can use comparative language.	**MNU 0-11a** • Share relevant experiences in which measurement of length, heights, mass and capacities are used, for example, in baking. • Describe common objects using appropriate measurement language, including tall, heavy and empty. • Compare and describe lengths, heights, mass and capacities using everyday language, including longer, shorter, taller, heavier, lighter, more and less. • Estimate, then measure, the length, height, mass and capacity of familiar objects using a range of appropriate non-standard units. I have experimented with everyday items as units of measure to investigate and compare sizes and amounts in my environment, sharing my findings with others.
Unit 7.2 **Estimating and measuring using non-standard units of measurement**	**An Introduction to Measurement A** Unit 7.4	• I can use non-standard units of measurement for object size, length, mass and capacity.	

Conceptual context

This concept is based on understanding the need for measure in relation to length, height, weight, mass and capacity. The children should develop the language involved and understand that measure can give them a better understanding of the properties of objects so they can make comparisons and decisions.

Initially children have a relational understanding of measurement and develop language to describe differences in size (big/small, etc.). Pre-school children use their perception to decide on size, who is taller or 'Can I fit under that table?' They then begin to use direct comparison. Direct comparison is a valid but limiting form of measurement. Only items that can be moved or lifted can be compared. Children often need help to understand that the ends must be lined up for a fair comparison. Often called matching, children can use this method to order items in size by direct comparison.

'Conservation' of measurement is a key aspect of a child's understanding. Young children often believe that if an object is moved its size changes. They need plenty of experience to observe that length, area, volume, mass and capacity can be conserved.

The next step that children take is to use a third object to compare two others. This can be achieved by using string or body parts, for example. This helps children to understand that measurement not only *compares*, but that this can be achieved by using some 'unit'. This is still a comparison skill, but children are building a repertoire of units or objects that they can use in the future to compare *and* estimate. By developing the skills of comparison, both direct and indirect, children are developing a conceptual sense that measurement helps to determine size.

Once we begin to use indirect measurement, we soon realise that there needs to be some element of uniformity in measurement to help us communicate with others. Non-standard units of measurement help children to compare objects and share their answers. The following key principles are introduced:

- Measurements are expressed by counting the total number of units used;
- The unit must not change during the measurement activity;
- The units of measure used are not absolute but are chosen for their appropriateness.

Children will become familiar with these measurement principles through activities such as baking and through play such as playing with toys in the water tray. Language should be introduced and developed during these activities. The children should be exposed to lots of comparisons, opposites and ordering as described in the following activities.

Resources

- Opposites sorting cards
- Texts involving measure such as 'Goldilocks and the Three Bears'
- Role play area and resources
- Baking resources
- Clothes and shoes – different sizes
- Balance scales
- Tubs – various sizes
- Funnels
- Spoons
- Water tray
- Dried rice
- Lentils
- Paper strips
- Ribbon/string
- Cubes
- Buttons
- Objects – various weights
- Natural objects
- Water tray
- Tuff tray

Digital

- Treasure Measure
- Capacity, Weighing

Record book

Pages 22–23

Key vocabulary

- **Length** – the size of an object or the distance between two points
- **Weight** – how much matter is in an item including the effect of gravity, therefore the weight of an object will change depending on its location
- **Mass** – how much matter is in an object, its mass will never change
- **Capacity** – the amount an object can hold
- **Estimate** – approximate, find something close to the right answer

- **Small**
- **Medium**
- **Large**
- **Bigger than**
- **Smaller than**
- **Holds more**
- **Hold less**
- **Longer than**
- **Shorter than**
- **Full**
- **Empty**
- **Half full**
- **Big**
- **Little**
- **Long**
- **Longer**
- **Short**
- **Shorter**
- **Tall**
- **Small**
- **Thick**
- **Thin**
- **Heavy**
- **Light**
- **High**
- **Low**

Misconceptions

Potential misconception	Addressing the misconception
Unit 7.1 The children do not understand the language of measure.	• When introducing measure highlight and explain the language used and take every opportunity to consolidate the language in daily experiences.
Unit 7.2 The children may notice one characteristic and not realise the specific characteristic linked to the type measure, for example, a tall and thin container may lead them to say it holds more than a short, wide container as they are focussing on height not capacity.	• Give the children lots of experience of different objects and keep their focus on the specific area of measure and their related characteristics.
Unit 7.2 The children do not use non-standard measures accurately, for example, for length and capacity.	• Teach the children the correct way to measure, for example, for length ensure the children measure the whole object from end to end. When moving the measure ensure they mark it correctly or for capacity ensure they fill the measure object each time. Discuss the fact it may not be an exact amount and therefore must use the word 'about'.

Teaching ideas

Unit 7.1 Using descriptive language for measurement

Stations

- **Order Strips** – Give the children cut up strips of paper. Ask the children to order them from the shortest to the longest. Discuss the strips using the words long, longer, longest, short, shorter and shortest.
- **Compare Objects** – Give the children some toys. Ask them to feel the weight. **Which one is heaviest and lightest? Which is tallest and shortest and which are biggest and smallest?** Ask the children to explain their choices and check with the group if everyone agrees.
- **Descriptions** – Give the children different objects to describe. These could be things like a tie, a scarf, a paper-weight, a feather, a jug, a fancy bottle and so on. Ask the children to describe the object, questioning them when necessary to ensure they are using the language of measure.
- **Investigation Stations** – set up tuff trays with a focus on length, weight or capacity. Have lots of objects and non-standard measures, tapes, rulers, balance scales, containers and so on for the children to explore. Pictures and feature words could also be displayed around the area.

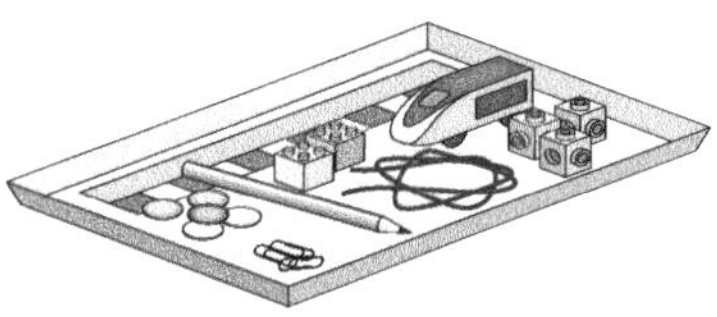

Station for length

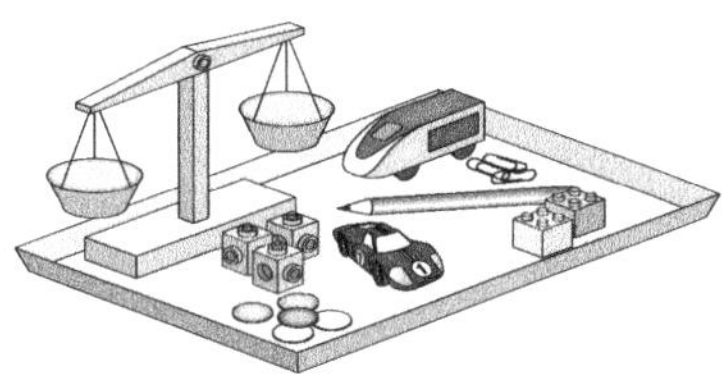

Station for weight

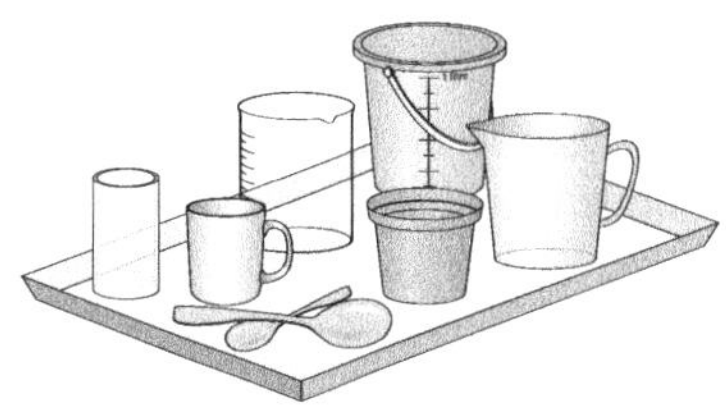

Station for capacity

- **Strings** – Give the children a ball of string and a pair of scissors. They cut a length of the string and then pass the ball of string on. Once all the children have a cut a piece of string the length they want, they can compare them to find out who has the longest piece and who has the shortest. **Are any strings the same length?** The children can then use their piece of string to go and find something in the room that is longer than their piece of string, something that is shorter and something that is the same length. **Is their piece of string longer or shorter than themselves?**

Unit 7.2 Estimating and measuring using non-standard units of measurement

Stations

- **Length** – Have a variety of objects on the table. Ask the children to order the objects from the longest to the shortest. Then use cubes to measure the length of each object and check to see if their estimates were accurate.
- **Different Measures 1** – Measure the length of a variety of objects using cubes. The children can guess and then measure to check how many cubes long the object is. The length of the same objects can then be estimated and measured using other non-standard measures such as buttons or paper clips. The children can use the cube measure to estimate the length with the other objects. Are they getting better at estimating?

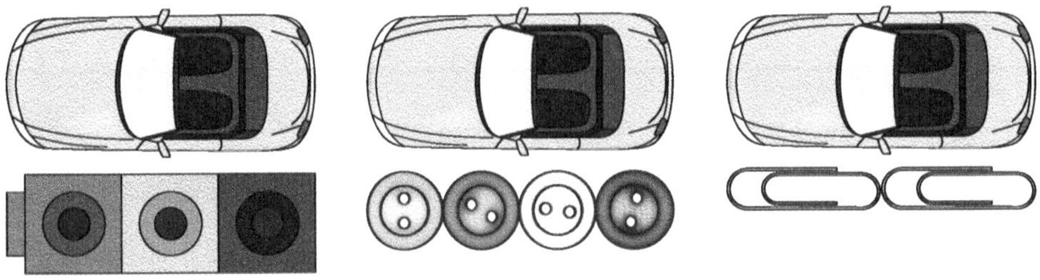

- **Height** – The children can use handspans to measure how tall one of the children is. When they have found how many handspans tall one person is, they can use that knowledge to estimate how many handspans tall other children in the group are. Measure the same child with a child's hand and an adult's hand. Discuss the difference in the answers. **Why do they think the measurements are different?** This will lead on to the children understanding the need for a standard measure. The children can create copies of their hands by drawing round them. Again, they can compare the different sized hands.
- **Big Measures** – Children love measuring big things. Give the children a stick (this could be a metre stick but don't draw attention to the unit just call it a stick) and let them measure the length and width of the classroom or an area of the playground.
- **Weight** – Have a variety of objects on the table. Ask the children to order the objects from the heaviest to the lightest. Then use cubes in the balance scales to measure the weight of each object and check to see if their estimates were accurate.
- **Different Measures 2** – Measure the weight of a variety of objects in the balance scales using cubes. The children can guess then measure to check how many cubes the object weighed. The same objects can then be measured with other non-standard measures such as buttons or lollipop sticks. The children can use the cube measure to estimate the weight using other measures. Are they getting better at estimating? Also, by comparing the measure found using different items to weigh the same object – for example, the car was 16 lollipop sticks, 5 cubes and 18 buttons – this highlights the need for a standard measure.
- **Capacity** – Give the children a number of containers of different shapes and sizes. Ask them to order them from the one that they think will hold the least to the one that will hold the most. Ask the children to explain their thinking. Then use cups of water/rice/lentils to find out if they were correct. Discuss the findings.
- **Different Measures 3** – Ask the children to choose a container. Show them a cup and ask them to estimate how many cups will fill the container. Then measure and discuss the accuracy of their guesses. Then measure the same container with spoonfuls and mugs. Remember to estimate first and discuss their accuracy based on the actual number of cups/spoonfuls/mugs it took. **How accurate were they?**

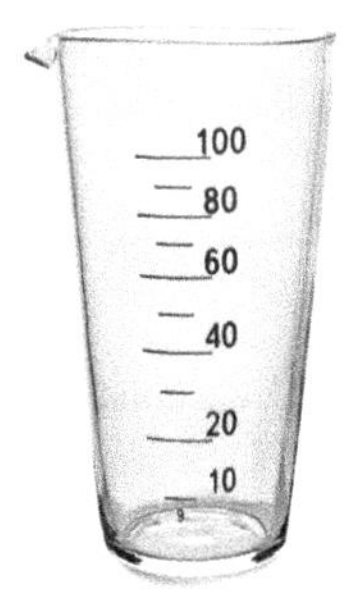

- **Capacity: Tall Versus Wide** – Show the children two containers, one tall and thin and one short and wide. Ask: **Which container will hold most? Why do you think that?** Ask the children how we could prove if we were right. From their ideas (hopefully), take the container the children thought would hold the least and fill it with water. Then pour the water into the other container. If they were right, it will not overflow. **Were they correct?**

What to observe

- Children who can use the language of measurement in relation to length, weight, mass and capacity.
- Children who can compare and order objects in relation to length, weight, mass and capacity.
- Children who can estimate and use non-standard units of measurement for objects size, length, mass and capacity.

8 Patterns and Relationships

Patterns and Relationships A

Section overview

Units of learning	Review units	Prior learning	Benchmarks/Es&Os
Unit 8.1 **Copying simple patterns**		• Children innately look for patterns in their world – as simple as bedtime comes after a story each day, Saturday always comes after Friday or visual patterns, e.g. the arrangement of bricks in a wall. Ensure that children have appropriate indoor and outdoor experiences where they can explore and observe in context. • Children should have experience of pattern in a range of forms – **colour**, **shape**, **object**, **action**, **movement**, **music**, **temporal**, **orientation and size.**	**MTH 0-13a** • Copies, continues and creates simple patterns involving objects, shapes and numbers. • Explores, recognises and continues simple number patterns. • Find missing numbers on a number line within the range 0–20. I have spotted and explored patterns in my own and the wider environment and can copy and continue these and create my own patterns.
Unit 8.2 **Continuing simple patterns**		• Children need to recognise what 'unit' is represented in the pattern in order to continue it.	
Unit 8.3 **Creating simple patterns**		• Children need to understand what a pattern is (repeating unit).	
Unit 8.4 **Developing the language to describe patterns**		• Being able to use/recognise vocabulary related to pattern.	

Conceptual context

A pattern always contains elements that repeat over and over again in a predictable order; a key element of any pattern should be a mathematical rule. Most early pattern experiences have a focus on shape, size or colour and involve using real objects. Children learn, through play, by observing and working alongside others. In order to develop an understanding of pattern, children need to have real-life experiences of pattern and hear the language of pattern; they need an enriched pattern experience to encourage the recognition, copying and creating of pattern. Children should be challenged to create, extend and change patterns. Remember making patterns need not involve lots of concrete objects – patterns can also be made using textures, or using markings in wet sand, mud and snow, and the fact that patterns occur in routines should be recognised too.

Children will begin to learn about patterns, not only across the curriculum but at home and in their wider environment. They will recognise that each nursery day has a pattern in the form of their routines, that pictures and designs have patterns and that our number system has many patterns. Patterns can be visual, auditory, temporal, involve movement and be numerical. Many everyday occurrences can be explored in a mathematical context – songs and poems can be learned by remembering the pattern of music or a string and beads can be used to make a necklace with a focus on repeating colours (even though the initial aim of the activity may have been to develop fine motor skills).

Pattern making is over-arched with a rich vocabulary of shape, colour, size, texture and positional language. Children require a 'hands-on' knowledge of manipulatives involved in pattern making and an awareness of key outside environmental objects that may be involved in pattern making. Remember, talking about vocabulary encourages children's reasoning skills. It is important to introduce new vocabulary (or to consolidate known vocabulary) in a context using visual and concrete materials, indoors and outdoors, or using stories, rhymes and songs. Vocabulary should not become an extra cognitive load.

Teacher directed rhythmic pattern is a very effective way to introduce children to the concept of pattern; it is not necessary for children to master a pattern before moving on. Children will continue to become more confident and improve their understanding if they are given a wide range of experiences and many opportunities to practise.

All practitioners working in the Early/First level stages, should check to ensure that all children are using a pattern to continue a sequence and not just copying the original. Careful questioning will assist in assessment.

Describing pattern is difficult for young children; they need lots of practical experiences scaffolded by supportive adults. Vocabulary is key to developing their mathematical knowledge. Once children tune into patterns that exist all around them many opportunities will arise to talk about patterns in meaningful contexts throughout each nursery day.

Resources

- Ribbons
- Rhymes and songs
- Un-tuned percussion instruments
- Music
- Chunky pens and mark-making materials
- Beads
- Buttons
- Coins
- Coloured pegs and peg boards
- Shapes
- Sand
- Shells
- Construction materials
- Stories (rhythmic patterns repeating chorus in stories, eg. *We're Going on a Bear Hunt* by Michael Rosen)
- Counters
- Cubes
- Nuts and bolts
- Pine cones, twigs and sticks
- Pebbles
- Bottle tops
- Photographs of pattern in the environment
- Textured papers/cards
- Washing line and pegs
- Leaves
- Petals on flowers
- Wrapping paper

Digital

- Leckie – Beads and Laces

Record book

Pages 24–25

Key vocabulary

- **After**
- **Before**
- **Continue**
- **Copy**
- **Create**
- **Different**
- **Make**
- **Matching**
- **Not the same**
- **Pattern** – an arrangement of objects or numbers which follow a rule
- **Printing**
- **Repeating**
- **Same**
- **Set** – a collection of items
- **Weaving**

Misconceptions

Potential misconception	Addressing the misconception
Units 8.1, 8.2, 8.3 and 8.4 Pattern is about colour and design.	• Ensure that children are exposed to correct vocabulary and a range of pattern 'experiences' where the repeating nature of a pattern is clear.
Unit 8.2 Inconsistency in copying a pattern – they may recognise the items in the pattern but don't consistently place them in order.	• Children need support in understanding proximity (nearness). Provide opportunities to copy and make patterns and talk about them using appropriate vocabulary. Use materials that the children are familiar with; if the child cannot accurately recognise colour, then use objects that they are familiar with.

Teaching ideas

Unit 8 Pattern (8.1, 8.2, 8.3, 8.4)

Whole class

- **Rhythmic Patterns –** Group the children so that they are standing or sitting in a circle, as appropriate for the physical pattern. Start with a pattern like:

  hands on shoulders clap **hands on shoulders** clap **hands on shoulders** clap

  Ask the children to copy the pattern. It is vital to check that they all understand what is meant by the commands. Engage the children in a range of physical patterns by acting out and saying the pattern. An extension to this (if the children are mathematically receptive and ready) is to name the parts of the rhythmic pattern using letters.

This is an example of an AB pattern:

hands on shoulders	clap	**hands on shoulders**	clap	**hands on shoulders**	clap
A	B	A	B	A	B

This is an example of an AAB pattern:

Clap, clap, stamp clap, clap, stamp clap, clap, stamp

This is an example of an AABC pattern

Clap hands, clap hands, tap knees, stamp feet, clap hands, clap hands, tap knees, stamp feet, clap hands, clap hands, tap knees, stamp feet

Ask the children to think of actions that we might use in pattern making. **Would they like to become a 'pattern maker'? Could they show their actions to the other children in the circle and think about how they might put their actions into a pattern? Could they stand and lead the children in the circle?**

Whole class

- **Story –** *My Mum and Dad Make Me Laugh* by Nick Sharratt. A humorous look at family dynamics where mum loves all things spotty, dad loves all things stripy and the little boy loves pattern-less grey. A great vehicle to discuss and provoke a search of spotty/stripy and pattern-less objects associated in our lives. **Are the spots actually a pattern or just a design? Are the stripes actually a pattern or just a design?**

  Ask key questions, e.g. 'Look at the front cover and discuss the picture.' **What do you think the book may be about? Why do you think mum likes spotty things? What else could she find that has spots? Where have you seen spots?** 'Dad likes stripes.' **Where have you seen stripes? Why do you think the little boy likes grey clothes?** 'The little boy likes elephants. They match his grey clothes.' **If your favourite animal was a penguin, what sort of pattern might you like?**

- **Exploring patterns indoors –** Use a range of patterned wallpaper/card or pictures depicting patterns; include some flock type papers so the children can feel as well as see the pattern. Before the activity, give the children opportunities to explore and familiarise themselves with the pattern examples to be included in the feely bag. The adult has the samples already hidden in a feely bag. The children should be seated in a circle for this 'Pattern pass the parcel' game. Choose a child to hold the bag until the music starts, then the bag will be passed round the circle from child to child until the music stops. When the music stops the child holding the bag pulls out a sample of paper or card and decides whether it shows a **pattern** or **not a pattern**. Ask: **Can you describe what you see?** This gives an opportunity to use pattern language in context. Sort cards into two piles: 'Pattern' and 'Not pattern'.

- **Indoor group activity** – Use a Tuff tray and place a selection of natural/tactile materials such as pine cones, conkers, small logs, beans, shells, nuts and bolts, etc., in it. It is good for the children to touch and experiment with the objects before making patterns.

  Provide the children with a pattern mat. Pattern mats can be made using a sheet of card and sticking a border of patterned hobby tape round the outer edge. They can be laminated, if required, to enable the use of dry wipe markers.

Craft tape

Craft tape applied to card to form a border

Allow free exploration so that the children can become familiar with the materials and the notion of using the mat to record their pattern on. Children can also be provided with pre-produced cards to copy or cards to extend (see Photocopiable resource 8.1).

Examples of pattern cards.

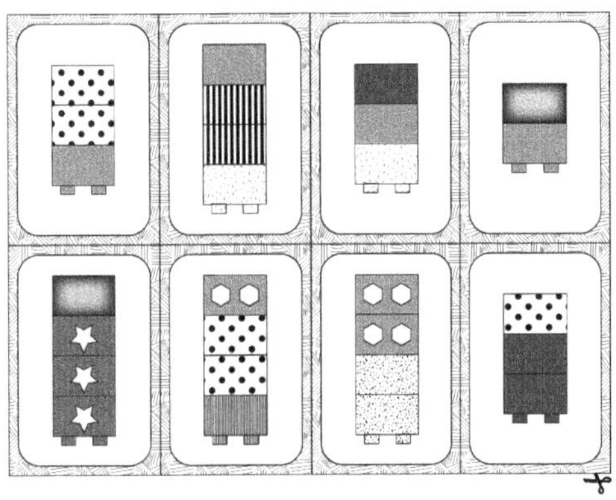

Photocopiable resource 8.1 – Pattern Cards.

When reflecting on the teaching of pattern think about opportunities for creating, describing and progressing and then moving to continuing a pattern. This provides an excellent opportunity for discussion, observation and assessment.

It is important to change materials and patterns regularly; children will quite happily indicate when they would like a change!

By providing a range of different materials there is an element of self-differentiation; the children can 'mix and match' materials coming up with their own categories of pattern and descriptions.

When having a group discussion on what type of pattern has been made, use effective questioning to find out what the children know and can demonstrate. Try to use open questions. **Tell me about the pattern you have made. What did you use? How many different types of objects did you need to make your pattern? Have you seen a pattern like this before? Can you tell Finlay how to make a pattern like yours?** 'Amman, look at Isla's pattern.' **Can you tell us a little about what she has done?**

Whole class

- **Outdoor Walk** – Take the children on an outdoor walk in the local environment/playground/outdoor area, encouraging the children to look for patterns in their local environment. Provide the children with a 'finding finger' or 'pattern picker'.

  Pattern pickers are made using pipe cleaners; you can use different colours of pipe cleaners wound together for finding coloured patterns or different shapes for finding shape patterns. Pattern pickers give the children an aide memoire: **What was it I was looking for again?** Photographs or paper samples can also be used to help find and match patterns – look in decorating stores for some wallpaper samples, e.g. faux brick wallpaper. *Finding fingers* can be used once the child is secure in pattern recognition. These can be used indoors and outdoors where children may be searching for patterns (walls/floor tiles/railings, etc.) or 'checking' patterns that other children have made. Outdoors, children could use sticks, stones, logs, toys, etc., to make repeating patterns.

Encourage the children to think about the patterns that they have spotted on railings, slabs, brick walls, roofs, etc., and record these patterns using photography, rubbings and so on. By using rubbings, the pattern is copied and can be looked at closely indoors. Some of the patterns that the children have seen can be replicated using water (making water lines and blobs using a water bottle), sticks, stones, flowers or other found materials.

Patterns can also be explored outdoors by making natural colour patterns. For this activity you will need to have prepared some white cotton rectangles. (Cut up some old white cotton sheets into pieces of about A4/A5 size) The cotton can be washed at the end of the activity and reused. Fold the cotton in half for the children and then open the cotton so that it resembles a book. Encourage the children to make a pattern with some natural materials (leaves or flower petals are ideal) on one half of the cotton. Once the children have completed their pattern fold over the other part of the cotton and press it down. The cotton now resembles a sandwich – the two pieces of cotton being the bread and the natural pattern materials making the filling. The children can (with supervision) hammer on top of the top bit of cotton, forcing the colours out of the plant material on to the bottom part of the cotton. This creates a lovely natural pattern but watch out for those thumbs!

Stations

The following ideas can be used as stations to engage and develop, using natural curiosity and interest. It is good to introduce the following ideas in a whole class situation so that children are familiar with the materials and the expectations; it is possible then to carry these activities in group sessions or as paired activities. Any materials used should not hinder the child's natural pattern development, e.g. ensure children know how to weave before expecting them to build a pattern round it.

- **Sounds/music** – Make a display of tuned and untuned percussion instruments such as castanet, tambourine, chime bar and drum. Let the children explore the instruments and find out how the sound is made using each one. **Do you need to tap this instrument or shake it? Can you make a loud sound or a quiet sound?** Seat the children in a semicircle so they can all see the pattern and all see you. The 'pattern' can be displayed using 2D shapes adhered to a white board, drawn on the whiteboard or reproduced on a Smart Board.

  The adult, with child input, selects an instrument, e.g. drum. Together they look at the pattern of triangles.

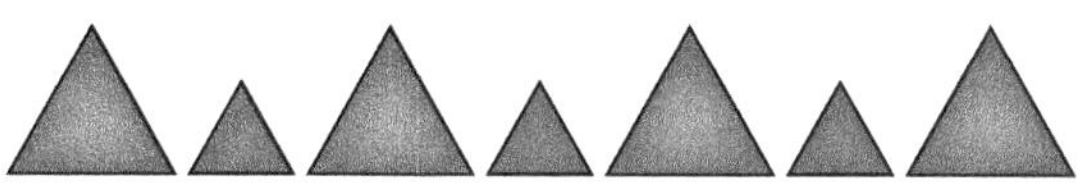

  This pattern is an AB pattern; big triangle, small triangle, big triangle, small triangle ... The adult tells the children that the big triangle signifies a loud sound and the small triangle signifies a quiet sound. The adult demonstrates how to play the sounds. **How many loud sounds do they think they will hear?** The adult uses the drum to 'play' the pattern. The children will hear the pattern – loud sound, soft sound, loud sound, soft sound ... Extend the activity by repeating the pattern using different instruments or with the children joining in using loud claps and quiet claps. The patterns and the instruments can be changed.

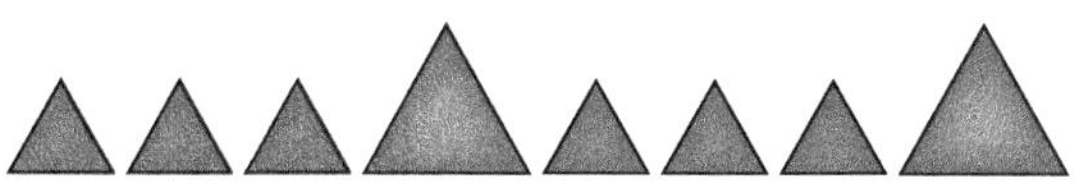

  This is an AAAB pattern. Ask: **What do you think this will sound like? Can you clap it?**

This activity can be developed and extended. More able children could make a pattern using 2D shapes for the adult to play. The ability of the children to draw accurately isn't required because they are substituting shapes for drawings, to show their pattern accurately.

When the children are confident with following the pattern and producing the correct 'volume' of sound, then the activity can become a group activity with the adult becoming the conductor. In the role of conductor, the adult points to each shape moving along the row of shapes slowly. The orchestra (the children with their instruments) follow the pictorial instructions and perform the desired action as the conductor points to the shape, e.g. soft, soft, soft, loud … This is fabulous fun and a great way to develop concentration.

- **Movement** – Ensure that, whatever movements you are going to ask the children to do, they have had experience of doing them and can recognise how to move their bodies into the required positions (see Photocopiable resource 8.2).

  Provide movement cards strips for children to follow on command; use the finding finger to point to the action.

  The physical pattern here is: arms up straight legs, arms up knees bent, arms up straight legs, arms up knees bent. Photograph the children doing actions and compile your own movement strips. Remember to reinforce vocabulary. Ask: **Can you can see a pattern? Are there any movements that repeat?**

Photocopiable resource 8.2 – Moving Body Positions.

- **Printing** – Explore patterns by printing with fruit and/or vegetables. It is also possible to hit leaves and flowers with a hammer onto old cotton. It is however less precise! See Angle, Symmetry and Transformation section for full details.

  In the creative area, place long strips of paper (wide enough to fit the vegetables on), paints and a selection of vegetables. Potatoes, broccoli and carrots tend to make good prints. Cut the vegetables into halves, quarters and slices for the children. Encourage the children to make patterns using different vegetables. Explore the vegetable patterns without paint and then apply paint so the children can make a permanent copy of their pattern. Talk to them about what they have made. Ask them open ended questions: **I wonder what might happen if …? What did you notice …?** 'Explain to me why …' **Could you make this pattern another way?** 'Tell me about it.' For those children unable to make their own pattern, have a vegetable print pattern for them to copy until they develop confidence. Remember to reinforce appropriate vocabulary.

- **Weaving** – This activity is best done outdoors on railings, fences or bamboo canes as these structures will provide examples of horizontal patterns. An adult should tie one end of the ribbon or string to the end post and then begin to demonstrate the weaving action of over and under (over the fence post, under the fence post, etc.); get the children to chant along with you. Children can then explore using their ribbons or strings; a range of colours, lengths and thickness should be provided. (A B A B pattern) Ask: **Can you see a pattern in your weaving?** The pattern is movement related – over and under – and not a colour related pattern.

  Using a chain link fence the children will be able to make horizontal, vertical and diagonal patterns.

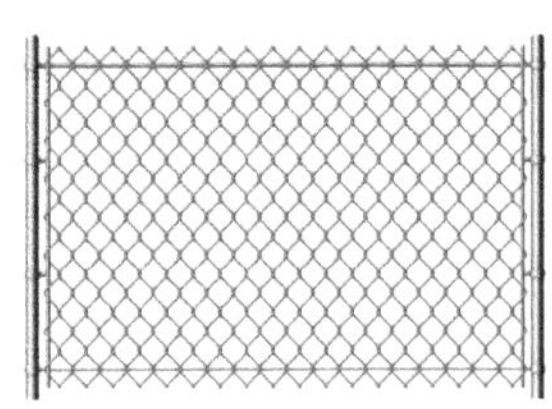

- **Dance Patterns** – A dance is composed of repeated physical patterns and sequences. Using the outdoor environment, it is possible to link objects to moves in a dance.

Take the children for a walk in the area you are going to create your dance and get them to look closely at everything that is in the 'dance arena', e.g. drain covers, grass, trees, bins, lampposts. Tell the children that they are going to create 'dance steps' but different 'dance steps' will happen when they move under a tree or move across a drain cover, for example.

Examples of dance steps:

passing by a lamppost – spin round once
walking under a tree – raise one arm in the air
going over the drain cover – clap hands once
walking on the grass – wave your hands in the air
passing by the bin – do a star jump

Children will follow the same 'steps' in order to make it a repeating physical pattern, i.e. lamppost, tree, drain cover, grass, bin: lamppost, tree, drain cover, grass, bin … For younger children, two or three dance steps may be appropriate. It is a fun, creative activity and becomes more of a Mexican wave dance as the children are not all doing the same steps at the same time although the steps are in the same order.

Discussion and effective questioning should underpin all the pattern experiences that you use.

What to observe

- Children recognising and taking an interest in pattern.
- Children who, when encouraged through effective questioning, will reflect on what they could try next/ how they would work it out and are able to respond appropriately both using concrete materials and with correct vocabulary.
- Children copying a sequence and being able to describe it (A B A B…).
- Children copying a pattern and being able to describe it (A A B C…).
- Children attempting to describe the pattern they made – ensure it is in front of them so they can talk about it.

Assessment opportunities will arise through discussion with the children; encourage them to use appropriate language.

Patterns and Relationships B

Section overview

Units of learning	Review units	Prior learning	Benchmarks/Es&Os
Unit 8.1 **Creating and describing patterns**	**Patterns and Relationships A** Unit 8.1	• Children should have had numerous opportunities (indoors and outdoors) to explore, discover and talk about patterns using appropriate vocabulary. They should be able to copy, describe, continue and create repeated patterns in a variety of contexts and using a range of materials. Opportunities should be provided for personalisation and choice involving design, manipulatives and objects chosen. • They should have an understanding of the terms copy, continue and create. • Children should have experience of growing patterns, through the context of stories.	**MTH 0-13a** • Copies, continues and creates simple patterns involving objects, shapes and numbers. • Explores, recognises and continues simple number patterns. • Find missing numbers on a number line within the range 0–20. I have spotted and explored patterns in my own and the wider environment and can copy and continue these and create my own patterns.
Unit 8.2 **Finding the missing number on a number line**	**Patterns and Relationships A** Unit 8.2	• Children need to be able to recognise numerals and be able to show what each represents e.g. 4 as a numeral, dot pattern, tally mark arrangement, number of legs on a donkey, number of tyres on a car etc.; children should also be able to identify numerals. • Children should be able to count confidently, saying numbers clearly. • Children should have had experience of using number lines and number tracks. • Children should be able to order numbers and be able to state the number before and the number after within a given range.	

Conceptual contexts

Mathematics is the study of pattern and pattern is a key mathematical idea; the search for pattern is an important tool for learning because it helps us to see relationships and make connections. In our teaching we should be trying to deliver mathematics by giving children lots of experiences in pattern in such a way that it allows them to see order and make connections. The ultimate aim is for children to be able to explore why patterns occur and how they can be used, rather than just simply spotting patterns. When children engage with patterns, such as those within our number system, they are engaging with important mathematical concepts.

Children engage with pattern from their earliest experiences and naturally explore it as part of their environment. Adults can help children understand pattern by talking with them about it. It is important that all adults working with children have an appropriate subject knowledge and understanding of the concepts and vocabulary associated with pattern. The adult has the responsibility to intervene in a sensitive manner; importantly this can produce opportunities to introduce and extend the child's understanding of the concept and the vocabulary associated with pattern.

Articulating a pattern, even informally, is a key part of understanding it, and this supports later, more sophisticated thinking such as making predictions about further items in the pattern, stating generalisations and beginning to reason about why these generalisations might be true. Recognising and making sense of patterns enables children to appreciate mathematics as a connected and beautiful subject.

Children should have experience of 3 types of patterns – repeating, growing and number. Young children first need to experience repeating patterns in rhythm, in colour, in design and in arrangement. It is from these experiences that the discovery of pattern in number originates.

Growing patterns are difficult for some children to recognise; they are often overlooked in early experiences. A lovely story to read that involves a growing pattern is *Rooster's Off to See the World* by Eric Carle. In the story, as rooster walks along, animals join him in groups that increase by one animal (two cats, three frogs, etc.). It is a simple introduction to the meaning of numbers and sets.

Children need to say what they see, make patterns for themselves, copy others and be able to predict how they will continue the pattern.

In order for children to embrace **number patterns** they need to be numerate and have a developing basic knowledge and skills in the following areas: counting accurately, remembering key facts, reasoning logically and be able to apply knowledge to derive new information. Children should be able to: recite number names in sequence (recitation), synchronise each number name with one object counted (1:1 correspondence), understand that the last number in a count represents the quantity of the set (cardinality), recognise that the last number in a count represents the place in a sequence (ordinality), understand that the position of a digit in a number affects its value (place value) and instantly recognise the number of objects in a small group without the need to count (subitising). Number line and numeral track activities rely heavily on the children being able to count, order, recognise and identify numerals, recite numbers forwards and backwards and being able to state the number before and the number after.

Resources

- Ribbons
- Rhymes and songs
- Untuned percussion instruments
- Music
- Chunky pens and mark making materials
- Beads
- Buttons
- Coins
- Pegs and peg boards
- Shapes
- Sand
- Shells
- Lolly sticks
- Construction materials
- Bread tags
- Feely bag
- Stories (rhythmic patterns repeating chorus in stories, e.g. *We're Going on a Bear Hunt* by Michael Rosen)
- Natural and produced manipulatives appropriate to pattern development
- Photographs of pattern in the environment
- Textured papers/cards
- Sticks
- Washing line and pegs
- Leaves
- Petals on flowers
- Wrapping paper
- Dish designs
- Books
- Calendars
- 100 squares
- Numicon
- Cuisenaire rods

Digital

- Leckie - Beads and laces
- Leckie - Pattern Tool

Record book

Pages 24–25

Key vocabulary

- **After**
- **Before**
- **Continue**
- **Copy**
- **Create**
- **Different**
- **Growing**
- **Make**
- **Matching**
- **Not the same**
- **Pattern** – arrangement of objects or numbers which follow a rule
- **Printing**
- **Repeating**
- **Same**
- **Set** – a collection of items
- **Weaving**

Misconceptions

Potential misconception	Addressing the misconception
Unit 8.1 Child confuses colours in patterns.	• This may be due to colour blindness. If this is the case, use shape pattern instead of colour pattern.
Unit 8.1 Difficulty with reversing patterns.	• Children lack left-right co-ordination until they are about 5 to 6 years old. Using a mirror to observe the reflection and careful pattern comparison their ability will improve.
Unit 8.1 Children often repeat what is being given to them rather than looking at the chunk that is being repeated.	• Lots of experience of making repeating patterns and talking about them; get the child to look closely at the pattern and describe the 'chunk' that is to be repeated.

Teaching ideas

Unit 8.1 Creating and describing patterns

Whole class

- **Story** – *Have You Seen My Dragon?* by Steve Light. This activity session involves the teacher using a story to interrogate the understanding a child has about pattern while weaving it into other mathematical contexts.

  Read the story together. This is a counting book but the quality of illustrations on each page are superb; there is incredible detail throughout the book. Patterns need to be looked for carefully but it is such a lovely book it would be a shame to miss it out for another book where pattern was the sole focus! Inside the front and back covers are 'maps' of New York showing the places where the boy visited while hunting for his dragon. Each place is numbered and is a great recall activity, though the map does give small pictorial clues beside the map location number. Look at the illustration of the dragon. Ask the children about the pattern of scales that make up the dragon's skin. Ask: **Have you seen a pattern like this in nature?** (Fish, pine cones, snakes, pangolins, overlapping roof tiles.) Discuss with the children the use of line drawings to create pattern, e.g. under the caption 3 Buses, point out the pattern: window, brick design, window, brick design … **What type of pattern is this?** (AB) On the pages relating to 16 carriages, ask for the children to think about pattern related to the trains. **What are the repeating units?** It provides a fabulous vehicle to highlight things which are pattern and which are not pattern; together scrutinise pictures and try to see if there is an element that repeats. If there is no repetition, it is not a pattern, it is a design. Manhole covers in the story exhibit cyclical patterns. Challenge the children to draw their own cyclical pattern for a manhole cover (provide circle-shaped card and black markers). This would introduce cyclical pattern in a relevant context and might link to work being done on rotational symmetry. This is a book that keeps on giving and can be read time and time again; each visit to the story brings new discoveries.

  Read *Pattern Fish* by Trudy Harris. This brightly coloured pattern book, cleverly written in rhyming prose, is about a range of creatures that live in the sea. The creatures inhabit a sea full of patterns; the patterns begin with AB patterns and develop to illustrate more complex ones. The illustrations express and reinforce the patterns of the words. The story is about six creatures moving in the sea and a rather naughty shark who scares them away. On each two-page illustration, every element from the decorative frame to the background characters share the same pattern. An end note adds some information about patterns but also some super activities to do with the book. This is a fabulous must-read book!

For example, the third sea character that is met in the story is a Seahorse. His rhyme is:

> chomp-chomp-munch-munch,
> chomp-chomp-munch-munch,
> A seahorse underneath the sea,
> Eats seaweed for its lunch,
> chomp-chomp-munch-munch, chomp-chomp-munch-munch, chomp-chomp-munch-munch…

The pattern suggested in 'chomp-chomp-munch-munch' is AABB. All the illustrations on these pages relate to this pattern: sea grass, the bubbles coming out of fishes' mouths, pebbles, markings on the arms of a starfish and the border surrounding the picture.

Trudy Harris has also written *Pattern Bugs* which is an equally worthwhile read. She again cleverly uses the sounds in the verses, the details of the pictures and the border colours to entice children to find and identify repetitive patterns.

Stations

- **People Patterns** – Group activity. Seat the children in a circle on the floor. Before starting the people pattern activity, check that each child understands the commands that may be used in the pattern, e.g. hands up/hands down; sit/stand. Begin a people pattern by having the children do a particular action. Point to each child in order as you say hands up, hands up, hands down, hands down, hands up, hands up … The first child adopts the pose of hands up and holds that position. The second child also adopts the pose hands up and holds. The third child is asked to model hands down and hold the pose. Ask: **Can you see what has happened? How would you describe this pattern in terms of ABCs?**

Other examples to use: stand and sit; front and back; stand, sit and kneel. It is important for many children to actually 'feel' and see the pattern.

Use Lego figures to demonstrate a Lego person pattern, e.g. arms up, arms up, arms down … For some children this may aid visualisation.

- **Growing Patterns** – This practical activity builds on the children's experiences developed in 'People Patterns'. The children will experience growing patterns using a physical activity. Sit the children in a line so they can all observe the front of the classroom and see the developing pattern from the same viewpoint. Ask a group of children to come and stand in front of the other children; they are going to demonstrate their pattern. Start the pattern for the children to respond by doing the appropriate action – stand, sit then stand, stand sit, then stand, stand, stand sit. In our line of children, we would then have a child standing, then a child sitting, then two children standing and one child sitting, then three children standing and one sitting. **Could they predict what would come next?** Another example may be carried out in the gym hall or playground. Have a series of hoops available, laid out in a row. In the first hoop there is one child, in the second hoop there will be two children in the third hoop three children, and so on. Ask: **Can you tell how many children there will be in the next hoop? How many in the tenth hoop?**

  'One elephant went out to play' is an example of a growing number pattern in a song.

- **Pegboard Patterns** – This activity is best carried out round a table with children seated. Ensure that each child has a peg board with a repeating pattern of coloured pegs along the top. The same sequence can be at the top of each board initially. Ask the children to describe the position of the pegs in the sequence. **Where is the red peg? Is the green peg after the yellow peg? Which colour of peg does the pattern start with/finish with?** Get them to copy the pattern on the row below.

  Ask the children to look away and alter the position of one of their pegs from the original sequence. Children now look at the pattern boards. Ask: **What has changed? Is the first peg still red? Which peg is now between blue and green?**

  The children can then explore their own patterns using the pegs or work in pairs to make patterns for their partner to copy or continue.

  Numicon base boards and pegs can also be used for this activity. Children who do not have good manual dexterity find the Numicon pegs easier to use.

Ideas of materials to use to develop patterns – sponge squares, pasta pieces, bottle tops, multi-link cubes, match stick patterns, coins, bread tags.

Unit 8.2 Finding the missing number on a number line

Whole class

- **Story** – *Absolutely One Thing* by Lauren Child. The story is based on Charlie and Lola going on a trip to the shop. Mum says they can pick one thing each at the shop. This animated story provides plenty of opportunities to practise counting and calculating simple operations, while corresponding numeric labels and number sentences make explicit the tallying and calculations. It is a wonderful stimulus to springboard into a maths lesson. Some critical maths knowledge is open for discussion particularly in terms of cardinality.
- **Reciprocal Counting** – Encourage the children to count clearly and carefully. Reciprocal counting can help children extend their range counting forwards and provide support when counting backwards. It can support children's counting while crossing decades (forwards and backwards). The adult and children take turns at saying numbers. The adult can say sequences of 1, 2, 3 or 4 numbers and the children follow on saying the next 1, 2, 3 or 4 numbers. Remember it is important to not always start at zero or one. Counting backwards should be included.

  Adult begins counting at 8, for example. The adult says 8, 9, the children in unison say 10, 11 then the adult says 12, 13. The children are getting a chance to hear the numbers in order. Children, when confident, can become the 'teacher' and lead the activity.
- **Looking for Patterns** – Display a 100 square on the whiteboard (Smartboard/Promethean board). Look for patterns. Numbers ending in zero; numbers starting with 5. Get the children to come to the board and circle or cover the number. Hide some of the numbers on the square. Ask: **Do you know what number is hidden?** Use the wipe clean 0–100 square in the pack, one for each child or one between two. Ensure that children recognise the terms 'column' and 'row'.

  What patterns can you see in the columns? In the rows? What number is above 16? Which numbers are on either side of 24?

  Ask the children to cover numbers ending in a zero or a 5, for example. **Can you see a pattern?**

  Children need to develop a fluency relating to 'number after' and 'number before' (one more than/one less than). This should progress to recognising 10 more than/ 10 less than.
- **Numeracy Memory Walk** – This activity can be introduced as a whole class lesson but can then be used as a group activity once the children can operate independently. The numeracy path shown here was made using good quality backing paper. Along the path there are pebbles with numbers written on them with a permanent marker. You can use number cards if required. My walk is from 1 to 10 but the number of pebbles can be decreased/increased as appropriate. The first task is to ask a child to put the pebbles along the path in order starting at 1. Next to each pebble the child then places an object – have a selection of interesting objects ready in a bag/box. You can see from the picture that number 1 pebble has a dog beside it and number 2 pebble has a coin beside it. The purpose of the activity is for the children to be able to recall the items in order. As a class, talk about the items you pass going from 1 to 10. Initially 1–5 might be sufficient! Cover the path and see what the children can recall. The activity can be different every time – think about changing the objects that are put beside the numbers (pebbles) and think about using different number ranges. The children could be going on a walk that starts at 9, a walk where they jump in 2s or 5s. This may alter depending on the children involved and what numbers would they need for the walk.

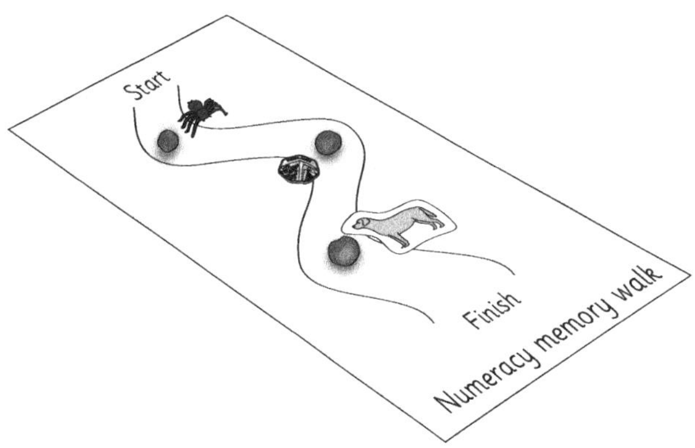

Stations

- **Making Number Patterns with Numicon** – In a group situation encourage the children to explore Numicon shapes. Then ask a series of questions. **Can you put the Numicon shapes in order (1–10 or 10–1)? Can you make a pattern (odd/even) using the Numicon shapes? Can you describe your pattern? Can you extend your pattern? Could you make a growing pattern?**
- **Number Lines and Number tracks** – A range of commercially produced number lines and number tracks are shown here. Children should be familiar with both lines and tracks and have experience of using them.

  A **number line** is a line marked with numbers that is a useful visual aid for calculations and showing relationships between values.

  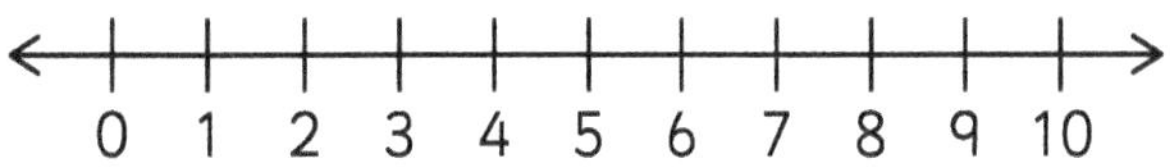

  A **number track** is a rectangular grid of squares which can be useful for counting, numeral identification and numeral order.

1	2	3	4	5	6	7	8	9	10

- **Counting On and Counting Back** – This game can be played with two or three children (two is best). Each child has a coloured cube. They place their cubes at 0. The children take it in turns to spin the spinner and move their cubes the corresponding number of jumps – it is vital that children are jumping on/back and not counting the number they are starting on. Winner is first to 10. If the number line is on a wipe clean board the children can extend the line marking in the numbers and play first to 15 or 20.

  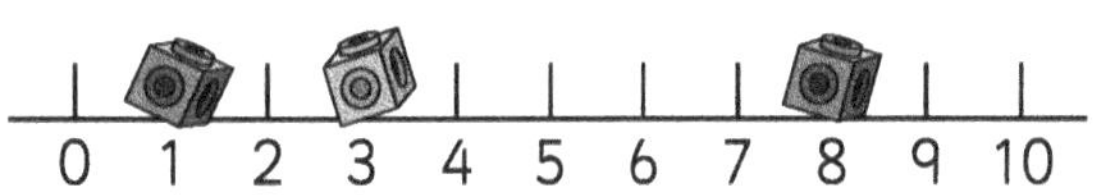

- **Ordering numbers** – A favourite resource is a book of raffle tickets. Instantly you have pairs of numbers of varying ranges. The can be laminated to preserve them. Useful for pairs, ordering, number recognition, place value, matching and sorting. For this activity the range is going to be 1–20. Draw a 0–20 numberline on backing paper or use a commercially produced one. Remove a number of tickets (1–20) from the book and place them in a feely bag. Ask the children to take turns at pulling out a ticket and placing it on the numberline. Ask questions about their ticket number. **Is your number before or after 6? What number comes after your number?** The activity can be made more challenging by having an empty numberline; the children need to estimate where their ticket can be placed. **Is your number nearer 20? Does your number come after 13? Where might it be placed on the line?**

What to observe

- Children who can construct their own patterns.
- Children who can explain their pattern.
- Children who can predict what will come next.
- Children developing an awareness of growing patterns.
- Children who have confidence in number order and who can state the number before and the number after.
- Children who can recognise and identify numbers.
- Children developing a sense of pattern involving numbers.
- Children using the correct vocabulary in the correct context.

9 2D Shapes and 3D Objects

An Introduction to 2D Shapes and 3D Objects A

Section overview

Units of learning	Review units	Prior learning	Benchmarks/Es&Os
Unit 9.1 **Describing the properties of shapes and objects using informal language**		• From their lives the children will have an ever expanding level of vocabulary and a range of personal experiences to draw upon.	**MTH 0-16a** • Recognises, describes and sorts common 2D shapes and 3D objects according to various criteria, for example, straight, round, flat and curved. I enjoy investigating objects and shapes and can sort, describe and be creative with them.
Unit 9.2 **Sort shapes and objects by describing features**	**An Introduction to 2D Shapes and 3D Objects A** Unit 9.1	• The children will have experiences of shape from their lives and have started to investigate this in their play using informal language.	

Conceptual context

Children will arrive at nursery or school with little knowledge of shapes and their properties, so they will need many opportunities to play with shapes and objects and learn the associated vocabulary. The world of young children is filled with 3D objects like building blocks, cardboard boxes and toys, so it is reasonable to start their shape work with 3D objects rather than 2D shapes. To make links between 3D objects and 2D shapes you can use the faces of cuboids and prisms as printing blocks to create shapes to discuss.

When sorting and classifying shapes we often focus on aspects that are the same rather than what is different, e.g. saying a rhombus is like a pushed over square. Use questions like **What is the same?** and **What is different?** to promote deeper thinking. When learners come to classify shapes and objects, they should have opportunities to use their own criteria *and* criteria specified by others. There are many criteria that we can use to classify shapes: shapes that are good for building, shapes that can tile, shapes that have curved lines. As the pupils get more experience, they should begin to include the following in their explanations:

- the similarities and differences between shapes;
- that the same shapes can be sorted in different ways;
- that classifying shapes has a purpose, i.e. we can tidy the equipment away by looking at the shapes of the objects.

As such this concept is based on the children being able to identify and name common 2D shapes and 3D objects. They will build up their understanding of the language involved in shape and use it to describe the properties and features. The children will be able to use this knowledge to sort, classify and match shapes.

Resources

- 2D shapes – different sizes, materials, colours and orientations
- 3D objects – both true shapes and natural 3D objects such as balls and tubes
- Sorting hoops/tray
- Paint
- Feely bag
- Lollipop sticks
- Egg timer
- Art materials such as coloured paper, glue, googly eyes, wool, junk and so on
- Play dough
- Shape cutters

Digital

- Animated song – Shape Song
- Game – Falling Shapes
- Tools – Shape set

Record book

Page 26

Key vocabulary

- **Circle**
- **Triangle**
- **Square**
- **Rectangle**
- **Oval**
- **Sphere** – a geometrical solid 3D shape that is perfectly round, three dimensional and circular
- **Cylinder** – a solid 3D shape with two flat circle faces and a curved face
- **Cone** – a solid 3D shape that tapers from a flat circular base to a point
- **Cube** – a solid 3D shape that has six square sides or faces
- **Cuboid** – a solid 3D shape with six rectangular sides or faces
- **Pyramid** – a solid 3D shape that has either a square or triangle base with triangle faces which come to a point
- **Faces** – the flat surfaces or sides of a 3D shape
- **Flat**
- **Curved**
- **Straight**
- **Round**
- **Corners**
- **Side**
- **Edges** – the line that joins two corners in a 3D shape
- **Pointy**
- **2D** – flat shapes having only two dimensions
- **3D** – three dimensional shapes

Misconceptions

Potential misconception	Addressing the misconception
Unit 9.1 The children do not understand the language of shape.	• When introducing shape and their properties highlight and explain the language used and take every opportunity to consolidate the language in daily experiences.

Teaching ideas

Unit 9.1 Describing the properties of shapes and objects using informal language

Stations

- **Investigate Shapes** – The children should be allowed to explore different shapes and objects in their play. Use this to extend their vocabulary introducing words like round, curved, straight, flat, edges and corners. Let the children lead their learning but maximise opportunities to ask them questions which help extend their learning and deepen their understanding.
- **Story Books** – Books are a lovely way to introduce shapes. Some titles to look for are: *Changes Changes* by Pat Hutchins, the Stanley books by William Bee, *Round is a Tortilla* and *Round is a Mooncake* by Roseanne Thong, *Colour Zoo* by Lois Elhert, *Shape Shift* by Joyce Hesselberth, *Shape by Shape* by Suse MacDonald, *Circle Rolls* by Barbara Kanninen, *Not a Box* by Antoinette Portis, *Apples and Robins* by Lucie Félix and *Shapes that Roll* by Karen Nagel.
- **I Spy a 2D Shape** – Say, 'I spy a square' (hold up a square so the children know what a square is, if this game is being played early in their shape learning). The children guess lots of square things they can see like a window, a book and so on until they guess correctly. It helps if they can then touch and discuss the item, comparing them to the square you held up. Repeat for other shapes.
- **I Spy a 3D Object** – Say, 'I spy a cube' (hold up a cube so the children know what a cube is, if this game is being played early in their shape learning). The children guess lots of cube shaped objects they can see like a tissue box, a dice and so on until they guess correctly. If helps it they can then touch and discuss the item, comparing them to the cube you held up. Repeat for other shapes.
- **Feely Bag 2D** – Put different 2D shapes in the feely bag showing them all to the children as you place them inside and naming each one. There can be multiples of the same shape in different sizes and colours. Limit it to two different shapes at first, then more different shapes can be added. The children then put their hand in the feely bag, pull out the shape and name it. Once the children are confident naming their shapes, they can try and guess the shape by feel alone before pulling it out of the bag and seeing if they were correct.
- **Feely Bag 3D** – Put different 3D objects in the feely bag showing them all to the children as you place them inside and naming each one. There can be multiples of the same object in different sizes and colours. Limit it to two different object types at first, then more different object types can be added. The children then put their hand in the feely bag, pull out the object and name it. Once the children are more confident naming the objects they can try and guess the object by feel alone before pulling it out of the bag and seeing if they were correct.

- **Compare Shapes/Objects** – Hold up two shapes/objects which are the same shape/object but not identical. Ask the children to describe what is the same and what is different. For example, 'They are both red but this one is pointy' or 'They both have straight sides but this one is smaller'.
- **Shape Walk** – Take the children on a walk looking for different shapes/objects in the environment. This can be quite general at first, looking for any shapes/objects the children know. The children might just say a brick, for example, and need help to name the shape. It can also be specific looking for a shape or object they have been learning about. The children can look for 2D shapes or 3D objects. If looking for a specific shape or object, the children can take an example of that shape or object with them and compare the shapes/objects they find in the environment encouraging the use of informal descriptive language. You can also discuss why that shape was used where it was; for example, **Why was the brick used to build the wall?** Making the children think about the properties of the shape.
- **Make a 2D Shape** – Outside using natural objects ask the children to make a 2D shape. Show the children the 2D shape they are being asked to make, name it and discuss the properties. Ask: **How many sides does it have? How many corners?** Count them together then let the children try to make their own shapes using materials they find.

- **Back Drawing** – The children sit in pairs. One child turns their back to the other child and shuts their eyes. The teacher holds up a 2D shape. This can be limited to a choice of two at first and extended to all the 2D shapes when the children are more confident. The second child, using their finger to draw, draws the shape the teacher held up on their partner's back. Then ask the child to turn around and open their eyes. They share what shape they think was drawn on their back and why, for example, 'I thought it was a triangle because it had three sides' or 'I think it was that shape', pointing to the appropriate picture 'because it was round'. Check if they were correct by holding up and discussing the shape the children were drawing. Swap over and repeat as many times as you like.

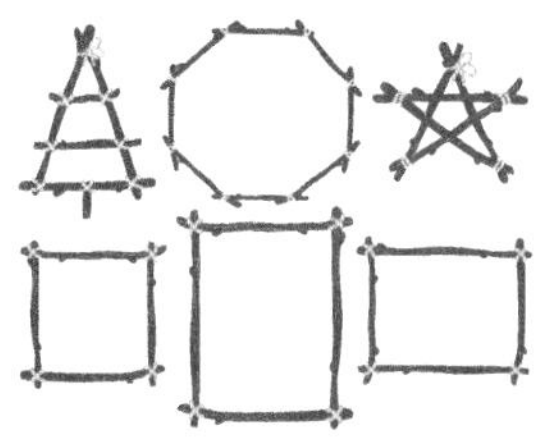

- **2D Shape Printing** – Choose a 2D shape. Look at and explore all the different types of the same shape. Discuss their properties and what is the same and different between the shapes, this is especially important with triangles which come in different forms. The children do not need to know the names of all the different types of triangles they just need to recognise them all as triangles and what they have in common – three straight sides and three corners. The children can use the shapes dipped in paint to print a page of the same shape. Encourage the children to do them all over the page and discuss how it is the same shape even if it is upside down or on its side.

- **2D Lollipop Shapes** – Pick a 2D shape and discuss it. Then help the children count out the correct amount of lollipop sticks to make the 2D shape. The children make the 2D shape using the lollipop sticks. At first the children can do this over the top of a laminated shape template.

- **2D Shape People** – Pick a paper 2D shape and discuss it. Give the children art resources to turn their shape into a shape person. Meet Mr Triangle and Ms Square.
- **Robots** – Give the children lots of junk. Ask them to make a robot. The children then describe their robot being encouraged to name some of the objects they used and discuss some of their properties. The junk can be collected as part of a recycling project.

- **Shape Hunt** – Pick a focus for the hunt, a 2D shape or 3D object. Show it to the children, name it and ask the children to describe some of the properties as they feel it. Then ask the children to hunt around the room to find as many of the same shapes/objects as they can and bring them back to the group. Turning a sand timer over to limit the hunt time makes it more exciting. Hiding lots of the shapes/objects before the hunt also adds interest, making sure lots of good examples are found. Once all the shapes/objects are gathered, discuss what they have in common with each other by asking: **What is the same? What is different?**
- **Shape Display** – Create a shape display on a table top with lots of words and pictures, shapes and objects (manufactured and natural). The children can also bring in shapes/objects from home to discuss and add to the display.
- **2D Play Dough Shapes** – Provide the children with 2D shape cookie cutters in different sizes to cut out 2D shapes. The children can name and play with the 2D shapes they make. Ask: **What shapes have you made? How do you know?** Encourage them to discuss the properties.
- **Shape Song** – To the tune of 'If You're Happy and You Know It', give each child a different shape to hold. Then sing the song encouraging the children to stand up at the appropriate time in the song.

  If you're holding up a square, stand up now,
  If you're holding up a square, stand up now,
  If you're holding up a square and you really want to show it,
  If you're holding up a square, stand up now.

Unit 9.2 Sort shapes and objects by describing features

Stations

- **Sort by Size** – Give the children lots of the same 2D shape or 3D object in different sizes. Ask the children to sort them by size, describing the features of the shape/object as they do.
- **Straight or Curved** – Give the children lots of 2D shapes. Ask them to sort the shapes into two hoops. One for shapes with straight edges and one for shapes with curved edges. Discuss each shape in turn and explain why they are putting the shape in that hoop.
- **Three or Four** – Give the children lots of 2D shapes. Ask them to sort the shapes into two hoops. One for three sided shapes and one for four sided shapes. Discuss the shapes in each hoop.
- **Two Hoops** – Give the children lots of 2D shapes or 3D objects and two hoops. Ask the children to sort the shapes/objects into the two hoops. Ask: **How did you decide to sort them? Do all the shapes fit their criteria? Were there any shapes left out? Why?** For example, these shapes are good for rolling.
- **Stack, Roll or Slide** – Give the children lots of 3D objects to explore. Ask them: **Which shapes could you use to build a wall? Which shapes would roll? When do they roll? How do they roll? Do any slide? Do any shapes fit more than one criteria?**
- **Sand Hunt** – Hide 2D shapes and 3D objects in the sand. The children hunt for all the items and put them into a 2D shape container or a 3D object container. Discuss their choices.

What to observe

- Children who can describe the properties of shapes and objects using informal language.
- Children who can sort shapes and objects by describing features.

Name and Discuss the Properties of 2D Shapes and 3D Objects B

Section overview

Units of learning	Review units	Prior learning	Benchmarks/Es&Os
Unit 9.1 **Recognising and naming common 2D shapes**	**An Introduction to 2D Shapes and 3D Objects A** Unit 9.1	• I can use informal language when describing and discussing shapes.	**MTH 0-16a** • Recognises, describes and sorts common 2D shapes and 3D objects according to various criteria, for example, straight, round, flat and curved. I enjoy investigating objects and shapes and can sort, describe and be creative with them.
Unit 9.2 **Recognising and naming common 3D objects**	**An Introduction to 2D Shapes and 3D Objects A** Unit 9.1	• I can use informal language when describing and discussing shapes.	
Unit 9.3 **Recognising the properties of shapes and objects using mathematical language**	**Name and Discuss the Properties of 2D Shapes and 3D Objects B** Units 9.1 and 9.2	• I can recognise and name 2D shapes. • I can recognise and name 3D objects.	
Unit 9.4 **Describing the properties of shapes and objects using mathematical language**	**Name and Discuss the Properties of 2D Shapes and 3D Objects B** Units 9.1, 9,2 and 9.3	• I can recognise and name 2D shapes. • I can recognise and name 3D objects. • I can discuss the properties of shapes and objects using mathematical language.	
Unit 9.5 **Using properties to describe, sort and/or match shapes and objects and justify choice**	**Name and Discuss the Properties of 2D Shapes and 3D Objects B** Units 9.1, 9,2, 9.3 and 9.4	• I can recognise and name 2D shapes. • I can recognise and name 3D objects. • I can discuss the properties of shapes and objects using mathematical language. • I can describe the properties of shapes and objects using mathematical language.	

Conceptual context

Children will arrive at school with little knowledge of shapes and their properties so they will need many opportunities to play with shapes and objects and learn the associated vocabulary. The world of young children is filled with 3D objects like building blocks, cardboard boxes and toys so it is reasonable to start their shape work with 3D objects rather than 2D shapes. To make links between 3D objects and 2D shapes you can use the faces of cuboids and prisms as printing blocks to create shapes to discuss.

When sorting and classifying shapes we often focus on aspects that are the same rather than what is different, e.g. saying a rhombus is like a pushed over square. Use questions like **What is the same?** and **What is different?** to promote deeper thinking. When learners come to classify shapes and objects, they should have opportunities to use their own criteria *and* criteria specified by others. There are many criteria that we can use to classify shapes, shapes that are good for building, that can tile, that have curved lines. But as the pupils get more experience, they should begin to include the following in their explanations:

- the similarities and differences between shapes;
- that the same shapes can be sorted in different ways;
- that classifying shapes has a purpose, i.e. we can tidy the equipment away by looking at the shapes of the objects.

As such this concept is based on the children being able to identify and name common 2D shapes and 3D objects. They will build up their understanding of the language involved in shape and use it to describe the properties and features. The children will be able to use this knowledge to sort, classify and match shapes.

Resources

- 2D shapes in different sizes, materials, colours and orientations
- 3D objects, both true shapes and natural 3D objects such as balls and tubes
- Sorting hoops/tray
- Paint
- Feely bag
- Lollipop sticks
- Art materials – coloured paper, glue, googly eyes, wool, junk and so on
- Play dough
- Pipe cleaners
- Camera
- Spaghetti
- Sticky notes
- Headbands

Digital

- Animated song – Shape Song
- Game – Falling Shapes Tangrams
- Tools – Shape set

Record book

Page 26

Key vocabulary

- **Circle**
- **Triangle**
- **Square**
- **Rectangle**
- **Oval**
- **Sphere** – a geometrical, solid 3D shape that is perfectly round, three dimensional and circular
- **Cylinder**
- **Cone** – a solid 3D shape that tapers from a flat circular base to a point
- **Cube** – a solid 3D shape that has six square sides or faces
- **Cuboid** – a solid 3D shape with six rectangular sides or faces
- **Pyramid**
- **Faces** – the flat surfaces or sides of a 3D shape
- **Flat**
- **Curved**
- **Straight**
- **Round**
- **Corners**
- **Side**
- **Edges** – the line that joins two corners in a 3D shape
- **Pointy**
- **2D** – flat shapes having only two dimensions
- **3D** – three dimensional shapes

Misconceptions

Potential misconception	Addressing the misconception
Unit 9.1 The children do not understand the language of shape.	• When introducing shape and their properties highlight and explain the language used.

Teaching ideas

Unit 9.1 Recognising and naming common 2D shapes

Stations

- **Name the 2D Shape** – Share all the 2D shapes with the children naming each one. Hold the shapes up quickly one at a time and the children shout out the name of each shape. Mix up the order the shapes appear in and hold up two of the same shape in a row to try and catch the children out.
- **Feel the 2D Shape** – Recap the name of each shape. Then ask a child to put their hands behind their back. Place a shape in their hand. The children feel the shape and try to name it.
- **2D Shape Book** – Take the children on a 2D shape walk. Discuss the shapes they see and take photographs of each one. Back in class, print off the photographs and create a shape book with a page for each different shape.
- **2D Shape Words** – Introduce the shape words and ask the children to match each shape to the word.

- **Print 2D Shapes** – Section a piece of A3 paper into areas with a shape name written in each section. The children find the appropriate shapes and print them in the correct sections.
- **Sort the 2D Shapes** – Give the children a selection of 2D shapes, commercial and/or natural. The children sort the shapes explaining what shape is in each group.
- **2D Fastest Finger** – Lay all the 2D shapes out in front of each child. Make sure there is one of each 2D shape. Ask the children to point to a named 2D shape. **Who found it first?**
- **2D Shape Bingo** – Give the children a base card with a selection of 2D shapes drawn on it (see Photocopiable resource 9.1). Call out a 2D shape. If the child has that shape on their board, they cover it with a counter. The winner shouts 'Bingo' when they have covered all their shapes. Check the board by asking the child to repeat back all their shapes.

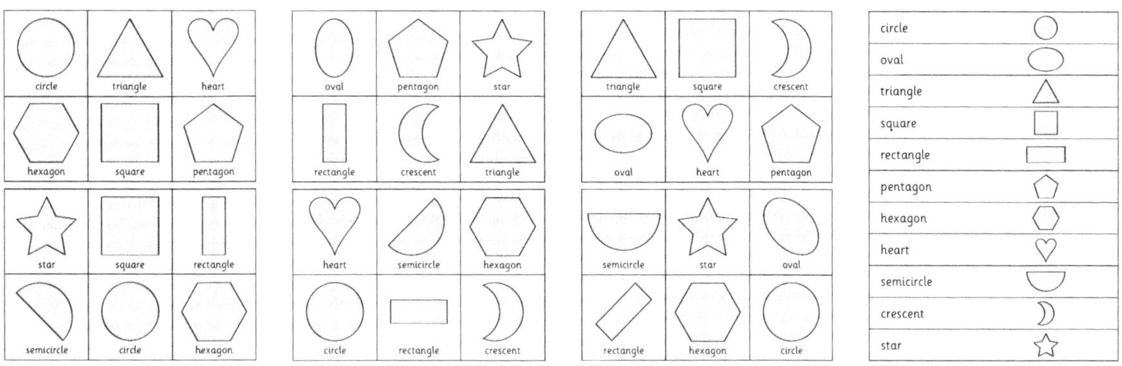

Photocopiable resource 9.1 – 2D Shape Bingo Board.

- **Draw a 2D Shape** – Give the children a whiteboard and pen. Ask the children to draw a 2D shape.
- **2D Shape Picture** – Give the children lots of cut out 2D shapes in different sizes and colours. Let the children use the 2D shapes to make a picture. The children explain what shapes they used in their creation. Ask: **What shape did you use for that part? How many circles did you use?** The children can also do this outside with natural materials or inside with commercially produced plastic shapes and then take a photograph to display and discuss.

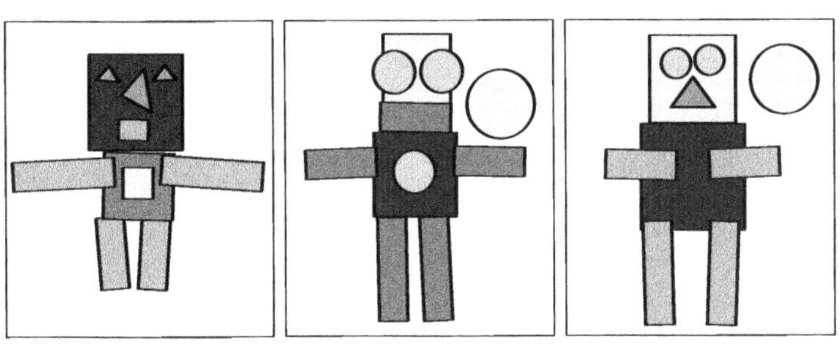

- **What's the Missing Shape?** – The children have a tray with one of all the 2D shapes on it. They shut their eyes and one child removes a shape. The children open their eyes and try to guess which shape is missing.

Unit 9.2 Recognising and naming common 3D objects

Stations

- **Name the 3D Object** – Share all the 3D objects with the children naming each one. Hold the objects up quickly one at a time and the children shout out the name of each shape. Start with only a few objects then as they become confident with them introduce more. Mix up the order the objects appear in and hold up two of the same shape in a row to try and catch the children out.
- **Feel the 3D Object** – Recap the name of each object. Then ask a child to put their hands behind their back. Place an object in their hand. The children feel the object and try to name it.
- **3D Object Book** – Take the children on a 3D object walk. Discuss the objects they see and take photographs of each one. Back in class, print off the photographs and create a 3D object book with a page for each different object.
- **3D Object Words** – Introduce the object words and ask the children to match each object to the word.
- **Sort the 3D Objects** – Give the children a selection of 3D objects, commercial and/or natural. The children sort the objects explaining what object is in each group.
- **3D Fastest Finger** – Lay all the 3D objects out in front of each child. Make sure there is one of each 3D object. Ask the children to hold up a 3D object. **Who found it first?**
- **3D Shape Bingo** – Give the children a base card with a selection of 3D object pictures stuck on it (see Photocopiable resource 9.2). Call out a 3D object. If the child has that object on their board, they cover it with a counter. The winner shouts 'Bingo' when they have covered all their objects. Check the board by asking the child to repeat back all their objects.

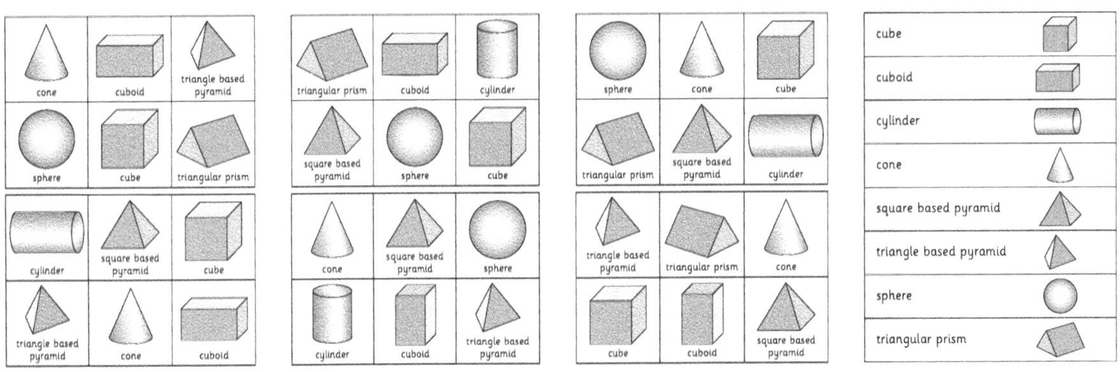

Photocopiable resource 9.2 – 3D Shape Bingo Board.

- **3D Models** – Give the children a selection of 3D objects and ask them to build a model. Ask the children what they built and what objects they used.
- **What's the Missing Object?** – The children have a tray with all the 3D objects on it. They shut their eyes and one child removes an object. The children open their eyes and try to guess which object is missing.

Unit 9.3 Recognising the properties of shapes and objects using mathematical language

Stations

- **I Am Thinking of a Shape** – Describe a 2D shape or a 3D object. The children guess the shape/object from the description. The children can either say it, draw it or hold up the appropriate shape/object.
- **Pipe Cleaner Shapes/Objects** – The children use pipe cleaners to create 2D shapes and 3D objects. They look at examples and notice the properties, the number of sides/edges and corners in order to recreate it. The children can work in pairs or team up with older children to help each other, especially when trying to make a 3D object.

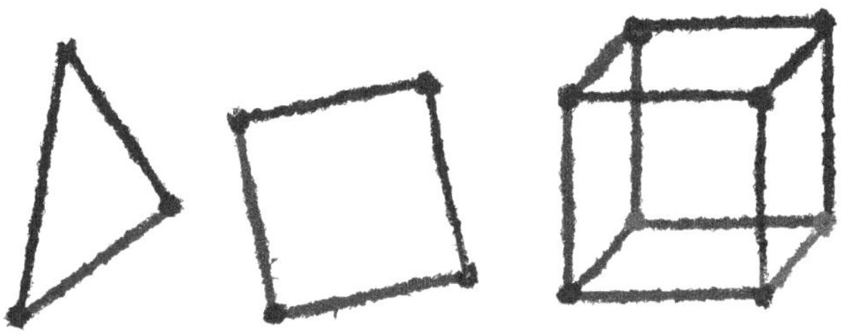

- **2D Shape Lollipop Puzzle** – Give the children different coloured lollipop sticks with the number of edges of the shape that can be made from them written on them; for example, four red sticks with a **4** written on each stick would make a square, three green sticks with a **3** written on each stick would make a triangle, six orange sticks with a **6** written on each stick would make a hexagon, and so on. The children sort all the sticks and make each 2D shape noting the number of sides each shape has.
- **3D Spaghetti Puzzle** – The children, working in pairs or team up with older children, use play dough and spaghetti to create 3D objects. The children must note the number of edges and corners to create the shape. They can have 3D objects available to discuss the properties of the shapes and compare them with their creations.

Unit 9.4 Describing the properties of 2D shapes and 3D objects using mathematical language

Stations

- **What is in the Bag?** – One child places their hand in a feely bag with a 2D shape or a 3D object in it. The child describes the shape they feel to the group. From the description the children name the shape/object.
- **Describe a 2D Shape** – In pairs, one child describes a shape. The other partner draws it on a whiteboard. Did they describe it well enough, paying attention to all the properties, so their partner could draw the correct shape?
- **Describe a 3D Object** – In pairs, the children have one of each 3D object on the table in front of them. One partner describes a 3D object. The other listens to the description and holds up the object they believe to be the one being described and names it.

- **Headband Shape/Object Game** – Create cardboard headbands with a picture of a 2D shape or a 3D object on the front. In pairs, one child chooses a headband and puts it on without looking at the shape/object on the front. The other partner describes the properties of the shape on the headband and then the child wearing the headband tries to name the shape on their headband. Then they take the headband off to check if they were correct.

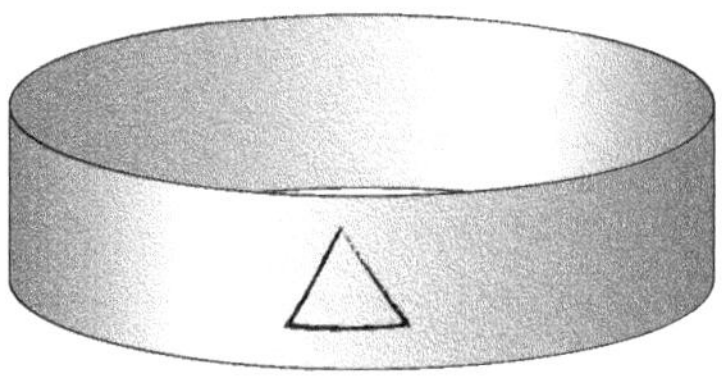

Unit 9.5 Using properties to describe, sort and/or match shapes and objects and justify choices

Stations

- **Sort 2D Shapes** – Give the children 2D shapes and two hoops and sticky notes. The children look at the shapes and sort them writing labels to show how they sorted them. Ask the children to justify their choices. Challenge the children to sort them another way.
- **Sort 3D Objects** – Give the children 3D objects and two hoops and sticky notes. The children look at the objects and sort them writing labels to show how they sorted them. Ask the children to justify their choices. Challenge the children to sort them another way.
- **How Did We Do It?** – One group are given 2D shapes or 3D objects or a mixture of both. They decide how to sort the items and then show their sorted items to another group. The other group then explain how they sorted them. **Can that group think of a different way to sort them?**
- **Screen Match** – Two children sit opposite each other with a box of 2D shapes, 3D objects or a mixture of both each. There should be a screen between the two pupils so they cannot see each other. One child lays out some shapes and/or objects behind their screen then describes what they have to their partner. Their partner listens and matches what they believe their partner has. The screen can then be lifted to see if the two collections do indeed match.
- **Create an Object** – Ask the children to create an object using given art resources to meet given criteria; for example, make an object that can stand by itself.
- **Matching 2D Shapes** – Fit 2D shapes together by turning if necessary. Discuss which shapes they chose to use and why.

What to observe

- Children who can identify and name 2D shapes.
- Children who can identify and name 3D objects.
- Children who recognise the properties of shapes and objects using mathematical language.
- Children who describe the properties of shapes and objects using mathematical language.
- Children who use properties to describe, sort and/or match shapes and objects and can justify their choices.

10 Angle, Symmetry and Transformation

Directions, Reflections and Symmetry A

Section overview

Units of learning	Review units	Prior learning	Benchmarks/Es&Os
Unit 10.1 **Giving basic directions**		• Children should be confident in movement and have had a range of experiences in moving themselves and objects through games and explorative play before there is an expectation of them to describe movements, give directions and follow directions. • Children should be given many opportunities to explore their environment (both indoors and outdoors) and to build a mental image of the space they inhabit; they should be able to locate familiar landmarks and remember short journey sequences. • The use of gestures and movements should be encouraged if children initially cannot describe a movement verbally, especially if they can perceive it, move and manipulate in response to it.	**MTH 0-17a** I have fun creating a range of symmetrical pictures and patterns using a range of media. **MTH 0-19a** • Understands and correctly uses the language of position and direction, including in front, behind, above, below, left, right, forwards and backwards, to solve simple problems in movement games. • Identifies, describes and creates symmetrical pictures with one line of symmetry. In movement, games, and using technology I can use simple directions and describe positions.
Unit 10.2 **Beginning to understand the language of position and direction, including in front, behind, above, below, left, right, forwards and backwards in movement games**		• Children should build on existing vocabulary through practical, directed tasks and become familiar with vocabulary that indicates where they are and their position in relation to others and the movement of objects. • Children should have experience of using their bodies in a range of ways e.g. stretching tall, running in straight lines/zigzags or circles and making shapes with their bodies. • Children should be able to respond to questions/directions and place.	

Units of learning	Review units	Prior learning	Benchmarks/Es&Os
Unit 10.3 **Reflections**		• Children should recognise what a mirror is and how it can be used to explore reflections (and symmetry). They should be beginning to consider the similarities and difference they observe. They should be aware of the term mirror image. • Children should have some knowledge of shiny objects which reflect light. They also should recognise reflections in nature and the outdoors. (Reflections in still water and shiny surfaces such as spoons or car doors).	
Unit 10.4 **Symmetry**	**An Introduction to 2D Shapes and 3D Objects A** Units 9.1 and 9.2	• Children should be able to recognise and name some basic 2D shapes and 3D objects; they should have had opportunities to play with shapes and explore their properties and how they fit together (tessellate). This may include the use of shape sorters which aids the early development of symmetry in a familiar context. • Through directed nature walks and exploration children should be able to identify objects which are symmetrical.	

Conceptual context

Children inherently develop spatial awareness as they begin to explore their worlds – for example, reaching for objects, crawling, climbing, running, manipulating small world toys, playing on bicycles, wheeling prams, bouncing balls and playing both physical and board games with others. All these activities need a commentary from supportive adults describing what the children are doing, using and supplying words to describe what is happening. Rich specific vocabulary needs to be used and developed in context and it is always very helpful to present 'position' words with visual and kinaesthetic support clues. By the time children are in the Nursery Class they should be beginning to locate items, deciding how to best get from one place to another, describing where objects are, using words and gestures or by making models. Some of these may not be perceived as being mathematical activities; however, they are foundational experiences for later mathematical work, particularly in geometry.

The Early Years learning environment should be bursting with opportunities to explore spatial relationships – children should have experience of stretching tall, running in straight lines/zigzags or circles and making shapes with their bodies, using language, using construction materials, using art materials and playing with puzzles and games. All these can happen indoors or outdoors. If children don't have maths in their hands and bodies, it will not get into their heads! Using investigations involving objects, e.g. things that turn (see Group Activity – Things that twist and turn) can help develop and consolidate the learning of position and movement as children will be able to see that they can create movement in other objects. This also provides a fantastic opportunity for children to use their newly extended vocabulary.

Children should be able to locate items (e.g. Can you find your jumper?), describe how best to get from one place to another and describe where objects are in the playroom or the outside play area. This information may be conveyed as a gesture, action or words depending on the confidence and experience of the child.

A wide range of opportunities, including the use of stories, art materials, songs and rhymes, should be provided to enable talking about, hearing about, drawing and modelling spatial relationships. Children should have opportunities for re-arranging and reshaping – putting things together and taking things apart. Children can be asked during tidying up to put the scissors in the holder on the shelf next to the sink or to put pencils in the pot on the shelf below the scissors. Is there enough space in the scissor holder or pot? When a child uses a shape sorter toy, they are developing their knowledge of symmetry, e.g. when a shape is rotated to enable it to fit through the hole, children can see a mirror image of the hole that the shape is then going to pass through.

Symmetry is used in creating patterns by **reflection** and by **rotation** and can be found both in nature and the built environment. If an object, e.g. butterfly or a snowflake, is exactly the same on both sides then it is said to be symmetrical. Children can be given large pictures to fold and check to see if it is the same on both sides, i.e. is symmetrical. It is helpful for young children to be given a picture that has the fold line already creased so that the only checking required by them is to see if the sides of the picture actually match and not whether they can fold along a line. In reflective symmetry, a shape or pattern is reflected in a mirror line or a line of symmetry and the reflected shape will be an exact mirror image of the original – it will be the same size, the same colour, the same shape and the same distance from the mirror line. In rotational symmetry, if the object is rotated or turned around a centre point, the object appears the same (e.g. a wheel or certain types of flowers). Opportunities should be provided to explore symmetry – folding paper shapes and pictures, experimenting with mirrors, building symmetrical patterns with bricks (Lego/Duplo/Stickle) or natural objects (pine cones/chestnuts/seeds) across a line, whether it be a fold or a drawn line or a bit of string or rope. These activities are explored in more detail in the Teaching ideas section.

Difficulties that may exist for supporting adults, especially if they do not know the children well, are:

- to know how and when to scaffold or push understanding;
- to know when it is appropriate to use ideas to focus the children's attention involving the 'trickier' parts of spatial concepts and;
- when to extend the learning and provide multiple opportunities to explore and manipulate.

However, the need for consistent rich vocabulary should never be in doubt.

Resources

- Blocks
- Small world figures
- Soft toys
- Art materials
- Sit and ride toys/bicycles
- Story books
- Music
- Play mats
- Toy vehicles
- Games
- Jigsaws
- Games such as positional pups game
- Movement games (Simon says)
- Mazes
- P.E. apparatus
- Parachute
- Mirrors
- Plastic/magnetic shapes
- Remote control cars
- Shape sorter toys
- Range of loose parts materials (e.g. pine cones, sycamore seeds, chestnuts, nuts and bolts, screws, buttons)
- Small world toys and objects

Digital

- My favourite toy (Animated Song)
- Ant crossing

Record book

Pages 27–28

Key vocabulary

- Above
- Backwards
- Behind
- Below
- Forwards
- In front of
- Left
- Next to
- Reflection
- Right
- Symmetry
- Underneath

Misconceptions

Potential misconception	Addressing the misconception
Unit 10.2 Children often struggle to understand and correctly use 'opposite' concepts for example before and after/in front and behind	• Teach/introduce one concept at a time; children should have lots of opportunities to explore and use the concept, e.g. **behind** before introducing **in front**. Initially children should be confident in using **behind** and **not behind** before they start to use **in front.**
Unit 10.1 The judgement of proximity. A child may say that the car is near the wall; if a motorbike then parks between the car and the wall, they will say that the car is not near the wall now. In real terms the car and the wall have not moved.	• Creating lots of experiences to explore using toys and themselves, and scaffolding their understanding will help develop their judgement.

Teaching ideas

Unit 10.1 Giving basic directions

Station

- **Treasure Hunt** – Turn the sand tray into a pirate treasure island. Dampen the sand and then bury treasure (coins, plastic jewels, etc.) in the sand; place small stones (useful to be coloured) and palm trees (lolly sticks which can be numbered) around and about the 'island'. Draw arrows and pathways. Put in small pirate figures in the sand and give the children directions so they can move the pirate to find the treasure. E.g. start at the yellow stone, move forward to the tree marked 1, turn (left) to the blue stone, etc. Get the children to bury the treasure and give directions to others.

Whole class

- **Making a maze**

1. Fill a tote tray or seed tray with compost. Mark out, using a pencil or a stick, a design of parallel lines or a spiral line making grooves in the compost and carefully sow cress seeds into the grooves. Water and watch the mini maze growing. Provide small world characters to move through the maze. Children can describe the directions that the toy must move to get to the end of the maze.
2. In the outdoor play area use chalk to draw a large maze. Children can move through the maze talking about the movements they make to enable them to get to the end. The children must realise that they cannot step out of the lines.
3. Children can practise using mazes found in puzzle books. They can provide a commentary of how to get to the end of the maze. Mazes can be laminated and be reused.

Unit 10.2 Beginning to understand the language of position and direction

Whole class/group

- **Story** – Read *Rosie's Walk* by Pat Hutchins. This simple story provides children with an amusing, visual introduction to positional language. The story is beautifully illustrated using autumn colours, and children are encouraged to follow the journey of the hapless hen, Rosie, as she moves around the farm, closely followed by a fox. Children can be involved in representing her movements using, for example, hand actions for movements such 'over' or 'round'. **How might they represent through or near?** Together with the children after reading the story, make a plan or map depicting the places Rosie visited. Encourage the children to talk about/give directions about how she found her way round the farmyard. Her path, once agreed, can be drawn onto the plan using a marker pen. Some children may be able to represent the story in a map-like form for themselves. The tale can then be further developed in the playroom/outdoors with the children deciding which apparatus/equipment might be best to represent the farmyard 'situations' that develop in the story, or the children can use Lego or small world materials to enact their own Rosie's walk. Effective open questioning and consistent use of positional language should accompany all this activity. Think about what you would like the children to represent in their enactment of Rosie's walk and model the actions with them so that they are confident and understand what is meant. The actions – over, across, around, past, through, under and up – could be modelled physically before representing them in their farmyard situations.

Whole class/group

- **Tidying-up** – This activity forms a natural part of every nursery/early years class/classroom. It can be appropriate for both group and whole class and requires no extra resources. Initially, the supportive adult directs the action but as children develop in their ability to carry out the actions and develop vocabulary, they then may become the 'director' of the action. They could wear a special hat to signify the importance of their role. The purpose of the activity (apart from getting materials put away tidily) is to develop an understanding of position and movement. The language used needs to be specific to sorting out the materials, if required, placing the materials in the appropriate containers and then providing directions to guide the children to the required area in the playroom/classroom. The directions can be straightforward – put the bucket **under** the sand tray – or more challenging – go **behind** the pop-up tent, go **in front** of the painting easel, **turn to face** the sand tray and walk towards it, now place the bucket **under** the sand tray. It is critical that the children are aware of the names of the equipment. The tidying up session will build on and consolidate the learning of the lesson or group activities and become a plenary. Below are typical activities that might be used as part of a group rotation; each will require a tidying up scenario.
    - **Block Play** – The children will need a range of 3D wooden/plastic construction blocks and books/pictures to stimulate children's construction ideas, presence of an adult at intervals to enrich and support, mirrors. The staff and children will be using the words beside, on top, under, next to, inside, outside, same, different, turn, move throughout this exercise. Children should have time and space to construct and deconstruct models, starting simply by making roads and enclosures and then bridges and castles. They need opportunities to place blocks next to each other, on top of each other, re-arranging and reshaping. They should be able to make and describe their model and then make a new model using the same bricks. By taking digital photographs the children can revisit the model. **Can they recognise the blocks they used? Can they build it from what they can see? Can they describe what they are doing to another?**

Next steps may be to look at the models made from different perspectives: digital photographs are useful for this. Simple mirrors can allow exploration of what an object looks like from a different angle. Using a selfie stick or attaching a mirror to a ruler allows children to see two views at once. **Can they match the picture of a model to the view from above or from the side?** Photographs of the models can be uploaded and slides produced that can be displayed on an interactive whiteboard to extend the activity. Children can discuss the similarities and differences. The challenge can be increased by having more than one correct answer flashed up on the screen – the children could talk about and justify what they see.

- **Things that Twist and Turn –** You will need screwtop jars, plastic bottles with lids, lids of varying sizes, door knobs, keys and locks, padlocks and keys, clockwork toys, clocks with keys, knobs, buttons and dials, wheels, nuts and bolts, screws and screwdrivers, twisty shaped pasta, ropes, shells, old style telephone cable, rotary whisk, spinning tops, football rattle, windmill, water wheel and wind-up music box, hula-hoops, sycamore seeds, propellers or helicopter blades. (Local locksmith or DIY store may donate some old locks and keys, ex-display door knobs and nuts and bolts.)

  The children are learning the key words of: turn, twist, on/off, up/down, switch, fix, screw, wind, spin, round, fast/slow.

  Leave the materials for the children to explore: watch for their wrists and fingers and hands twisting and turning as they play, and listen for talk as they play. Take photos. Get children to demonstrate the actions of twisting and turning and share ideas with other groups in plenary session.

  This activity can be taken outdoors where children can engage in more physical activity relating to twisting and turning – spinning on the spot, hula hooping, spinning tops, rope swings, roundabouts, ring games (e.g. 'Clock in the Tower'; 'Duck, Duck Goose'; 'Here we go around the Mulberry Bush').

  Next steps would be to see if the vocabulary and actions manifest themselves in role play, e.g. knobs on the cooker, turning keys in the ignition of their car/rocket, etc.

Unit 10.3 Reflections

Whole class

- **Go on a 'Mirror' Spotting Walk** – Take the class on a reflective walk and encourage the children to look carefully for reflections. Plan your route and ensure that it will provide opportunities/situations that will allow children to observe their reflections: window panes (at a height the children can see their reflections), puddles or calm water, mirrors on car doors and shiny paint surfaces on vehicles.
- **Follow My Leader** – Follow My Leader is a form of reflective movement. It can be played in open areas. The leader of the line calls out an action for the others behind to copy. This may involve lifting an arm **up** in the air, walking **backwards**, bending **down** to touch the ground, etc. Children can take turns at being the leader and thinking of movements.

Stations

- **Investigating Station: Looking for Reflections** – You will need safety mirrors (range of sizes and shapes), plastic mirror sheet, tall mirror, handbag mirror, CDs, spoons, foil, polished metal trays/objects, stainless steel trays, string/tape/drawing materials/small objects, periscopes, kaleidoscope, pictures, dentist's small mirror.

IDEAS: To enable children to see what objects look like from different viewpoints, place mirrors around the playroom (under tables, behind toys, etc.) for children to find and explore.

Encourage children to look in large mirrors and explore their reflections to reinforce the term mirror image – sticking tongue out, finger on nose, etc. Talk to children about what they see; encourage them to look closely and describe what they see, thinking about their actions and what is happening.

Outdoors, place mirrors on the ground and talk about what can be seen from this different view. Lie on the ground and look upwards. **Can they see the same things?** Encourage and support vocabulary development and careful looking. Add a rear facing mirror to wheeled toys (on bicycle or scooter handlebars) to add a different dimension.

Indoors use mirrors to explore objects. Two small mirrors can be hinged together using masking tape, providing different viewpoints to familiar objects. Stand the mirrors upright (like a small book) and in the angle where the two mirrors meet place a small world figure. Encourage the children to slowly open the mirrors and look at the reflections they can see as the angle on the mirror widens. Leave a range of figures/objects for the children to explore and talk about. Draw the children's attention to the number of objects they can see in their mirrors.

Unit 10.4 Symmetry

Whole class

- **Movement cards** – This activity builds on from children observing their reflections (Unit 10.3). In this activity all the children are standing, facing the adult who stands at the front of the class/group. Ensure all children have a clear view of you and that you as the adult can see them. Music is played and the children can dance/sway, etc., on the spot. The adult stops the music and holds up a large movement flash card (see Photocopiable resource 10.1). The children look carefully at the card and try as quickly as possible to get their bodies into the position shown on the card.

  Initially, the adult can provide a commentary, e.g. 'Yes, that is great Finlay, you have put both your arms up in the air'. Check that all the children are making the same action. Children are then encouraged to resume dancing/swaying when the music restarts and continue to do so until the music is again stopped and another movement card (see Photocopiable resource 10.1) is held up. This activity can progress to no verbal guidance by the adult and the children when asked can state what movement they have made. Picture guidance can be removed and the adult may just say the action, e.g. arms up. The adult may then show the flashcard and children can self-monitor whether they got the action correct. This is a quick, but valuable, assessment activity.

Photocopiable resource 10.1 – Movement Cards.

Group

- **Creating Mirror Line/Symmetrical Paintings** – Materials: piece of heavy-duty paper (sugar paper) folded down midline, ready mixed paints (selection of colours), paint brush.

  Ask the child to put blobs of paint on one side of the folded paper. When this has been successfully done, assist the child in folding the paper across the midline and press gently on the now half-size (doubled) piece of paper. Carefully open the paper to reveal a symmetrical blob painting. The paper can then be shaped, using scissors, to make a symmetrically coloured butterfly, etc.

- **Creating Mirror Line/Symmetrical Pictures** – Materials: piece of paper/card minimum A4 size or pieces of plain material (old heavy-duty parasol material is good to cut up for this activity, Lego/Stickle Brick base-boards can also be used with thick elastic bands down the middle to create the midline), selection of objects (including nuts, bolts, screws, pine cones, stones, sticks, chestnuts, beans, coins, cubes, small toys, Lego, Stickle Bricks, Duplo, buttons, etc.).

  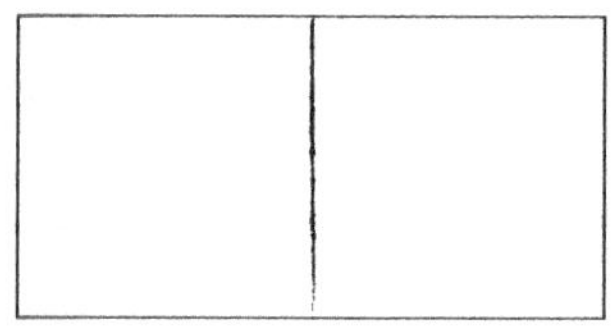

  Ensure that the paper/material is folded in the middle – the mirror line (mid line) or line of symmetry – and is marked with a heavy-duty pen.

  Children are told that any object that they place on one side of the line they then must place an identical object in exactly the same position on the opposite side of the line. A symmetrical picture is then composed. This is a good individual activity but can progress to a pairs activity when the children are competent. In the pairs activity, one child makes their side of the picture and the other child makes its reflection.

What to observe

- Children show an interest in shape and space by playing with shapes or making arrangements with objects. They may use 3D objects to build castles or houses or use 2D shapes to create faces (symmetry) and pictures; listen carefully for the use of positional language in context or look for gestures and movement that provide suggestion (tall tower may be signified using stretched arms) Can they copy a created model?
- Children flipping or turning pieces to find a fit when using jigsaw puzzles.
- When children are using mirrors or making symmetrical patterns can they describe the patterns, pictures or shapes? What can they see in the mirror? Can they predict what their 'fold and blot' picture will look like?
- Listen for children using positional language whilst playing in the playroom, outdoors or in PE hall. Can they describe their actions or the actions of others?

Directions, Reflections and Symmetry B

Section overview

Units of learning	Review units	Prior learning	Benchmarks/Es&Os
Unit 10.1 **Using language to describe the relative position of objects**	**Directions, Reflections and Symmetry A** Units 10.1 and 10.2	• Children should have had a range of experiences where they can talk about, model and draw spatial relationships (e.g. using building blocks); the input of a knowledgeable adult is critical to ensure rich, precise vocabulary is used and developed. • Children should have experience of simple mapping activities, drawing and sketching, playing with tangrams and shape sorters, using a range of jigsaws and games (both physical and board). • Children should be developing the ability to decentre and to see things from different perspectives, i.e. what an object looks like from above, or from the other side. • Through daily classroom activities children will have physical and visual opportunities to develop vocabulary, appropriately in context.	**MTH 0-17a** I have fun creating a range of symmetrical pictures and patterns using a range of media. **MTH 0-19a** • Understands and correctly uses the language of position and direction, including in front, behind, above, below, left, right, forwards and backwards, to solve simple problems in movement games. • Identifies, describes and creates symmetrical pictures with one line of symmetry. In movement, games, and using technology I can use simple directions and describe positions.
Unit 10.2 **Creating and identifying symmetrical pictures or arrangements of concrete materials**	Link to shape **Name and discuss the Properties of 2D Shapes and 3D Objects A** Units 9.1 and 9.2 **Directions, Reflections and Symmetry A** Units 10.3 and 10.4	• Children should have opportunities to become familiar with three types of mathematical movement – translation, reflection and rotation through informal activities. • Children should recognise names of 2D shapes and 3D objects, understand how to fold a paper shape in half to investigate the symmetrical nature of the shape, be confident in the use of a mirror and recognise that an object which is symmetrical is the same on one side as the other. • Children should have had numerous opportunities to observe and explore symmetry using story books, toys, natural objects and classroom apparatus.	

Conceptual context

Most young children will have had a range of experiences relating to shape, position and movement before entering Nursery education and indeed Primary 1; it is important to recognise their previous learning but also to realise that they might have developed some misconceptions due to exploration without support or guidance, particularly in terms of vocabulary; this can be remedied by adults providing opportunities to explore while scaffolding learning through effective interaction. Shape, position and movement are significant in a child's mathematical journey.

Many of the suggested teaching activities, included in the Teaching ideas section, are designed to give the children foundational experiences that are essential for developing their understanding for future work in Geometry (including Angles) and Mathematics. Children should become familiar with the concepts of transformation, rotation and translation by experiencing them in appropriate practical contexts.

Transformation – when a shape is moved it changes position but it is still the same shape. Children should be encouraged to think about what is different, what has changed and what remains the same. For example, when building with different blocks, a cylinder is still a cylinder even though it is representing the base of a tree or part of a window frame. See *Changes, Changes* by Pat Hutchins.

Rotation – an object is moved in a circular way around a fixed point. For example, think of the movement of the hands on a clock face. Tangrams can provide a visual way to explore rotation of shapes.

Translation – an object is slid/moved from one position to another without turning. For example, when moving objects across a line of symmetry or children moving on a track/path; they may go backwards/forwards/left/right but keep facing the same way.

Reflection – a shape or object is flipped over a mirror line.

Symmetry – an object is **symmetrical** when it is the same on both sides. A shape has **symmetry** if a central dividing line (a mirror line) can be drawn on it, to show that both sides of the shape are exactly the same. Symmetry is used in creating patterns by reflection and rotation; it can be found in nature and the built environment.

Visualisation – children should be able to summon a mental image in their head. Can they 'see' a number with their eyes closed and use this information to prompt the writing of the number using the correct orientation? More complex visualisation relies on making connections and not just memory, so experience and discussion are vital.

Landmarking – this is a term used to work out the position of something relative to something else or somewhere else. For example, when giving directions it is often easier to say turn left at the church rather than turn left at Hill Street. Landmarking relies on our ability to visualise where familiar objects are to guide our way.

The development of spatial thinking relies on visualisation and relational understanding (being able to recall that the pencil tub is on the shelf next to the tub for the felt pens) occurs through experience, discussion and familiarisation. It is important to build upon children's thoughts and ideas, using them as a catalyst to develop their understanding.

By using small world objects, children will begin the development of their ability to decentre (to see the model/object from someone else's perspective); this is developed through rich experiences and the use of appropriate language. Think about the scenario where a child places a cube and a cylinder side by side.

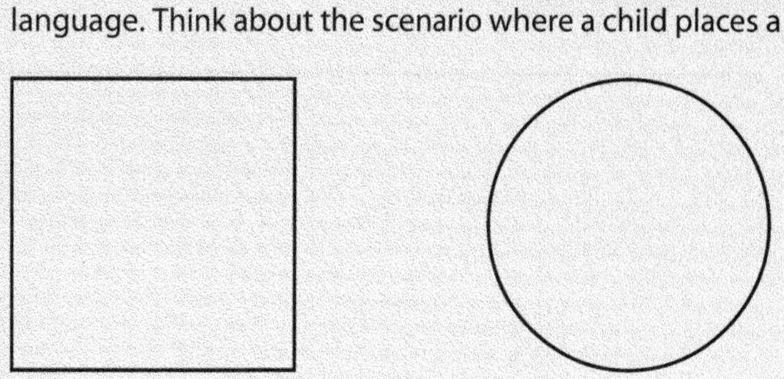

Depending on where other children view the 3D objects from, the cube may be behind the cylinder or the cylinder may be behind the cube or next to the cube, and so on. This is further complicated if the objects used are small world figures (e.g. a fireman, a cat, etc.) and they have features, such as faces. If these figures are placed side by side; the definition this time depends this time on the way the figure is 'looking'. Children require support and opportunities to discuss what they see.

The ability to imagine the effects of movements begins with children turning themselves round, then applying this knowledge to turning shapes around and over to test how they might fit, e.g. into the corresponding space in a shape sorter toy. Further experience can be gained by doing jigsaws where they are trying to get pieces to fit together or using tangrams where the child is trying to fill a template or copy a picture using the given shapes.

Some thought should also be given to providing opportunities to begin preliminary steps towards angle development – angle often is not fully understood as being a measure of the size of a turn. It is important to ensure that children experience physical rotation initially as turning through part and whole turns. Children should have physical activity before trying to measure angles with a right angle or angle measurer or whether they are turning clockwise or anticlockwise. Provide opportunities to explore turning; these activities are easy to do and can become part of the classroom/P.E./outdoor routine. A simple activity might be to stand in the classroom with everyone facing the same *wall* (everyone facing the same way). On command, all turn to face the *window* (90-degree turn); moving through another 90 degrees has us facing in the opposite direction in order to face the *painting table,* then a further 90 degrees to face the *Smartboard* and then back to facing the *wall* and a full 360 degrees or whole circle has been turned. Look at numbers on a clock face and try to make connections with the hands moving. Watch a trundle wheel turn and listen for the click when the wheel has done one full revolution. In the playground, draw a chalk circle and mark lines on the perimeter at ¼, ½ and ¾ or relate the positions to the numbers on a clock, 3, 6 and 9 (an 'X' marked at the top of the circle to indicate the starting/ending point). Mark the lines in three different colours: red for ¼ (3), blue for ½ (6) and green for ¾ (9). Ask the children to face the 'X' and see if they can spin round and stop at the red mark, etc. The examples above are set out using vocabulary for adult understanding. It is important to keep vocabulary simple at this point until the understanding of the children develops.

Resources

- Blocks and construction materials
- Small world figures
- Soft toys
- Art materials
- Chalk
- Stories
- Music
- Play mats
- Toy vehicles
- Games
- Jigsaws
- Positional pups game
- Stories
- Mazes
- P.E. apparatus
- Parachute
- Stories
- Tangrams
- Board games
- Mirrors
- Plastic/magnetic shapes
- Programmable toys
- Mazes
- Range of loose parts materials
- Mirrors
- Scissors

Digital

- Rotate and Reflect
- Tangrams

Record book

Pages 27–28

Key vocabulary

- Above
- Across
- Backwards
- Behind
- Below
- Diagonal
- Flip
- Fold
- Forwards
- In front of
- Left
- Mid-line
- Next to
- Position
- Reflection
- Right
- Symmetry
- Tangram
- Turn
- Under

Misconceptions

Potential misconception	Addressing the misconception
Units 10.1 Jigsaw pieces may be put into place by trial and error.	• Children need lots of opportunities to complete a range of jigsaws; a supportive adult should engage in discussion with the children whilst they are solving the puzzles eliciting from them appropriate language and supporting their understanding about flipping, turning or rotating pieces.
Units 10.2 Many children cannot orient a matching pair of shoes and locate them on the correct feet.	• We are expecting such children to be able to rotate shapes, locate jigsaw pieces and recognise fine differences in objects. Many children are unable to recognise left and right consistently; this ability doesn't develop until they are at least 6 years old. By providing lots of discussion about similarities and differences children can be supported on this journey.
Units 10.2 For many children this is the only representation of a triangle they know. 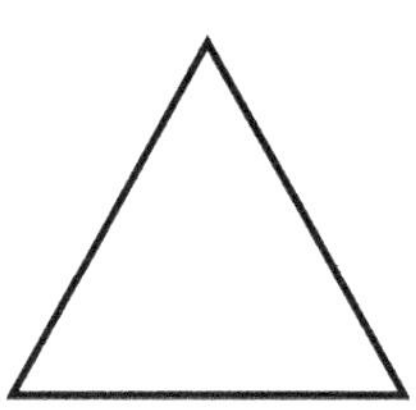	• Children need to experience transformation and rotation of shapes. 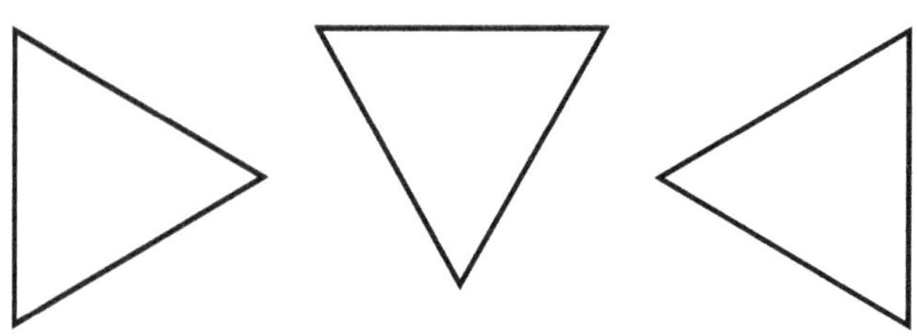 This can be achieved using tangrams or free exploration of wooden, plastic and magnetic shapes.

Teaching ideas

Unit 10.1 Using language to describe the relative position of objects

Whole class

- **Follow My Leader: Revisit from 10A Unit 10.3** – Follow My Leader is a form of reflective movement. It can be played in open areas. The leader of the line calls out an action for the others behind to copy. This may involve lifting an arm **up** in the air, walking **backwards**, bending **down** to touch the ground, etc. Children can take turns at being the leader and thinking of movements.
- **The Detective Game** – This game is an extension of Follow My Leader. Children should be seated in a circle on the floor/carpet. One child is selected to be the detective; they leave the circle and then move away so they cannot observe the circle of children. From the seated children, one child is chosen to become the leader. The detective returns to the circle and stands in the middle. The leader starts to make an action (e.g. patting shoulders) and everyone in the group copies this action. The leader can change the action when they wish. The detective has to work out who is the leader. Once they have discovered the leader, a new leader and detective are chosen and play resumes.
- **Simon Says** – The game of Simon Says is a fun game that helps exercise listening skills. The object of the game is to reduce the numbers of children playing and end up with one or two super listeners. Children stand facing the adult who takes on the role of 'Simon'. The children should only respond to commands that begin with 'Simon Says'. For example, if the adult says 'Simon says put your hands in the air', then the children should respond and indeed put their hands in the air. A child not responding would be out. Equally, children who respond if the command was 'Put your hands in the air' would also be out because 'Simon says' didn't precede the command.
- **Hunt the 'Thimble'** – An object (thimble) is hidden in the room and the children ask questions to find out where it is. For example, **Is it under the chair? Is it behind the box?**
- **I Spy** – The children must locate an object with a given letter. For example, 'I spy with my little eye, something beginning with B'. The children ask questions: **Is it on the floor? Is it on the wall? Is it near the door?**

Stations – group/paired activities

Where Shall I Put Finlay? – It is a great paired/small group activity (see Photocopiable resource 10.2) where the children gain confidence in moving the character and saying where he is or asking a friend to place him in a particular place: 'Put Finlay beside the dog'.

Photocopiable resource 10.2 – Where Shall I Put Finlay?

- **Where is Isla?** – See Photocopiable resource 10.3. Cut the cards out. Place cards, face down, in a pile. Children pick a card from the top of the pack and describe where Isla is. If correct, they can keep the card. Challenge can be added by encouraging the children to draw their own cards or look for positional pictures to add to the game.

Photocopiable resource 10.3 – Where is Isla?

- **Where is the Ladybird to Go? Paired activity** – See Photocopiable resource 10.4. Cut the ladybird out. The children gain confidence in moving the ladybird and saying where it is or asking a friend to place it in a particular place: 'Put the ladybird on the table'.

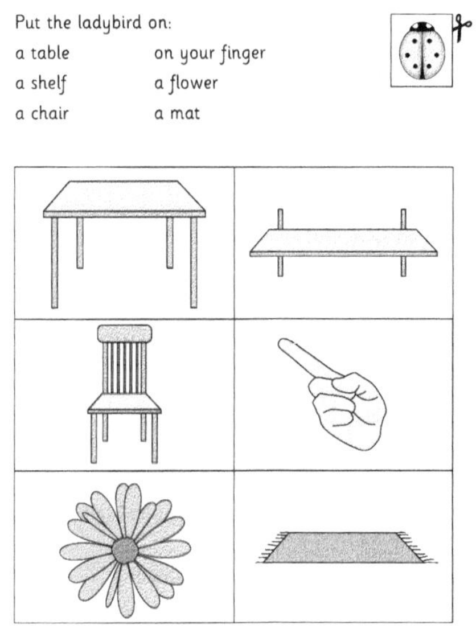

Photocopiable resource 10.4 – Where is the Ladybird to Go?

- **Tangrams: Group** – Materials**:** tangram sets, accompanying challenge cards.

  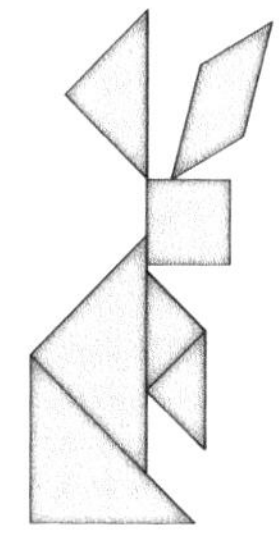

  Provide each child with a tangram challenge card (see Photocopiable resource 10.5) and ask them to use the shapes provided to create the given figure e.g. a rabbit.

  Ask the children to look closely at the tangram rabbit. Can they point to the large triangles? What shape is used to make the head? Tangrams are available for iPad/smart board (see Digital resources).

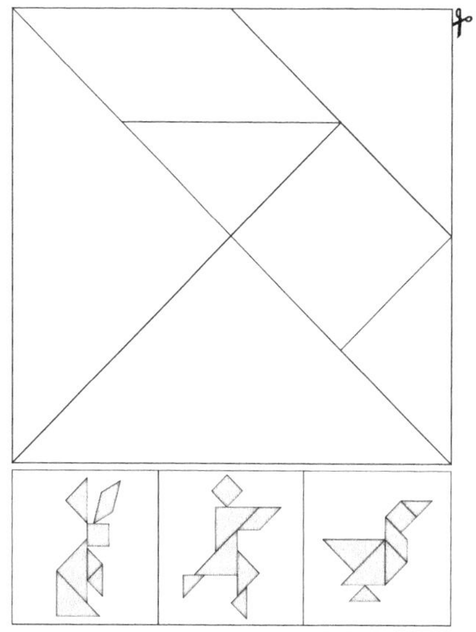

Photocopiable resource 10.5 – Tangram Challenge.

Group stations

- **Following Instructions: Outdoors** – You will need; dice (spots or numbered), coloured chalk (a different colour for each child in the group), a series of direction cards (diagonal lines, straight forwards/backwards/left/right) arranged face down in a pile (dot in top corner so the orientation of the card is always correct) (see Photocopiable resource 10.6), a child to take charge of dice rolling and directing the other children. All children start at the same spot. The leader child rolls the dice and turns over the movement card. Child 1 moves in the direction suggested and using the correct number of steps from the dice. When child 1 has reached their new position, they mark a cross on the ground using their colour of chalk. This continues for all the children in the group. The children all roll the dice three times. **Where do they all finish up?** The route they travelled can be seen by joining up the chalk crosses; the children have made a simple map of their movements.

  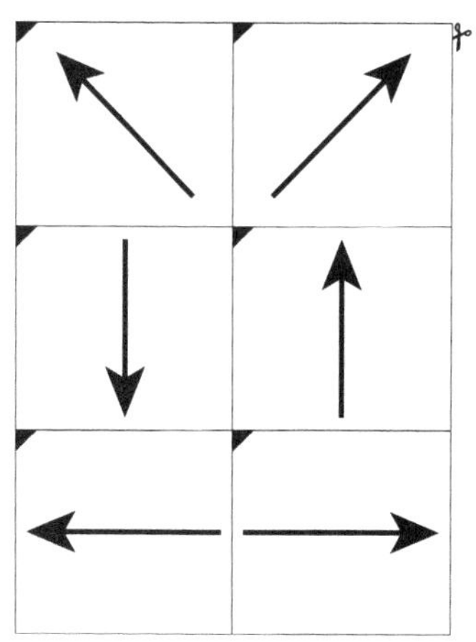

  Photocopiable resource 10.6 – Arrow Cards.

  For example:

  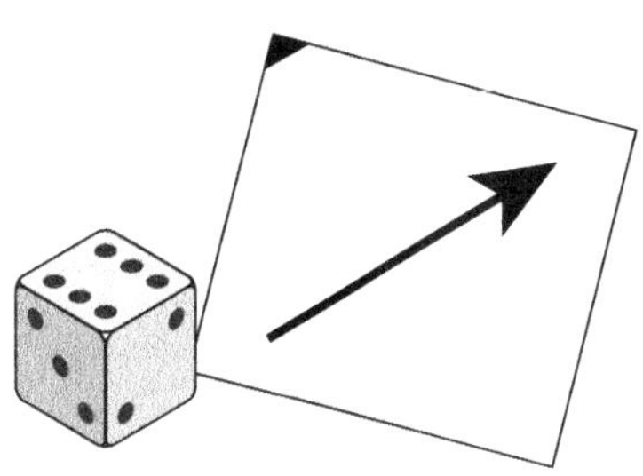

  The child moves 6 paces, diagonally.

Increase the challenge by using two dice.

The activity suggested above where the children can record their movements using a simple 'map' links to action stories such as *Rosie's Walk* (10A Unit 10.2) where a class map can be made plotting the movements of Rosie. Children can be challenged by thinking of their own version of the story and plan a series of directions to help their Rosie find her way. The children can re-enact their story in the hall; support may be required to aid the understanding of positional and directional terms because, for example, left and right depend on the direction you are facing.

- **Giving instructions to a programmable toy** – You will need sticks or chalk to make/draw a road way with junctions and programmable toys cars.

  In the activity 'Following instructions' the children were physically moving in response to directions/distance, but in this activity, they are being encouraged to use their knowledge of direction and distance to control the programmable toy.

  Children should be given opportunities to explore the programming of these toys. The previous game of following instructions should have provided an insight into direction and movement. Encourage them to put in a direction and a number and see what happens. **Where does the car go?** They quickly learn the need for accurate programming and soon learn that adjustments of numbers are needed when programming angles and distances.

Stations relating to block play

- **Follow my instructions** – In order to play this game effectively, children should be confident and able to name common 3D objects usually used in block modelling. Confidence in recognising and naming objects can be increased by playing 3D Kim's game. In Kim's game the objects being used are looked at, talked about and then are placed on a tray/table; the children then cover their eyes. The adult removes one of the objects from the tray and children are asked to say what is missing. They are encouraged to use the correct name for the object. The correctly named object is placed back on the tray/table in exactly the same space and the game continues with another object.

  Follow my instructions is an activity for two. You will need two sets of blocks and a screen. The first few times you play this game, provide only a small number of bricks for each child ensuring that each child has the same type of bricks. Challenge can be introduced by giving a greater pool of bricks so that the vocabulary and the directions need to be more specific.

  Say: 'Finlay has built a model behind the screen without Amman seeing. Finlay has to describe his model in such a way that Amman can build exactly the same one.' Remove the screen. **Are the models the same?**

- **Story** – Read *Changes, Changes* by Pat Hutchins. This story follows the building of objects to suit circumstances but has no text. It provides a fabulous opportunity for the children to get really involved and expand the story. For example, **Can you remember what the *cylinder* was used as/for in the story?**

  Some quick practical follow on activities relating to the story:

  - Give children 10 bricks to build a model. Photograph or sketch the result. **Can they build a completely different model using the same ten bricks?** Get them to look at it from above and from different sides. Ask them to describe what they see.
  - Give the children a photograph/sketch of a model and challenge them to build it.

Whole class

- **Listen and respond** – Use a set of position cards and enough cubes for children to have one each ensuring that there are a range of colours.

  Children should be seated in a circle on the floor/carpet area. Each child should have a coloured Unifix or equivalent cube placed on the floor in front of them. Cubes may be given out in a pattern (red, blue, yellow, green, red, blue, yellow, green, …) so that no child has the same colour cube as children on either side of them. The teacher has a stack of movement cards (see Photocopiable resource 10.7) placed in front of them. The teacher turns over a card (e.g. under your chin) and asks for everyone with a green cube to stand up and hold their cube under their chin. Quickly glance around and offer support to those needing it. Praise. Ask children to sit down and the process starts again with different cube colour and different position card. You can make some additional position cards of your own relevant to your setting.

Photocopiable resource 10.7 – Movement Words Cards.

Whole class/group

- **Jigsaw making** – Materials: selection of old Christmas/ birthday cards, retain the picture part, scissors.
- Children begin by choosing a card. Encourage them to make a single cut across the card. The cut can be diagonal, horizontal, vertical, jaggy, curvy, etc. Ask the children to move the pieces apart and then fit them back together. Encourage discussion about:
  - type of line – **Is it easier if it is a straight line rather than a jaggy line?**
  - **What happens if a piece gets turned over?**
  - **What movements are needed to solve problems.**
- Encourage the children to make a few more cuts and then put the picture back together again.
- Provide opportunities for the children to complete a range of jigsaw puzzles.

Unit 10.2 Creating and identifying symmetrical pictures or arrangements of concrete materials

Whole class

- **Revisit 10.4 Symmetry from 10A** – Recap the game and play until children are confident in making the body shape as a reflection (when they see the picture) but also able to make the body shape from a verbal description. The game can then be further extended:
  - Show/flash the picture once the music stops – **Is the body shape you are making symmetrical? If it is not symmetrical, how could you make your body shape symmetrical?**

  For example, children with one arm raised could make their body symmetrical by either lifting one arm so they have two arms raised, or by lowering one arm so they have two arms down. This involves the children thinking and using positional language.

- **Make a Bingo Game** – Use movement cards (see Photocopiable resource 10.1) and bingo base boards of stick people representing the possible actions (see Photocopiable resource 10.8) to play a simple matching game, where children locate the correct body card and place it on the base board. Progress to playing in a bingo format (using the movement cards as a set of bingo body cards) where the first person to describe the action, e.g. arms up, will win the card. Think about giving no visual clues and have children responding to questions, e.g. **Who has the picture with one arm up? Can you model the action to win a point?** Further extend this activity to children becoming the bingo caller or operating the cards and checking others.

Photocopiable resource 10.8 – Movement Bingo Boards.

Group activity (pairs)

Revisit symmetrical activities (10A Unit 10.4) and then build upon these experiences using some of the following suggestions.

- **Find 5** – You will need a piece of A4 card (A3 backing paper/sugar paper) with a mid-point line marked on it, selection of objects (small world, natural, loose parts, cubes, etc.). Card/paper can be placed either landscape or portrait.

  Child 1 chooses five objects (the number of objects can be altered to suit the children involved) and places them on the A4 card. Child 2 provides the mirror image on the opposite side of the card. The children then swap roles and the activity starts again. Challenge can be added by Child 1 placing the objects so that they are not in a uniform line. Child 2 makes the symmetrical pattern.

  Further challenge can be added by the children having a piece of paper each and a screen between them. Now they are arranging the items on the whole piece of paper (no midline marked on paper) and child 1 is required to describe the items used and position of these items on his/her whole piece of paper.

  Find 10: Ten Frame Symmetry – Materials: Make a large tens frame – draw one on sugar paper/on a roll of backing paper or make one using masking tape stuck to the floor, selection of objects (child's choice – small enough to fit more than one item into the boxes of the frame).

  The tens frame can be either portrait or landscape orientation; the children will be reflecting or showing symmetry across the mid line. Child 1 arranges the items in either 5 boxes in the first column or the first row. Child 2 then puts items in place to match Child 1's, making a symmetrical representation.

Class/Group activity

- **Story** – Read *Have You Seen My Dragon?* by Steve Light. While reading the story with the children, encourage them to look for symmetry in the illustrations (e.g. rotational symmetry in drain covers; symmetry in window arches and designs). This is a fabulous opportunity to discuss what is symmetrical and what is not and why.
- **Face Drawing** – Give the children half a 'picture' of a human face with a line down the mid-point. Ask the children to complete the other 'half' by drawing in features (eyes/nose/mouth/ears/hair) to make it identical on both sides. Ask them to talk about the positions of what they have drawn.

- **Symmetrical Pictures** – Can the children make symmetrical pictures using sticky shapes or 2D shapes? See examples here of the type of picture. Ask questions such as: **What shapes can you see? How many circles can you see? How many different shapes are used for the reindeer?**

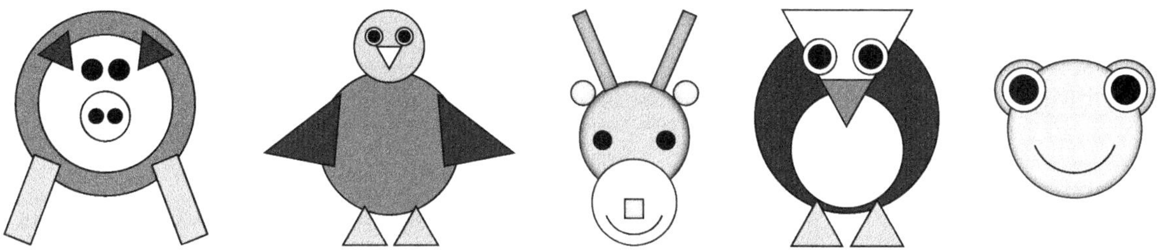

- **Ladybird Symmetry** – Provide the children with the outline of a ladybird; can they make their ladybird into a symmetrical ladybird? Can they draw spots on it to do so? Looking at pictures and photographs of ladybirds adds authenticity to this task. Add challenge by suggesting the ladybird needs to have a particular total of spots, e.g. 6 or 8 or 10 (this may reflect a knowledge of doubles or the ability to share 'out' spots). **If we have six spots, how many are there on each side of the ladybird?**

What to observe

- Children use appropriate mathematical vocabulary to describe how they made building 'constructions' using blocks/3D objects. They can make a different model using the same pieces. They can describe the model from different perspectives (from above, side view).
- Children describe how to move to a specific location or give directions to get a toy from one location to another (including programmable toys).
- Children can follow instructions that use position and movement vocabulary and can make observations from different viewpoints (e.g. looking down from the climbing frame or looking through a tunnel) when outdoors or in the hall, and when using apparatus. They can use the vocabulary of position and movement.
- Children can describe what they can see in a picture when reading stories or looking at pictures or drawings? For example, in the story of 'Goldilocks and the Three Bears', they can describe if mummy bear was *behind* daddy bear or if the bears were going *into* a forest or coming *out* of a forest?
- Children, using mirrors or making symmetrical patterns, can describe the patterns, pictures or shapes. 'What can they see in the mirror?' When exploring symmetry through art, they can predict what their 'fold and blot' picture will look like. They can predict where the coloured blobs will appear.
- Children who point and say 'it's there' when asked about the position of an object, may be unsure about the positional language and will need more support.

11 Data and Analysis

Data and Analysis A

Section overview

Units of learning	Review units	Prior learning	Benchmarks/Es&Os
Unit 11.1 **Collecting objects and sorting them**		• Children should have had a range of early experiences in collecting objects/items and sorting them both outdoors and around the home. • Children need to have recognition of colour/s to enable them to complete some sorting and matching tasks. • Children should have had opportunities in exploring everyday objects e.g. by texture – hard/soft/rough/smooth – to further develop an awareness of same and different (not the same).	**MNU 0-20a** I can match objects, and sort using my own and others' criteria, sharing my ideas with others. **MNU 0-20b** • Asks simple questions to collect data for a specific purpose. • Collects and organises objects for a specific purpose. • Applies counting skills to ask and answer questions and makes relevant choices and decisions based on data. • Contributes to concrete or pictorial displays where one object or drawing represents one data value, using digital technologies as appropriate. • Uses knowledge of colour, shape, size and other properties to match and sort items in a variety of ways. • Interprets simple graphs, charts and signs and demonstrates how they support planning, choices and decision making. I can collect objects and ask questions to gather information, organising and displaying my findings in different ways.
Unit 11.2 **Matching objects and talking about the similarities and differences**		• Children should be able to recognise when objects are the same/not the same. They should have had experience in matching a range of objects or pictures. Jigsaw puzzles provide experience in matching as they are trying to build the puzzle to match the accompanying picture. Everyday items including patterns, shoes, gloves should be included. Playing pairs electronically on an iPad/tablet and/or in a card game format provides other exposure to same/not same. A supportive adult and appropriate vocabulary input will aid early development.	
Unit 11.3 **Recognising signs and charts in a familiar context**		• Children should be confident in daily schedules and recognise that signs give us information.	
Unit 11.4 **Collecting data by grouping items/objects together**		• Children need to have had practice in making collections; looking closely at objects and thinking about similarities and differences. • After being to the supermarket the child can sort foods – tinned foods, fruit, things that go in the fridge	

Units of learning	Review units	Prior learning	Benchmarks/Es&Os
Unit 11.5 **Displaying data by creating a picture using the concrete materials collected**		• Children should have had lots of practical 1:1 experience, e.g. one fork to one knife when setting the table, matching up pairs of socks after they have been washed and dried; they should be able to talk about what they have done.	
Unit 11.6 **Asking simple questions to collect data**		• By listening to others asking appropriate questions the children should begin to develop their abilities in asking questions. • Children should realise that by asking questions they get answers to something that they want to know.	
Unit 11.7 **Recognising signs and charts outside the immediate environment**		• Children should have had exposure to environmental print.	

Conceptual context

Handling data is an integral part of real life and of the everyday activities in a Nursery setting; it is a key tool for solving problems. Children should be engaged in sorting and data handling tasks which are related to their own experiences; where possible the initial question should be their own and they should be involved in all stages of the data collection and representation in order to be able to interpret the data.

While tidying up, children are sorting toys into the correct storage containers, rearranging the correct items onto the shelf in the shop corner but in their daily routine, they also are being given opportunities to collect information for real world purposes, e.g. How many children are here today? How many plates will we need for snack? Supportive adults need to model, introduce and consolidate vocabulary which is important to these experiences.

Sorting and matching are important mathematical skills in the early stages of handling data; to be able to sort and match objects, children need to have a concept of the attribute they are using for comparison. Sorting is different from matching as it involves reorganising a whole collection (set) into two subsets. Children in Nursery should be able to move objects from one group to another to sort them and to say why objects belong in each of the groups. They are matching items that are fully alike; matching at this early stage is object to object but later when dealing with pictograms and block charts the children will be required to match the object to a word or sign. At this stage much of this sorting is purposeful to them as individuals and it can be difficult to sort according to an attribute chosen by someone else. For example, Nuria may sort a collection of ribbons into the ones she likes and the ones she doesn't like but may be unable to sort when asked to put all the patterned ones together. At this stage in their development, children may find it difficult to interpret someone else's sorting and to interpret graphs created by others. These are points to bear in mind when planning activities.

Children should progress, with enough experience and support, to being able to sort by one attribute. If the activity involves 2D shapes, it is important to be certain that they recognise the similarities and differences presented by 2D shapes, especially if they are being asked to find all the triangles. A different activity with shapes may involve the child being asked to look for shapes with three sides. It is important to check that they know what three is. Children can be asked to sort a collection of footwear so that all the socks are together. This collection of things worn on feet may include wellies, sandals, slippers, socks, shoes, flip-flops. I always think, where possible, it is valuable to include a 'mystery object' – in this case a slipper sock – to promote discussion and to add some challenge. Is it a sock or is it not a sock? How does it differ from other socks? By using objects that the children have experience of, supports early development, or by creating a story round the objects to be sorted intrigues the children further.

Pre-school children often find it difficult to answer 'how do you know' questions which seek them to give justification. e.g. 'How do you know that more people like fish fingers?' and often will not respond. With the assistance of a supportive adult who models and encourages and shows the children how to 'analyse' the graph/pictograph, etc., children will gain confidence and can begin to formulate their own answers. Young children also find difficulty in asking questions; many are unsure what a question actually is. When the questions are related to a pictogram, for example, the children will need many occasions where an appropriate range of questions are modelled. Young children vary in their abilities to use spoken language; some will be highly verbal and able to explain their thinking whereas others may have emergent language skills. It is very important to take this into account and pay particular attention to what the children are showing you rather than what they are telling you.

Representing data can be time-consuming when done by hand; however, it is very important for young children to be involved in and engage with the process of representing their data. It develops more of an understanding of the process, supports ownership of the data and enables meaningful discussions to ensue with the supportive adult. In the earliest stages of representing data, young children use themselves to show the numbers in each category, e.g. eye colour, favourite ice-cream flavour, etc. The children constitute a frequency table when they line up beside the sign or symbol which indicates their eye colour or favourite ice cream flavour. The next step for young children is to use a pictorial representation of the data. A typical context might be favourite pets. Children, after being questioned about what their favourite pet might be, will choose a picture of their favourite pet. The children may be asked to hold their picture up if their favourite was a dog and a child may be invited to count. It becomes apparent that this can be a tricky task for the child trying to see who has chosen a dog. Children should be guided to see that if the pictures of all the dogs were brought to the carpet area or numeracy table and then arranged in a row/line beside a sign for dog, they would be easier to count and interpret. This becomes a pictogram. Pictures can be replaced with blocks when children can cope with the increased level of abstraction involved, and thus Block Graphs are introduced.

The most important thing that children can learn about data and analysis is why we actually do it. We are challenged to ensure that children see the purpose of collecting data to answer questions when answers are not immediately obvious. With scaffolding and the support of a thoughtful adult, young children will be able to follow the steps involved in data analysis. Once they have several different experiences under their belts, they begin to see that data needs to be represented so it can be interpreted and that how data is gathered depends on the question. Children should be involved in thinking about what the 'big' question might be. It is important that they understand that the big question is what drives the whole investigation; we are striving to get children to notice what is happening and then reflect and wonder what might happen next.

Resources

- Sorting hoops
- Mark making materials and paper
- Commercially produced sorting toys
- Cotton reels
- Buttons
- Clothes pegs
- Sweets
- Pasta
- Coins
- Socks
- Gloves
- Shells
- Leaves
- Pictures/photographs
- Ribbons
- Sea glass
- Pebbles
- Seeds
- Loose parts materials
- Feely bag
- Stories – *My Mum and Dad Make Me Laugh* by Nick Sharratt, *Handa's Surprise* by Eileen Browne, *Socks* by Nick Sharratt and Elizabeth Lindsay

Digital

- Carroll diagram
- Venn diagram
- Pictogram

Record book

Pages 29–30

Key vocabulary

- **Chart**
- **Collect**
- **Data**
- **Different**
- **Display**
- **Match**
- **Not the same**
- **Pairs**
- **Same**
- **Set**
- **Sign**
- **Similar**
- **Sort**

Misconceptions

Potential misconception	Addressing the misconception
Units 11.5 and 11.6 Children may not be able to count accurately how many are in a set	• Practice is required in 1:1 correspondence, number order and using appropriate vocabulary.
Units 11.2, 11.5 and 11.6 The language of maths can cause difficulties as children handle data	• Practitioners need to model and promote useful and appropriate vocabulary in all contexts – not just during guided teaching activities but also when playing, carrying out daily routines and talking generally about things happening in daily lives.

Teaching ideas

Unit 11.1 Collecting objects and sorting them

Whole class

- **Same and Different Game –** Seat the children on the carpet in a circle arrangement with a large blanket in the middle. Place several objects on the blanket such as soft toys, cars, toy trucks, building blocks, shapes, pebbles, buttons, dolls, toy animals, pencils, pens.

  The adult chooses two objects from the blanket and then asks the seated children: **What is the same about these two objects?** Once a few answers have been given, the adult then asks: **What is different about the objects?**

  'I chose a toy car and a pony. The *similarity* between these objects is they are linked by 'four', four wheels on the car and four legs on the pony. The *difference* is that one is red and the other is not.' Ask each child to select two objects. Let the children know that you are excited about what they might choose and you are wondering about what they might say. You can start by only looking for a similarity. Once the children have more experience, introduce the concept of difference.

Stations

- **Pasta Sort –** Provide the children with a bag/tray of mixed pasta pieces. Encourage the children to explore the pasta and ask them to think about how they might sort it. Colour? Shape? Size?
- **Money, Money, Money** – Provide the children with a collection of money. Use real currency where possible. **What might be the big question that the children will think of? How can the money be sorted? Paper and metal? Shiny and dull? Silver and bronze? Size? Shape? Numbers (1s, 2s, etc.)?** Older children may wonder what value each set has once the sorting is complete. They may need support to add up the values of the coins.

Unit 11.2 Matching objects and talking about the similarities and differences

Whole class/group

- **Pairs Card Games** – Lots of commercially produced pairs games exist; children can match picture to picture, picture to word or word to word, dot pattern to number, dot pattern to dot pattern and number to number. Start off with a small number of cards until the children's concentration improves then gradually increase the number of cards used. Set the cards out in a rectangular array arrangement, face down and encourage the children, to take turns, and to turn over two cards. Try and ensure that the cards are turned over where they are in the array. This helps recall and visual memory. The children need to concentrate and remember where each of the items were. If a child turns over two cards that match, (for older children, ask: **Can you identify what is on the card?**) they can then keep that pair of cards. The winner is the person with the most cards at the end.
- **Paper Plate Pairs** – This is a variation on a pairs card game. Paper plates are bigger and easier to turn over and they are a blank canvas for you to put on them whatever is pertinent to the children. The range of numbers used can start with 1–3; variations include matching number to number or number to regular dice pattern within 6. Think about age and stage and what the children need to practise.
- **Feely Bag Fun** – Lots of quick and easy activities can be based on the use of a feely bag and a pass-the-parcel type routine. For example, seat the children in a circle and scatter some numerals (alter the range of numerals displayed to suit the needs of the children) in the middle of the circle. Remember that the children may not see the numerals on the floor in the middle of the circle in the correct orientation, as they will be upside down to some of the seated children; however, each child can stand and walk freely round looking at the numbers to match an item from the feely bag if they are unsure. Place a selection of Numicon shapes in to the feely bag. Children should be familiar with Numicon shapes before carrying out this activity. Pass the bag round the circle, accompanied by music if appropriate. When the music is stopped, the child holding the bag is asked to remove one Numicon shape from the bag. The child then stands up and walks around looking at all the numerals scattered inside the circle. Once located, they place their Numicon shape beside the numeral. The seated children can be asked to verify if the match is correct. Dice pattern cards, plastic animals and insects and finger pattern cards can be substituted for Numicon. When selecting plastic animals and insects look for ones with differing numbers of legs, e.g. penguin, crab, dog, beetle.

Stations

- **Find My Match** – Children are given a single sock and bag of socks. They search in the bag to find the sock which is an exact match for the original sock.
- **People Sort** – This activity can be done outside. Draw two large chalk circles and a chalk line. Children stand on the chalk line; they are asked a variety of questions and have to listen carefully. For example, 'If you are wearing blue, step into the chalk circle that has a blue cone beside it. If you have nothing blue, move to the other chalk circle.'
- **Leaf Sort** – Collect leaves in autumn. Initially they can be sorted by colour. Encourage the children to think about what might happen to the collections when we try to sort them by shape? Support and guidance will be required to help children re-sort in this scenario.

Unit 11.3 Recognising signs and charts in a familiar context

Whole class

- **Signs** – Children will benefit from daily interaction with signs and the information they present. This can be achieved by creating a playroom where: key furniture is labelled; storage trays/tubs are labelled, usually with both a picture and word combination; there is pictorial and numerical information, such as visual timetables, registration and weather charts, snack time choices (and the quantity) or how many can play at a particular area is shown.

  Using signs to label key equipment means that children can assist in tidying up. For example, glue sticks can be numbered with a sticky label on the base and a corresponding number on a sticky label on the lid so a matching exercise is carried out when asked to tidy. The children will also become familiar with common words through this approach.

Stations

- **Washing Line** – Set up a washing line. Provide a basket of pegs and cards with symbols, e.g. socks. The children peg the sock card to the line and then search through the items in the bag selecting all the socks. The children then peg the socks to the line. This activity can be repeated using other items.

Unit 11.4 Collecting data by grouping items/objects together

Whole class

- **Scavenger Splash** – Tell the children that the current activity in the water table needs to be changed. The plan is that it is going to be set up as a float and sink discovery zone. Explain that you need their help to make it very interesting. Ask the children to think of objects that are in the playroom that might be good to investigate. This is a good time to make a list of inappropriate objects, with the children, that might not be welcome in the water tray – no personal items of clothing, worms from the garden, biscuits from the snack table, etc. Organise the children, as appropriate, and ask them to find an interesting object. Objects can be gathered in a large tub which is accessible to all. Once all the items have been collected the tub can be emptied out and the contents studied. Provide three large sorting hoops – labelled *float*, *sink* and *don't know* – and encourage the children, one by one, to place their item into the correct hoop. The contents of each hoop can be transferred to 3 separate marked trays and taken to the water table. The children can engage in checking if indeed the object used did float or sink. What about the objects that were in the *don't know* hoop/tray? Where would they now be placed? If the predictions were wrong, the materials may need to be resorted.

Stations

- **Box Collection** – Provide each child with a smallish box (smaller than a shoe box is best). Send them to the outdoor area to find objects that will fit inside the box. When the box is full or nearly full ask them to bring their box to a table, e.g. in the maths area. The children should be encouraged to tip out their box and see if they can find similarities and differences in the objects they have found. Support them to label and talk about their collections.

- **Magnet Walk** – Give a group of children a small magnet and set them off on a magnet walk (accompanied by a supportive adult with a bag) to see what they can pick up using their magnet. Encourage discussion while checking objects. Small objects that stick can be transferred to the bag for later investigation. Once the activity is complete (indoors and outdoors), empty the bag out and look together at all the items. **Why were these objects picked up by the magnets? Can anyone think of objects that didn't stick to their magnet?**

  Set up a tray of objects, some of which were picked up by the magnets but also include some that were not (tins, nails, screws, clips, sticks, pebbles, small plastic toys, wooden items); encourage the children to explore and then to sort the items into two hoops labelled *stick* and *non-stick*.

Unit 11.5 Displaying data by creating pictures using concrete materials collected

Whole class

- **Duplo Pictogram** – Tell the children they are going to make a Duplo pictogram. Ask each child to select a piece of Duplo from the Duplo tub and then sit all the children in a circle on the carpet. Most Duplo tubs contain a selection of Duplo bricks (Lego), people and vehicles. This is ideal. Have paper (to create labels), felt pens and hoops with a couple of Duplo pieces in the middle ready.

  Going around the circle, ask the children one-by-one to bring their Duplo piece into the middle of the circle; they place their Duplo piece in a hoop beside something that is the same or similar. Continue until everyone has had a turn. The sorting criteria depends on how the first few children place their pieces in hoops as they make decisions that decide how the sorting will be defined. It may be that the sorting is by colour. All the red bricks together, yellow bricks together, etc. Labels could be made showing the brick colours and the bricks can then be lined up above the label to form a pictogram. Children can then engage in counting.

  The sorting can be more complex and vary each time the game is played. Within your Duplo/Lego box you will have a variety of different pieces. Think about encouraging the children to tip the box out and look for commonalities. Encourage the children to explain what they have done. Some ideas that the children might sort by: Duplo vehicles – wheels and no wheels; Duplo/Lego people – people and bricks. Carry out an investigation of the number of circles on top of each brick – 2s, 4s, 6s, 8s, etc.

  These all represent 4. Check that the children can see why the items in the picture represent 4. Ask the children to find some Lego that might represent, e.g. 2. **What other numbers might be represented if other bricks were used?** The children will need to have had opportunities to sort the bricks on previous occasions. Can they remember what each brick's value was without counting the dots on top? (Conservation of number.)

Stations

- **Snatch and Grab** – Materials: a tub with counters, a small bowl or dish for each child, name card for each child in the group. Children take it in turns to grab a handful of counters from the tub (using only one hand) and place the counters into their own bowl. Once every child has had a turn, the adult will place each child's name card in a row. Children then can lay out their counters, from their bowl, in a column above their name. **Who has the most/least?** For older children the difference between two children's counters could be found.

Unit 11.6 Asking simple questions to collect data

So that children can ask questions they need to understand what a question is and have opportunities to hear questions being asked. Children need to hear and see adults modelling appropriate vocabulary and using effective questioning in all areas of their daily routines.

To encourage the development of a child's ability to ask questions, we can:

- Create an environment where children feel secure and have the confidence to speak out;
- Read story books together and encourage participation by using effective questioning, e.g. *The Most Magnificent Thing* by Ashley Spires (fiction) and *A Chicken Followed Me Home* by Robin Page (non-fiction);
- Encourage curiosity;
- Demonstrate questions;
- Value the questions that children ask.

Stations

- **Play Question Ping Pong** – As with lots of the activities already described, effective questioning can take the learning to a higher level. Initially this game is played with an adult leading. The child and the adult (in this example) collect several small items in a box (indoor and outdoor) and then tip out the contents, sorting it, for example, into similar and different or by colour or by using their own criteria.

  The adult may ask: **I wonder why you have put this stone with the smooth pebble?** The child should be able to justify their criteria while responding to the question. The child then takes a turn at asking a question to the adult; they may ask the adult: **Why have you put this twig beside the fir cone?** The adult then justifies their choice. This is not an activity to be hurried and can be used for any age.

Unit 11.7 Recognising signs and charts outside the immediate environment

Whole class

- **Devise and Use Environmental Print Matching Activities** – Make a game of road sign bingo, lotto or pairs. The children should have opportunities to talk about and look at signs in their environment. Examples of typical signs that the children may see are shown here. Check if they can state what each sign 'says' and in this case what they mean. They may be able to tell you where they have seen such signs. **How many red signs can you see on the way to Nursery? Why might the sign be red?**

- **Symbols and Logos** – Use symbols and logos that the children are familiar with. Stick a set of matching logos on to paper plates and play the matching game.

- **Make a Collage of Wrappers** – Ask the children to collect wrappers of their favourite cereal or favourite snack and bring them to nursery to make a collage/display. After collecting the wrappers and making their display, encourage the children to think about how they might go about disposing of the paper/cardboard that their wrappers were made of, in a responsible way. This may lead to sorting the papers and wrappers. **Is the wrapper recyclable? How would they know or find out?** The children could use magnifying glasses and check for the recyclable symbol. This will promote another use of signs! It definitely is a big question and something we all need to be responsible about. **Have they noticed how litter in their local environment is recycled? Have they any suggestions for their nursery?**

What to observe

- Children who show an interest in the attributes of objects.
- Children who can sort objects using one attribute; they can talk about how they have sorted the objects.
- Children who can to match objects.
- Children who are beginning to make purposeful collections.
- Children who can identify similarities and difference between objects.
- Children using questions.
- Children who can collect data and display it or participate in the display.
- Children who can count how many items are in a set.

Data and Analysis B

Section overview

<table>
<tr><th>Units of learning</th><th>Review units</th><th>Prior learning</th><th>Benchmarks/Es&Os</th></tr>
<tr><td>Unit 11.1
Using signs and charts to plan, choose and make decisions</td><td>Data and Analysis A
Unit 11.3</td><td>• Children should be familiar with everyday charts and signage which are linked to daily routines. E.g. weather charts, snack charts, who is here today at school; all providing data which can be investigated by the children.
• Children should have experience in different types of charts/graphs and be beginning to think which one will display their data in the best way.</td><td rowspan="4">MNU 0-20a

I can match objects, and sort using my own and others' criteria, sharing my ideas with others.

MNU 0-20b

• Asks simple questions to collect data for a specific purpose.
• Collects and organises objects for a specific purpose.
• Applies counting skills to ask and answer questions and makes relevant choices and decisions based on data.
• Contributes to concrete or pictorial displays where one object or drawing represents one data value, using digital technologies as appropriate.
• Uses knowledge of colour, shape, size and other properties to match and sort items in a variety of ways.
• Interprets simple graphs, charts and signs and demonstrates how they support planning, choices and decision.

I can collect objects and ask questions to gather information, organising and displaying my findings in different ways.</td></tr>
<tr><td>Unit 11.2
Collecting data by counting one-to-one</td><td>Data and Analysis A
Units 11.1 and 11.2</td><td>• Children should have had experience of working with tally marks; experience in counting in fives is useful.
• Children should have had a range of experience to develop 1:1 correspondence to improve the accuracy of their counting.</td></tr>
<tr><td>Unit 11.3
Displaying data by creating a picture or using technology</td><td>Data and Analysis A
Units 11.5

An Introduction to 2D Shapes and 3D Objects A</td><td>• Children should have had experience in making graphs with real objects but also using pictures to making pictographs. Opportunities should be available for children to interact with data handling programmes on the Smartboard allow for the manipulation of objects in different contexts.</td></tr>
<tr><td>Unit 11.4
Asking simple questions to collect data and making suggestions how to do this</td><td>Data and Analysis A
Unit 11.6</td><td>• Through experiential learning and scaffolding in Nursery, children should have developed an understanding of the processes involved in data handling. They need to understand what a question is to be able to ask a pertinent one but also have a grasp of the vocabulary required.
• Children need to have sufficient experience to enable them to identify the attributes of objects.</td></tr>
</table>

Conceptual context

Children live in a world full of data, presented to them as pictures, music, spoken word, numerals and objects. Data handling enables the development of ideas, discussions and investigations across a range of subject areas within and beyond mathematics. Children's questions and observations can be explored using data handling and this mathematical approach forms a framework for practical problem solving and talking about their experiences. A child's use of data needs to be purposeful, meaningful and relevant to them.

Children's understanding of data handling can be traced back to their earliest experiences of sorting everyday items. As their understanding develops, they can represent information in increasingly abstract ways and answer more complex questions. Unfortunately, many of the collections that are provided in schools for children are multi-coloured plastic sorting sets which are purchased from educational catalogues; these can be sterile and uninspiring for children with a fascination for learning. Suggested materials which are versatile and interesting to sort are contained in the resource section; look carefully at what you are providing. Ask yourself, how might I sort these materials? Can I think of another way of sorting them? Look for resources which are flexible.

There are four critical questions and steps which children should learn to answer/use to develop their data-handling knowledge and skills.

1. Posing a question: the big question – What do I want or need to know?
2. Collecting the data – How am I going to get the information that I need?
3. Representing the data – How am I going to organise and represent/display the information?
4. Analysing/interpreting the data – What did I find out?

Children need to experience scenarios where, once the data has been interpreted in terms of the original question and the problem still exists, then a new question may need to be formulated to enable answers to be found.

Children develop through progressive stages in data and analysis development:

- Sorting by one attribute – Isla can sort the shapes. She was given the shape tray and asked to find all the yellow shapes. She has put all the yellow shapes in one sorting ring and left all the others together in the tray. She cannot explain that the shapes remaining in the tray can be categorised as being 'not' yellow.
- Sorting by more than one attribute - Nuria can sort using two attributes at the same time; given a tray of shapes she is able to sort the shapes that are both red and square. However, when presented with Amman's shape sorting, she was unable to work out his criteria for sorting.
- Stating the rule – by this stage Nuria can state the criteria she used to sort her shape collection but can now also state what criteria Amman had used to sort his.

A wide range of scaffolded sorting experiences will enable children to progress through these stages. It is important to remember that there always should be a reason for collecting data.

Once preschool children are confident in sorting and matching (see Section A) they should be exposed to a wide range of supported experiences in representing data using pictograms and to a lesser degree, block diagrams. As children progress through Primary 1 they should have opportunities to represent data using Block diagrams, Carroll diagrams and using 1:1 correspondence in the form of tally marking.

Carroll diagrams are very useful for recording the results of a sorting problem. In Carroll diagrams a label is given to each section to show whether the items within each section meet or do not meet the given criterion. An understanding of the language of not meeting a criterion is vital for the Carroll diagram. Carroll diagrams can be used to represent sorting by one criterion or for two criteria.

One criterion

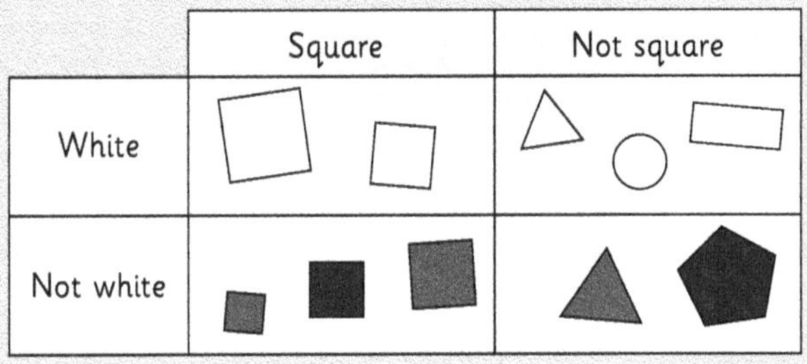

Two criteria

In using a block diagram, the children are substituting a block for the picture previously used to display information in a pictogram. When appropriate, blocks can be replaced by interlocking cubes and the 2D representation becomes a 3D block graph. At this level of development, for most block graphs and pictograms, it isn't necessary to have a vertical scale (y-axis) as the information is found by counting the blocks or pictures. The horizontal category (x-axis) is necessary to keep each line of blocks or pictures level to allow for comparison. Signs and labels for each category are located on the x-axis.

Children should naturally progress to a bar chart where data is represented as a continuous bar. Now each individual piece of information is no longer instantly identifiable and a vertical axis (y) must be labelled to allow the reading of the measurement at the top of each bar. Labelling of the vertical axis resembles a number line, normally starting with zero.

Another way of displaying data in a 1:1 correspondence is a tally chart. In a tally chart each piece of information is represented with a single mark. A frequency table is used to show how many times each data category has occurred/been chosen. Tally marks are arranged in 5s; they are very useful for counting larger numbers. Counting in 5s is a more efficient way of counting rather than in 1s.

Children in Primary 1 should have opportunities to further develop their skills involving the use of both fact-finding and preference surveys. Ideas for group and whole class activities can be found in the Teaching ideas section. When carrying out surveys, having an appropriate question to ask is vital; the question is essential in driving the data investigation, collection and interpretation.

Resources

- Sorting hoops
- Mark making materials and paper
- Commercially produced sorting toys
- Cotton reels
- Buttons
- Clothes pegs
- Sweets
- Pasta
- Coins
- Socks
- Gloves
- Shells
- Leaves
- Pictures/photographs
- Ribbons
- Sea glass
- Pebbles
- Seeds
- Loose parts materials
- Boxes
- Cars
- Ribbons
- Toy animals
- Dice

Digital

- Carroll diagram
- Venn diagram
- Pictogram

Record book

Pages 29–30

Key vocabulary

- **Chart**
- **Classify**
- **Collect**
- **Common**
- **Criteria**
- **Data**
- **Different**
- **Display**
- **Estimating**
- **Favourite**
- **Graph**
- **Group**
- **List**
- **Match**
- **Not the same**
- **Order**
- **Organise**
- **Pairs**
- **Same**
- **Set**
- **Sign**
- **Similar**
- **Sort**
- **Bar chart** – a graph using bars to represent quantities and numbers
- **Block graph** – a graph which uses blocks to represent items; unless specified one block equates to one item
- **Carroll diagram** – used to sort numbers and objects; the numbers and objects can be categorised as belonging or not belonging, e.g. red or not red
- **Frequency table** – a table used to organise data; it lists a set of scores and their frequency
- **Tally marks** – used to record the frequency
- **Pictogram** – a pictorial representation of data on a graph or chart

Misconceptions

Potential misconception	Addressing the misconception
Units 11.2, 11.3 and 11.4 The language of maths. When a child is asked, e.g. how many more than - the child gives the total number that is in the 'most' group as their answer.	• Work on 1:1 correspondence Activities involving 'Finding' the difference between numbers. (Counters, correspondence grids, number lines or number tracks.)
Units 11.3 and 11.4 Understanding the 'not' criteria; children may continue to put data that 'is' into the 'not' set	• Lots of practical activities: ○ Start by physically removing the items that do not belong (red and not red – take away all the items that are blue/green/yellow, etc. leaving only red. ○ Make symbolic cards – a card symbolising not red would be red in colour but would have a large cross through it. ○ Use in-class routines – who is **not** having a school lunch today.
Unit 11.2, 11.3 and 11.4 Children may struggle with counting – counting how many are in the set and how many altogether.	• Teach strategies to enable effective counting (1:1, cross off, arrange objects in a ten frame, or similar, to avoid re-counting) and encourage counting on rather than counting from 1.

Teaching ideas

Unit 11.1 Using signs and charts to plan, choose and make decisions

Whole class

- **Missing Property** – This whole class activity centres around the Lost Property Box (Lost and Found Box). A real problem exists in all schools with clothing, toys and footwear being found in classrooms at the end of the school day and in corridors, cloakrooms and in the playground after break. A solution is very apparent to all teachers – if the clothing was labelled then this wouldn't be an issue.

  The Missing Property Box is a relevant problem to engage children in mathematical thinking, talking and learning.

  The children should be engaged with the teacher in thinking about identifying a *big question* and planning what kind of data they might want to collect; the method of collection should emerge naturally.

  Typical *big questions* could be: **What does the Lost Property Box look like? How would we know which box it is? What items are in the Lost Property Box? Do items have names on them? Are all the items things you might wear? Are there pairs of items in the box? Do children in P1 lose more items than children in P7? Do boys lose more items than girls?**

The first step would be to locate the box labelled Lost Property (usually in the Janitor's office). Children could be involved in opening the box and making an inventory of the contents with adult support. Sort the items into categories and count them. If the question selected was about boys and girls and who lost more items, then two categories would be required. Items would need to be sorted into a collection for boys and a collection for girls. Each collection would have an appropriate sign. Depending on the number of items in each collection, the children may need to use tally marks to count the objects or they may use the signs (boys and girls) and lay the objects in a tower above each sign for ease of counting. The objects could then be substituted by blocks to make a display. The data in this case could also be represented using a Carroll diagram. This Investigation could lead to further studies if the contents of the box are re-sorted using different criteria. This activity could lead on to the pupils making 'I notice …' statements and even a further 'I wonder …' question that could extend the investigation.

Whole class

Using daily activities such as recording the weather, which is usually displayed alongside day, date and month information, with a little adaptation, this information could be collected for a month, for example, to provide useful data for the children to display. They could investigate the information that they gather and hopefully come up with several questions (depending on their interest), e.g. **How many wet days were there in March? How many wet Mondays were there in November? Were there any sunny days in October?** Remember to use the steps when guiding children to carry out their investigation:

1. Ask the *big question* – **What are we finding out?**
2. Make a plan – **What do we need to know?**
3. Collect the data – **How do we do this?**
4. Present the data
5. Interpret the data – *I notice* and *I wonder.*

- **Using a Pictorial Stimulus:** Dinosaurs and other things – Imagine a scenario where a child has been given, for example, a collection of dinosaur stickers. As the adult I might ask the child what they notice about the stickers and could they tell me a little about what they can see. **Are the stickers all the same? Could they sort them? Would colour be a good way of sorting them?** I would provide encouragement to get them to sort the stickers into collections or sets. Through discussion, the child has decided the sorting they did (see picture) should be titled 'Dinosaurs and other things'. They asked to make a sign to accompany they picture. The picture shows what the child chose to do. They said they noticed the legs on the dinosaurs; some had four legs and some had two. They sorted the dinosaurs into four legs and two legs and then left the other items, entitling them 'other things'. They didn't know what to do with the dinosaur with wings, although it did have two legs; they hadn't heard of flying dinosaurs and didn't feel it should be sorted with the other legged dinosaurs. They justified their decision making. The picture of 'Dinosaurs and other things' was then presented to the whole class. It was left on display and the other children in the class were encouraged to look at it and wonder how they might have sorted the stickers. **Did they agree with what had been done?** The children shared their thoughts and ideas. The original sorting was photographed and the stickers were jumbled up. Another child can then be chosen to re-sort and thus the investigation is extended. Children could be encouraged to 'wonder' what they might do.

Stations

- **Sorting Stations** – Encourage the children to think about signs for their collections when engaged in sorting activities. This may be linked to using Carroll diagrams or pictograms.

  Key questions to ask are: **What signs might you need to write for this sorting? If you are not able to write the captions, how could you display the captions or signs?**

  For the Numicon collections shown here, you can ask: **Can you tell from looking at these two collections what Isla's criteria for sorting her Numicon pieces was? How would you label them?** Isla asks Finlay to sort Numicon in another way and she will try and work out what he did.

Unit 11.2 Collects data by counting one-to-one

Whole class

- **Estimating Activities** – Quickly show a picture; do not allow children to count the items in the picture. Ask the children to estimate, in this example, how many dots they see.

  Try and encourage everyone to have a go at estimating. Once everyone has had their guess, the class all join in and count the dots as the teacher points to them, touching each spot as it is counted. Ask the children how many dots there are altogether. The aim is to get the children to become closer in their estimation but it is a good opportunity to practise 1:1 counting.

Stations

- **Encouraging 1:1** – Children need as many opportunities as possible to ensure their 1:1 counting is accurate. Using the one illustration, e.g. the cartoon dog with eight dots, it is possible to provide small group activities to meet the varying needs of children in the group.
- **Cartoon Animal Dots** – Nuria has a picture of the cartoon animal. She finds it difficult to count the dots and seems to count some of them twice. Nuria can recite her numbers in order but gets confused when she sees lots of items to count. To support Nuria and help her see the dots she has already counted, Nuria can be encouraged to get a pencil and cross off the spots as she counts them. This makes the task much easier. Next steps for Nuria would be to develop her subitising ability. **What dot patterns can she see when looking at the spots?** Most people see an arrangement of 5 dots and 3 dots. Others see 3 and 3 and 2. This would help her to focus on the dots.

  Other children may need the image of the cartoon dog enlarged. This will enable them to physically cover the spots with counters rather than crossing off the spots. Finlay isn't always reliable at reciting numbers in order and crossing off dots still resulted in the wrong total. Finlay is provided with a bowl of counters and is encouraged to cover each spot on the cartoon dog with a counter. Finlay then can take the counters off the spots one at a time and arranges them in a line. He counts them as he puts them down. He then can check his counting to see if he was correct. He may need to place them on a suitable number track or number line.

Isla has the same picture as Finlay and has been asked to place a counter on top of each spot. Isla, however, has been provided with a blank tens frame to record her information on. She places the counters on the spots; she can either fill her tens frame arranging her counters in a five wise or pairs wise pattern. Isla has a knowledge of tens frames so can visualise the answer once it is in one of the dot patterns she can recall.

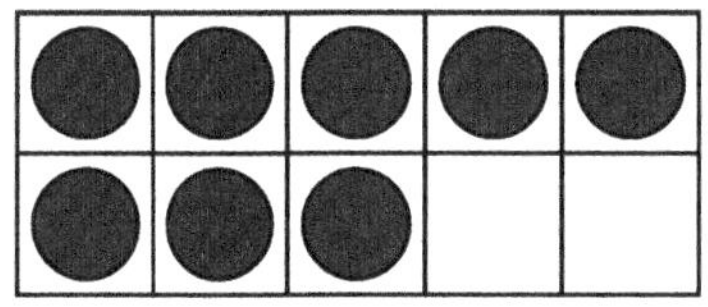

Five wise pattern for 8 on a tens frame.

Teacher led group

- **Movement and Counting** – Ask the children to watch carefully the movements you are about to make. You may jump several times or hop or perform a chopping movement with your hand without making a sound. The children are then asked to make the same movement the same number of times and to count with you.

  To extend this activity the children are asked to listen while you make a slow steady beat by clapping. The child counts them and tells you the amount. Playing Clock in the Tower helps to reinforce listening to the number of claps.

- **Clock in the Tower Game** – This game is best played where there is space to move around. Arrange the children in a circle and ask them to remain standing. Give each child a number (1–6). Ask a child to stand in the middle of the circle; this child becomes the clock. The clock child shouts 'The clock in the tower strikes the hour of …' The child does not say the number but claps the chosen number of times. All the children who are that number set off running around the outside of the circle and back to their own space. They run through their space and rush into the middle of the circle. The first child to touch the clock is the winner. They then become the clock.

- **Counting objects** – Have to hand a number of counters or other small objects (1p coins are good). The counters should be in reach of all the children in the group. The teacher or supportive adult are looking for a child's ability to exhibit 1:1 correspondence in their counting but also their ability to say numbers in order. Ask a child to remove a given quantity from the whole collection, e.g. 9. Watch and listen carefully as they count out the objects. **Do they have a counting strategy? Do they arrange them in a row? Do they slide them and say the number name as they do so? Do they miss any numbers out? Do they stop at the correct number?** This is a very useful assessment task.

- **Exploring Tally Marking** – Children will need to be competent with the concept of tallying and have experience of how to represent numbers using tally marks. Cotton buds, strips of paper, sticks and lollipop sticks are good resources to use to build tally numbers.

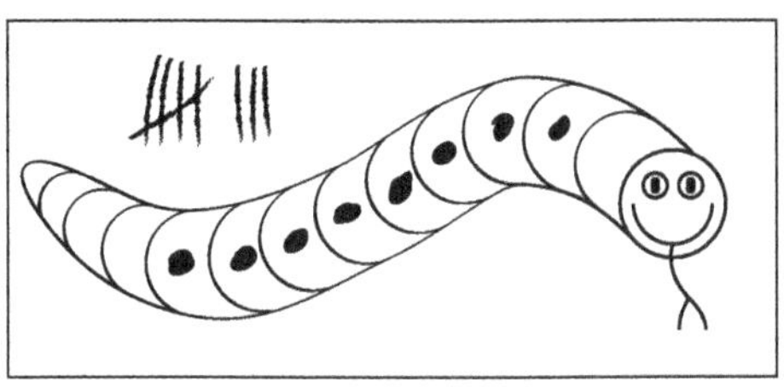

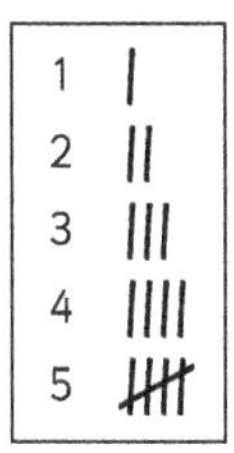

Any visual counting activity which gives opportunities to count in fives is advantageous:

- **Rekenrek** – There are 10 beads on each row of the Rekenrek and they are arranged in 5s: five red beads and then five white beads. It presents the children with a visual image of 5. Children can move the beads along the metal rod to practise counting in 5s. initially they may count in 1s but should be encouraged to push along groups of 5.

- **Hands** – Having five fingers on each hand is a useful aid to counting in fives.
- **Bead strings** (5 red and 5 white) – Children can move the beads along the string to practise counting in 5s. Initially they may count in 1s but should be encouraged to push along groups of 5.
- **Jumps along a number line** – The children can perform the jumps physically on a chalked number line too. Children can start exploring number patterns and see that if they start on 0 and begin to jump in fives, the numbers landed on end in a 0 or end in a 5.

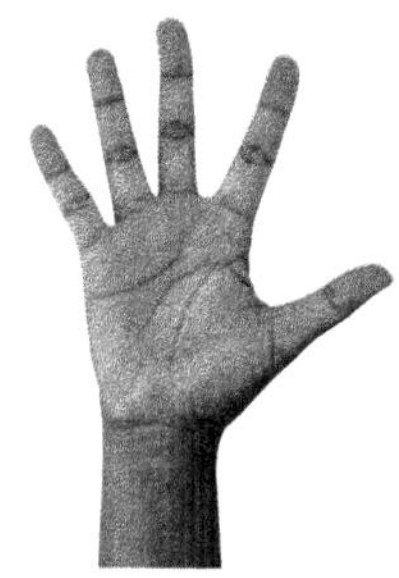

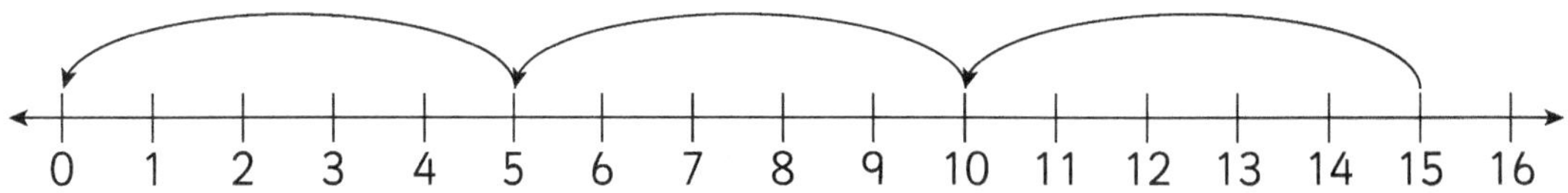

Unit 11.3 Displaying data by creating a picture or using technology

Whole class

- **Mr Chilly's Ice Cream Display** – Present the children with a scenario. Mr Chilly has an ice cream van. He has had a problem with two of the freezers in his ice cream van and as a result, Mr Chilly can only sell one flavour of ice cream. He normally sells vanilla, chocolate and strawberry. What flavour should he pick from these three to sell? Ask the children to think of their favourite flavour of ice cream from the list of chocolate, strawberry or vanilla. Provide the children with paper cone shapes. Two options are available here; either give blank cone shapes and the children colour their ice cream brown, pink or leave it white or provide the children with cone shapes which have the words chocolate, strawberry or vanilla printed on them and they have to select the one with the correct flavour on it.

  Ask the children: **How would you make a display showing the flavour that Mr Chilly should sell?** The children now have the question, a means of collecting the data, so they now need to think about how best to display it. Ask the children what should be done. Guidance and encouragement may be required. The children may need help to organise the production of a pictogram – creating signs and labels, sorting their cones into flavours, arranging them in towers, counting how many in each column and then making a label to state which ice cream flavour Mr Chilly would sell until his freezers are fixed.

Stations

- **Bottle Top Display** – Provide the children with a tub of plastic bottle lids to explore. On closer inspection the lids are different. Provide a magnifying glass so the children can look closely and see some of the small details, e.g. colour, size (diameter and height), writing on top, lids with flaps.

Ask the children to think about how they might display the lids once they had sorted them into their chosen criteria. A Carroll diagram may be useful if they chose lids with writing on and lids with no writing on as their criteria. The children may come up with other ideas depending on what they want to know. They might tip out the jar to see how many lids there are and want to find out how many of each colour there are. **Would they need to create a frequency table or could they just then arrange the lids in a pictogram and count how many?** Lots of opportunities to explore and assess how well the children can respond to a challenge.

- **Using Technology** – A number of data handling packages are available for children to use to create a display. For example, the Carroll Diagram digital activity (see the Digital resources).

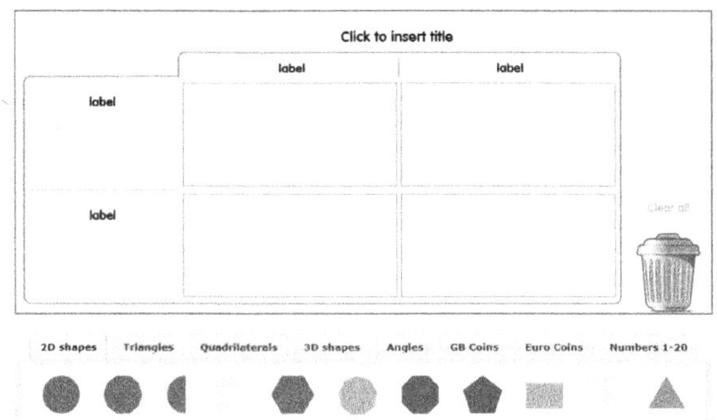

Unit 11.4 Asking simple questions to collect data and making suggestions how to do this

Whole class

- **What Makes a Shop a Shop?** – Engage the children in discussions about *what a shop is?* **Are all shops the same? What is/are the common factor/s that may link different types of shops?** Help the children to compile a list of criteria that they think all shops should have in common (model effective questioning when engaging with the children). The list may take the form of a visual check list or a written check list. Take the children to visit a number of different types of shops in their local environment and encourage them to look for items pertaining to those on the check list. Mark items as required on the list. Once back in school the lists for different shops can be compared. **Do they all have the same features? What is the same and what is different? Are differences apparent when comparing the findings from a supermarket visit and a small local store visit?**
- **Surveys** – Engage the children in creating surveys. Surveys usually take one of two forms: fact-finding or preference. In a fact-finding survey the children may engage with other children, by asking questions, to find out the how the children get to school (scooter/bike/car/walk, etc.), or they may do some research, looking at books to find out about the types of birds that might visit bird feeders in their local area. Their question might be how many different types of birds visit the bird feeder in the school garden. In a preference survey, the children will be given the opportunity to select, from two or more options, their favourite.

Stations

- **Fact finding** – The examples of fact-finding surveys below involve different techniques; in the bird watching example the children observe a scenario to gain the data whereas in the shoe fastenings survey the children will be required to ask questions of others to gather data. They should be involved in exploring in both scenarios.
- **Bird Watching in Groups** – One group of children will compile a list (pictorial) of common garden birds. These pictures can be organised into a tick chart which the children will then use to gather information on the numbers of these birds visiting the bird feeder, using tally marking. More able children could write the name of the bird alongside the picture.
- **Shoe Fastenings Group Survey** – Another group of children will be engaged in thinking about the types of shoe fastening that children in the class have on their shoes. This fact-finding survey can be extended to include the whole school. The fastenings may include: Velcro/laces/buckles/slip-on. The children will be required to devise a form of tick chart to gather the data.
- **Preference Surveys** – One important aspect of preference surveys is that children learn to appreciate that preferring one thing over another is not a question of right or wrong, good or bad; it is evidence that we each have a unique blend of sameness and difference with others.
- **Ice Cream Toppings** – A group of children could compile a list of ice cream toppings, e.g. chocolate sauce, raspberry sauce, flake, sprinkles; they can make and use pictorial or written labels for their tally chart. Children will be asked to choose their favourite type of topping and the data will be recorded in a tally chart.
- **Favourite Play Apparatus** – Another group of children could compile a list of favourite apparatus in the play park, e.g. swing, chute, roundabout, climbing frame. Again, pictorial or written labels can be used for the tally chart. Children will be asked to think about the park apparatus and give their favourite. The data will be recorded in a tally chart.

What to observe

- Children can sort and talk about a collection; they can describe how they have sorted their collection. They can think of their own criteria for sorting. They can sort the objects using more than one attribute.
- Children can count how many are in a set and compare the groups using language such as more/less/most/least/fewer/popular.
- Children can give ideas for collecting, sorting and presenting their findings.
- Children can record their 'maths' in their own ways, which they can explain and interpret.
- Children who can explain why two or more things are the same/are different (not the same).
- Children who can explain why things in a group go together.
- Children can use appropriate vocabulary to explain their thinking.
- Children who can sort, match and compare quantities in play and daily routines.
- Children who can ask questions.
- Children using signs and charts.